Learning from Dynamic Visualization

Richard Lowe • Rolf Ploetzner
Editors

Learning from Dynamic Visualization

Innovations in Research and Application

Springer

Editors
Richard Lowe
Curtin University, Perth, Australia

University of Burgundy, Dijon, France

Rolf Ploetzner
University of Education
Freiburg, Germany

ISBN 978-3-319-56202-5 ISBN 978-3-319-56204-9 (eBook)
DOI 10.1007/978-3-319-56204-9

Library of Congress Control Number: 2017941552

Printed on acid-free paper

This Springer imprint is published by Springer Nature
The registered company is Springer International Publishing AG
The registered company address is: Gewerbestrasse 11, 6330 Cham, Switzerland

Acknowledgments

The development of this book was in part supported by two research projects granted during the years 2011–2014 to Richard Lowe (Curtin University, Western Australia) and Rolf Ploetzner by the German Academic Exchange Service (DAAD) and the Australian Technology Network (ATN) within the Australian–German Joint Research Cooperation Scheme.

Contents

About the Contributors and Editors

About the Contributors

Sandra Berney received a Ph.D. in psychology from the University of Geneva, Switzerland, where she specialized in cognitive and educational psychology. She is an associate researcher in the Department of Psychology and Educational Sciences, and a member of TECFA, a research and teaching unit in the field of educational technology at the University of Geneva, Switzerland. In her research, she focuses on the interplay between spatial abilities and dynamic visualizations in learning contexts.

Mireille Bétrancourt is Professor of Learning Processes and Information Technologies at Geneva University, Switzerland. After undergraduate studies in psychology and a master's degree in occupational psychology, she completed a Ph.D. in cognitive science at Grenoble University, France. Her research focuses on the design of computer-supported instruction and its impact on learners' strategies and cognitive processes. She has been particularly interested in multimedia instruction and animated visualization, investigating the interplay between delivery features, learners' characteristics and instructional factors. Her goal is not only to contribute to academic knowledge but also to provide evidence-based recommendations for the design of multimedia instruction.

Jean-Michel Boucheix is Professor of Cognitive and Experimental Psychology at the University of Burgundy, France. After receiving a Ph.D. in cognitive and developmental psychology at the University of Poitiers, he initially worked for government agencies. Since 2000, he has been working at the University of Burgundy in the Laboratory for the Study of Learning and Development – National Center for Scientific Research (Laboratoire d'Etude de l'Apprentissage et du Développement, LEAD-CNRS). His research focuses on the cognitive psychology of learning and training technologies with respect to the processing of multimedia documents. He has a particular interest in the comprehension and processing of interactive animations and the use of online methods such as eye movement analysis for investigating multimedia and visual scene processing.

Bianka Breyer initially studied psychology in Marburg, Germany, and then worked as a research assistant at the University of Education in Freiburg, Germany. In her research, she participated in a number of projects concerned with learning from animations and investigated strategies, including drawing, that were designed to enhance such learning. Since 2015, she has been working at the Leibniz Institute for the Social Sciences in Mannheim, Germany, where her focus is upon on scale construction and open-access measurement instruments in the social sciences.

Jodi L. Davenport is the director of the National Center for Cognition and Learning at WestEd, USA. Davenport received a Ph.D. in cognitive science from the Massachusetts Institute of Technology, specializing in visual cognition and computation. As a postdoctoral fellow at the Pittsburgh Science of Learning Center at Carnegie Mellon University, she applied her background in basic research to study multimedia learning. Her current research and development efforts focus on leveraging technology to improve student learning and engagement in science and mathematics.

Benjamin Fillisch initially studied sociology at the University of Trier, Germany. After completing his MA, he received a scholarship from the German Youth Institute (DJI) in Munich, Germany. Since 2012, he has been conducting Ph.D. research at the Institute for Media in Education at the University of Education in Freiburg, Germany. The focus of his research is the use of strategies designed to enhance learning from animations.

Halszka Jarodzka holds a master's degree and a Ph.D. in psychology, both from the University of Tübingen, Germany. She is an assistant professor at the Welten Institute of the Open University of the Netherlands and chair of the research group Processes of Learning and Expertise Development in Information-Rich Environments. Her research focuses on the use of eye-tracking in educational psychology, and she is a visiting scholar at the eye-tracking laboratory of Lund University, Sweden.

Jodie Jenkinson is an assistant professor of Biomedical Communications at the University of Toronto, Canada, and principal investigator of the Science Visualization Lab (www.sciencevis.ca). She holds a Ph.D. in education from the University of Toronto, specializing in cognition and learning, with a focus on technology in education. Her research focuses on the role that visual representations play in learning. This includes investigation along various lines of inquiry, including the efficacy of visual media within different learning contexts, the design of visual representations for optimal impact, and the development of standards of visual communication in the scientific visualization community.

Björn B. de Koning is an assistant professor of Educational Psychology in the Department of Psychology, Education, and Child Studies at Erasmus University Rotterdam, the Netherlands. His research projects focus on optimizing learning from dynamic visualizations, designing effective multimedia instruction, using cognitive strategies and mental model construction in reading comprehension, and understanding the influence of collaboration on learning processes and performance in academic learning environments. His work is characterized by a multidisciplinary approach that aims to provide insights for improving educational practice by integrating cognitive (neuro-) science, developmental and educational psychology research perspectives.

Lucia Mason is Professor of Educational Psychology at the University of Padova, Italy, where she also received her Ph.D. Before returning to the University of Padova, she was an associate professor at the University of Lecce, Italy. In addition to being a former editor-in-chief of *Learning and Instruction*, she is currently an editorial board member for several high-ranking journals. She has carried out studies on conceptual change and epistemic beliefs. Her studies also investigated search, evaluation, and comprehension of online information on controversial issues. In more recent research, her focus has been on process data, in particular that obtained from eye-tracking, as a means of investigating cognitive processing in learning from science texts and graphics, as well as the connections between cognition and affect.

Richard E. Mayer is Professor of Psychology at the University of California, Santa Barbara, USA. His current research involves the intersection of cognition, instruction, and technology, with a special focus on multimedia learning, computer-supported learning, and learning with games. He is past president of Division 15 (Educational Psychology) of the American Psychological Association and past vice president of the American Educational Research Association for Division C (Learning and Instruction). He received the E. L. Thorndike Award for career achievement in educational psychology, the Sylvia Scribner Award for research contributions in learning and instruction, and the Distinguished Contribution of Applications of Psychology to Education and Training Award from the American Psychological Association. He is ranked No. 1 as the most productive educational psychologist in the world by Contemporary Educational Psychology. He currently is on the editorial boards of 12 journals mainly in educational psychology. He is the author of more than 500 publications, including 30 books, such as *Multimedia Learning* (2nd ed.), *e-Learning and the Science of Instruction* (3rd ed., with R. Clark), and *Applying the Science of Learning*.

Gaël G. McGill is faculty and Director of Molecular Visualization at Harvard Medical School (USA), where his teaching and research focus on visualization design methods in science education. He is also the founder and CEO of Digizyme Inc. (USA), a firm dedicated to the visualization and communication of science to a variety of audiences both in and outside the classroom. He recently co-authored and served as digital director for E.O. Wilson's *Life on Earth* iBooks biology textbook, and is the creator of the scientific visualization portal Clarafi.com and the Molecular Maya (mMaya) software. After undergraduate studies in biology, music and art history, he received his Ph.D. in biological chemistry and molecular pharmacology from Harvard Medical School (USA) and completed his postdoctoral fellowship in cancer biology at the Dana-Farber Cancer Institute (USA).

Frank Papenmeier is interested in the cognitive processes involved during the processing of dynamic visual information, with a particular focus on attention and working memory. During his Ph.D. in psychology at the Leibniz-Institut für Wissensmedien (IWM) in Tübingen, Germany, he studied the representation of dynamic spatial configurations in visual working memory. Since then, he has been working in the research group Applied Cognitive Psychology and Media Psychology at the University of Tübingen. His research covers the processing of dynamic objects (such as attention tracking), memory for three-dimensional object structure, and the processing of dynamic scenes (such as real-world events in movies and grasping actions).

Edys S. Quellmalz is Director of Technology-Enhanced Assessment and Learning Systems in WestEd's Science, Technology, Engineering, and Mathematics Program. She received her Ph.D. in educational psychology from UCLA's Graduate School of Education. She leads SimScientists projects funded by NSF and the US Department of Education related to simulation-based science curricula and assessments for formative and summative assessments that can serve as components of balanced state science assessment systems. She was Associate Director of the Center for Technology and Learning at SRI International and the Director of Assessment Research and Design. She served on the faculty at the Stanford School of Education and as research faculty in the UCLA Graduate School of Education.

Christopher A. Sanchez received a Ph.D. in cognitive psychology from the University of Illinois at Chicago and is currently an assistant professor of Psychology at Oregon State University. His research broadly focuses on the assessment and evaluation of individual differences in cognitive abilities, with the goal of understanding how these abilities influence and impact the effectiveness of learning environments. By understanding this interaction, his research hopes to better match the delivery of learning materials to the specific needs of each individual, thus maximizing learning outcomes and the successful attainment of educational goals.

Katharina Scheiter is Professor of Learning and Instruction at the Leibniz-Institut für Wissensmedien (IWM) in Tübingen, Germany. She completed her Ph.D. at the University of Tübingen in 2003 with research located at the intersection of cognitive and educational psychology. From 2003 to 2009, she was a postdoctoral researcher at the Department of Applied Cognitive Psychology and Media Psychology at the University of Tübingen. Since 2009, she has been heading the multimedia learning lab at the IWM, where with her team she investigates cognitive processes underlying learning with textual and graphical representations (such as diagrams and animations) and means of supporting this learning.

Wolfgang Schnotz is Professor Emeritus of General and Educational Psychology at the University of Koblenz-Landau, Germany. He received his Ph.D. from the Technical University Berlin, was a researcher at universities in Tübingen and Vienna as well as Full Professor of Educational Psychology in Jena, before he moved to the University of Koblenz–Landau, Germany. He served as chief editor of the international journal *Learning and Instruction*, was a member of the Reading Expert Group of PISA, and served as chair of the Multimedia Research Group and chair of the Graduate School of Teaching and Learning Processes at the University of Koblenz–Landau. His primary research interests are currently in the field of cognitive visualization with a special focus on text-picture integration.

Stephan Schwan received his doctoral degree in psychology from the University of Tübingen, Germany. Since 2004, he has been a principal research scientist and full professor for research on teaching and learning at the Leibniz-Institut für Wissensmedien (IWM) in Tübingen, Germany. His primary research focus is on perceptual and cognitive processes that occur during learning with three-dimensional and interactive visual presentations. He also investigates the role played by digital media in the way visitors to museums and exhibitions process information and acquire knowledge.

Mike Stieff is Associate Professor of Learning Sciences and Chemistry at the University of Illinois – Chicago. He holds a Ph.D. in learning sciences and an MS in chemistry from Northwestern University. He is an expert in the design and evaluation of learning environments, particularly those in undergraduate STEM (Science, Technology, Engineering, and Mathematics) courses. His research addresses three specific goals: the identification of the task-specific use of visuospatial strategies in scientific problem solving, the characterization of the interaction of visuospatial ability and scientific expertise, and the development of visualization software for teaching high school and college chemistry. His current research explores new theoretical frameworks regarding the role of visualization and diagrammatic reasoning at multiple levels of science instruction and the design of simulations and animations for teaching secondary science.

Inga Wagner initially studied psychology at the University of Koblenz-Landau, Germany. Upon her graduation, she undertook a Ph.D. on learning with animations and static graphics, which she completed in 2013. She then participated in a research project funded by the Deutsche Forschungsgemeinschaft (DFG) that dealt with the integrative processing of texts and graphics. Currently, she is conducting two research projects that are concerned with the acceptance of learning assessments and with the evaluation of school inspections.

Jennifer Wiley is a professor of Psychology at the University of Illinois at Chicago with a long-standing interest in research at the intersection of education and cognitive psychology. She received her Ph.D. from the University of Pittsburgh, where she was trained at the Learning Research and Development Center. She continued her training as a postdoctoral scholar at UMass Amherst, where she further explored factors that impact reading and comprehension processes, and developed expertise in eye-tracking methodologies. Her research interests include exploring how individual differences can affect comprehension and meta-comprehension, and how to facilitate conceptual learning and reduce illusions of understanding in multimedia contexts.

About the Editors

Richard Lowe is an Adjunct Professor at Curtin University, Australia and a Visiting Researcher at the University of Burgundy, France. After his undergraduate studies in science, he worked as an industrial chemist, science educator and textbook writer, then completed a Ph.D. in educational psychology at Murdoch University, Australia. His research focuses on the effectiveness of explanatory graphics with a particular emphasis on learner processing of complex technical animations. He has a special interest in how the findings from such research can be applied to improving the design of explanatory dynamic graphics used not only within educational settings but also across society more generally.

Rolf Ploetzner received a Ph.D. in psychology from the University of Freiburg, Germany, where he specialized in cognitive and educational psychology. Across his research career, he has participated in various research projects at the University of Freiburg, the Learning Research and Development Center (LRDC) at the University of Pittsburgh (USA), and the Knowledge Media Research Center (KMRC) in Tübingen, Germany. Since 2002, he has been a full professor of Media in Education at the University of Education in Freiburg. In his current research he focuses on strategies for learning from dynamic and interactive media.

Introduction

Dynamic visualizations in general and animations in particular have become a ubiquitous feature of today's technology-based learning resources. Nowhere is this more evident than in the multitude of educational apps to be found on electronic tablets and similar devices. Further, dynamic visualizations rather than written or spoken text are increasingly given major responsibility for presenting explanations. This reverses the traditional explanatory roles of text and graphics in education where graphics were typically used as mere adjuncts to the text. However, educators' enthusiastic embrace of dynamic visualizations has a downside. This motivating and potentially powerful form of representation has not proven to be the educational magic bullet that many assumed it would. Rather, as is the case with other more established forms of representation, the educational effectiveness of dynamic visualizations depends on how well they are designed, used and supported.

This book brings together leading international experts in the use of dynamic visualization for fostering learning. In one sense, it presents a review of the current state of play. However, rather than providing a comprehensive coverage of this area, the book's chapters offer selective snapshots that highlight some of the more innovative contributions to the field. Our hope is that this approach will suggest possible directions that could shape the field's future growth. This book is not about the application of research to practice. Instead, it deals with innovations in both areas of endeavor that have the potential to improve learning from dynamic visualizations. The book's title and the choice of contributors reflect our conviction that formal scholarly research is not the only source of valuable insights about this topic. Equally important are the accumulated wisdom and grounded craft knowledge of expert practitioners who daily grapple with the multitude of challenges involved in developing effective dynamic visualizations. We hope the book will stimulate a closer and mutually beneficial dialogue between these two communities of experts.

The diverse perspectives of the book's authors ranging across various theoretical, empirical, and practical orientations mean that individual chapters can differ markedly in their approaches. Further, because of our selective emphasis on innovation in the field, a number of the contributions contain aspects that are somewhat speculative in character. These features of the book are intended to expand thinking about dynamic visualizations and learning beyond the presently prevailing orthodoxies. While it is possible to make numerous links across the different chapters, it is clear that many gaps remain to be filled.

The first major publication that could be considered as specifically devoted to dynamic visualizations and learning was *Learning with Animations*, a book edited by Lowe and Schnotz (2008). It reflected a growing realization that learning from animations is not as unproblematic as had been generally assumed. In particular, it drew attention to the challenges that the dynamic character of animations may pose for learner processing. The most influential theoretical model used to frame research and practice in the field at the time was Richard Mayer's Cognitive Theory of Multimedia Learning (Mayer, 2001). However, this framework is primarily concerned with media in *combination* rather than a detailed consideration of individual types of media such as dynamic visualizations. In the final chapter of their book, Lowe and Schnotz (2008) foreshadowed some of the major considerations that would need to be taken into account for a theoretical framework that dealt specifically with learning from animations. These ultimately stimulated the development of the Animation Processing Model (APM; Lowe & Boucheix, 2008). The APM features in a number of chapters in the present publication, which in a sense can be seen as a successor to Lowe and Schnotz (2008).

Because dynamic visualizations are widely used within multimedia learning resources, there has been substantial investigation of their contribution to learning when combined with other media. However, rather than being concerned with such combinations, the focus of this book is primarily upon dynamic visualizations in their own right. Its main concern is with dynamic visualizations such as animation (rather than video) that are authored rather than captured (cf. Ploetzner & Lowe, 2012). Our reasons for deciding upon this more selective focus include the increasing explanatory responsibility given to dynamic visualizations, the need to better understand the distinctive contribution these visualizations make to learning, and the accumulating evidence that the spatiotemporal characteristics of dynamic visualizations can have a profound effect on how learners process the presented subject matter.

There are four main sections in this book, each addressing a theme of importance to learning with dynamic visualizations. Each section is preceded by a brief introduction that provides an overview of the main themes covered and how they are related across the constituent chapters. Part I, *Innovations in Representation and Design*, includes chapters that range from a theoretically motivated proposal for improving the design of dynamic visualizations to consideration of the approaches expert developers use in their design practice. Part II, *Innovations in Assessment*, focuses on how to collect better evidence of the educational outcomes that may result from learning with dynamic visualizations. Part III, *Innovations in Scaffolding*, introduces general approaches for improving learning from dynamic visualizations by supporting how learners process the information presented. Part IV, *Innovations in Learner Engagement*, complements this support theme by examining the potential of a number of specific intervention strategies intended to improve learner processing. We are extremely fortunate to have two outstanding commentators responding to the issues raised in these sections (Katharina Scheiter, Parts I and II; Richard Mayer, Parts III and IV). Our sincere thanks to you both.

We envisage this book as being useful to a broad spectrum of readers – those who design dynamic visualizations for learning, those who use them for instruction, and those who investigate ways to improve their educational effectiveness. On one hand, the book offers theoretical perspectives and empirical evidence that can provide a principled basis for how the design, use, and support of dynamic visualizations might be optimized. On the other, it contains invaluable insights into opportunities and constraints that shape how dynamic visualizations are developed in real-life contexts. Our hope is that by including these complementary aspects, the book will foster productive interchanges between the various communities who deal with these fascinating representations and help to advance this fertile field of activity.

References

Lowe, R., & Boucheix, J.-M. (2008). Learning from animated diagrams: How are mental models built? In G. Stapleton, J. Howse & J. Lee (Eds.), *Diagrammatic representations and inference – Proceedings of the Fifth International Conference, Diagrams 2008* (pp. 266–281). Berlin: Springer.

Lowe, R. K., & Schnotz, W. (Eds.). (2008). *Learning with animation: Research implications for design*. New York: Cambridge University Press.

Mayer, R. E. (2001). *Multimedia learning*. New York: Cambridge University Press.

Ploetzner, R., & Lowe, R. (2012). A systematic characterisation of expository animations. *Computers in Human Behavior, 28*, 781–794.

Part I
Innovations in Representation and Design

How a representation portrays to-be-learned information is perhaps one of the most fundamental issues influencing its educational effectiveness. At one level, we can distinguish between representations on the basis of the different senses (e.g., hearing, vision, touch, etc.) that are required in order to extract information from them. Within each of these categories, various ways of representing the information can be employed. Our particular concern in this volume is with visual representations that are depictive rather than descriptive and dynamic rather than static. The dynamic nature of these depictions confers on them special characteristics that can have a profound influence on how they are processed by learners. A key challenge for those who design dynamic visualizations intended to foster learning is to recruit these characteristics in ways that help rather than hinder learning. Essentially, their task involves devising designs that optimize the match between the characteristics of dynamic visualizations and the capacities of learners. The chapters in this part address not only the distinctive characteristics of dynamic visual representations and their psychological consequences, but also ways in which the interaction between those characteristics and human information processing may be managed to produce educationally effective designs.

This first part of the volume presents a diverse range of contributions that both reflect the current state of the field and look forward to how it might develop in the future. Readers whose main interest is in research may wish to work through the chapters in sequence, whereas others may prefer to start with the last two more practically-oriented chapters in this part before delving into the research. There are a number of common threads that link several of the chapters in this part. One of these is the effect that perceptual characteristics of dynamic visualizations can have on how effectively individuals process the presented information. Another is the use of the Animation Processing Model as a theoretical framing. Nevertheless, the field is still very much in its infancy when compared with far more established areas such as learning from text so there are many gaps in our knowledge of how to design effective dynamic visualizations.

Lowe and Boucheix (2017, this volume) invoke the Animation Processing Model to propose a principled basis for expanding the somewhat limited range of current approaches to designing dynamic visualizations. Their *Composition Approach* is intended to complement existing design approaches, particularly where the to-be-learned subject matter is complex and unfamiliar to the target audience. It addresses the fundamental mismatch between how animated depictions often present information and the capacity of humans to process it effectively. This approach takes particular account of the powerful but largely neglected perceptual effects that animations can have on where learners direct their visual attention and how that influences their extraction of key information.

Schwan and Papenmeier (2017, this volume) continue the focus on the design of dynamic visualizations in their consideration of 2D versus 3D representation of the subject matter. In response to the increasing tendency of designers to use 3D rather than 2D animations, the authors canvas a range of new opportunities for fostering learning that can be provided by using 3D representations. Like Lowe and Boucheix (2017, this volume), Schwan and Papenmeier identify the key role that perception can play in the processing of animations. While suggesting that 3D animations may offer psychological benefits such as improved direction of attention, they also acknowledge the distinctive processing challenges that these representations may impose on learners.

Ploetzner and Lowe (2017, this volume) report an empirical investigation of how presenting multiple animated segments simultaneously within the same display affects learning. The aim of this novel form of animation design is to enhance learners' relational processing of dynamic subject matter by supporting linkage of different pieces of spatiotemporal information through facilitation of compare-contrast activities. At first glance, this approach which involves *increasing* the complexity of the visual display, may seem to be at odds with the efforts of Lowe and Boucheix (2017, this volume) to *reduce* learner processing demands. However, this apparent disparity encapsulates the real-life challenges animation designers face in trying to find an appropriate balance between (a) limiting the demands learners must deal with in terms of the display characteristics and (b) assisting learners to compose the internal relationships that are the hallmark of a high quality mental model.

Wagner and Schnotz (2017, this volume) also highlight the role that perceptual characteristics have on how visual displays are processed by learners and how this may ultimately influence the cognitive consequences of instruction. Learning from static and animated displays of the same subject matter was investigated with respect to spatial and temporal aspects of the to-be-learned content. From their finding that sequential presentation was a crucial factor with respect to instructional effectiveness, the authors call for a reconsideration of the usual simplistic distinction between static and animated graphics. They suggest that when making decisions about how best to represent the targeted instructional information, designers should instead foreground spatial versus temporal aspects of the subject matter.

Jenkinson (2017, this volume) offers a distinctly different but complementary perspective from those presented in the previous chapters. From her viewpoint as a practicing animation designer and university teacher of biomedical illustration,

Jenkinson uses examples to give an insider's account of the processes involved in designing dynamic visualizations of specialized scientific content. She maps the broad-scale design phases that need to be carried out in order to develop an effective dynamic visualization. In contrast to most other chapters in this volume, Jenkinson emphasizes the key role that accumulated craft knowledge and expert professional judgement plays in this design process. Her chapter makes it clear that a considerable gap remains between the findings of animation researchers and the activities that animation designers engage in on a daily basis. Bridging this gap successfully will mean addressing the very different priorities and professional praxis of the practitioner and research communities.

McGill (2017, this volume) provides further insights into the work of a practicing animation designer and the very real constraints that operate in this professional space. A key focus is the potential tension that can arise within the client-designer relationship due to their different perspectives and priorities. Formulating and implementing a satisfactory response to the client's design brief under these circumstances can pose significant challenges for the designer. These extend beyond a consideration of the perceptual and cognitive effects that particular presentational characteristics may have on the target audience. In principle, today's visualization technology allows animators to produce almost anything that can be imagined. The author uses examples from his own practice to demonstrate how the vast array of representational possibilities available to designers must be intelligently marshalled in a way that not only satisfies the client but also provides end users with a psychologically appropriate depiction of the content.

References

Jenkinson, J. (2017). The role of craft-based knowledge in the design of dynamic visualizations. In R. Lowe & R. Ploetzner (Eds.), *Learning from dynamic visualization – Innovations in research and application*. Berlin: Springer (this volume).

Lowe, R., & Boucheix, J.-M. (2017). A composition approach to design of educational animations. In R. Lowe & R. Ploetzner (Eds.), *Learning from dynamic visualization – Innovations in research and application*. Berlin: Springer (this volume).

McGill, G. (2017). Designing instructional science visualizations in the trenches: Where research meets production reality. In R. Lowe & R. Ploetzner (Eds.), *Learning from dynamic visualization – Innovations in research and application*. Berlin: Springer (this volume).

Ploetzner, R., & Lowe, R. K. (2017). Looking across instead of back and forth: How simultaneous presentation of multiple animation episodes facilitates learning. In R. Lowe & R. Ploetzner (Eds.), *Learning from dynamic visualization – Innovations in research and application*. Berlin: Springer (this volume).

Schwan, S., & Papenmeier, F. (2017). Learning from animations: From 2D to 3D? In R. Lowe & R. Ploetzner (Eds.), *Learning from dynamic visualization – Innovations in research and application*. Berlin: Springer (this volume).

Wagner, I., & Schnotz, W. (2017). Learning from static and dynamic visualizations: What kind of questions should we ask? In R. Lowe & R. Ploetzner (Eds.), *Learning from dynamic visualization – Innovations in research and application*. Berlin: Springer (this volume).

Chapter 1
A Composition Approach to Design of Educational Animations

Richard Lowe and Jean-Michel Boucheix

1.1 Introduction

Animated graphics have recently become ubiquitous components of modern learning materials (Ploetzner & Lowe, 2012). However despite the rapid ascendency of animations, their designers lack authoritative guidance on how to make them educationally effective. This is not the case for other longer-established forms of visualization such as traditional textbook illustrations. When it comes to including such *static* graphics in educational resources, developers of these materials can draw on hundreds of years of practical experience as to which design approaches are most likely to be effective (Tversky, Heiser, Mackenzie, Lozano, & Morrison, 2008). In contrast, those who wish to include *animated* graphics have a very much shorter legacy of animation-specific design experience to call upon (see Jenkinson, 2017, this volume). Further, there is relatively little guidance in the research literature on how to design educationally effective animations. To date, most design-oriented research has been within the multimedia tradition where animations are used as adjuncts to text-based information (e.g., Ayres, Marcus, Chan, & Qian, 2009; Mayer & Anderson, 1991, 1992; Mayer & Moreno, 2002; Spanjers, Wouters, van Gog, & van Merriënboer, 2011). The main focus of that research has understandably been on the effectiveness of text-animation combinations and not on the design of animations *per se*.

Findings from the relatively few studies that have singled out animations for investigation in their own right suggest that animations may not be the educational

R. Lowe (✉)
Curtin University, Perth, Australia

University of Burgundy, Dijon, France
e-mail: r.k.lowe@curtin.edu.au

J.-M. Boucheix
University of Burgundy, Dijon, France
e-mail: Jean-Michel.Boucheix@u-bourgogne.fr

R. Lowe, R. Ploetzner (eds.), *Learning from Dynamic Visualization*,
DOI 10.1007/978-3-319-56204-9_1

panacea many believe them to be. On the contrary, it seems that some animations can actually be problematic for learners, particularly if the topic portrayed is complex and unfamiliar to the target audience. This has prompted a quest for ways of modifying animations in order to improve their educational effectiveness. For example, some researchers have explored the utility of principles such as visual cueing (De Koning, Tabbers, Rikers, & Paas, 2009) that have been borrowed from approaches long used in designing static educational graphics. Unfortunately, the findings from these studies have been inconclusive (see De Koning & Jarodzka, 2017, this volume). Although animated and static graphics have much in common (Schnotz & Lowe, 2008), there are also key differences in how they present information and how their distinctive forms of presentation are processed by learners (cf. Sanchez & Wiley, 2017, this volume). Merely recruiting design principles that have been found effective in the realm of static graphics may not provide sufficient guidance for designing sound educational animations. This is due to the additional and dominating influence that the dynamics of animated graphics can have on how learners process visual information. A consequence of the current lack of design guidance is that today's educational animations tend to be the result of designers' intuition and opinion rather than a more principled approach. Given the increasing reliance that learning materials place on animations for explaining to-be-learned content, this is a matter for considerable concern. However, recent developments in our understanding of how learners process animated graphics point to new possibilities for designing effective educational animations.

An essential first consideration for designing an educational animation is to be clear about what it is hoped to achieve in terms of learning. For many animations, the overarching design goal seems to be largely affective – their main purpose is to pique learner interest and provide on-going motivation. While these affective aspects are undoubtedly important, they are beyond the scope of the present chapter. The main focus here will be on how an animation might directly foster deep understanding of the subject matter it presents. Our criterion for an animation to be educationally effective is that it results in a target learner constructing a high quality mental model of the to-be-learned subject matter (see also Lowe, Boucheix, & Fillisch, 2017, this volume). By high quality, we mean that the mental model is (i) appropriate for the task that the learner will be required to perform, (ii) an accurate internal representation of the referent subject matter, and (iii) sufficiently comprehensive to encompass all essential aspects of the required task performance. When the subject matter portrayed by an animation is complex, its understanding requires learners to develop mental models that properly represent such complexity. There are several key issues to be considered when contemplating how to design educationally effective animations. These include the nature of animation as a depictive genre (i.e., how animated graphics represent their referent subject matter), the proclivities of the human information processing system as well as its inherent capacity constraints, the particularities of the subject matter that is to be depicted, and the type of learning task being targeted.

This chapter begins by considering how information that is both complex and unfamiliar to learners is presented via conventionally designed animations. We have chosen this particular focus not only because animations are now commonly recruited for explaining such information, but also because their challenging nature is likely to expose issues of fundamental importance to animated explanations more generally. The chapter then considers some challenges that such presentations may pose for learners. This is followed by an overview of the Animation Processing Model (APM; Lowe & Boucheix, 2008a) that provides an account of processes involved when learning from animations. Next, a principled APM-based alternative to current animation design approaches is outlined that aims to facilitate the learner's composition of an internal representation by adopting a different perspective on the characteristics of the animation as an external representation. The chapter concludes by describing how this Composition Approach has been implemented, providing evidence of its effectiveness, and considering implications for future research.

1.2 Animated Presentation of Complex Dynamic Information

Most animations currently used for instructional purposes are behaviorally realistic in that they depict the dynamics of their referent subject matter in an essentially faithful manner. This *temporal* resemblance between depicted and actual dynamics tends to be a fundamental design feature, irrespective of whether or not there is a high degree of *visuospatial* resemblance between the depicted entities and their real-life referents (Lowe & Schnotz, 2014). Even when animations depict the appearance of their subject matter in a highly abstract and manipulated manner (e.g., diagrammatically), they nevertheless tend to mimic the referent's dynamics very closely. We describe these conventionally designed animations as 'comprehensive' because of their relatively complete portrayal of the events being represented. Animation designers' privileging of behavioral realism is perfectly understandable and seems intuitively reasonable – one of the most compelling arguments for using animations rather than static graphics is that it is only when graphics have been animated that they are able to show explicitly and directly how the subject matter changes over time. Animated graphics provide full representational continuity, a property not available in their static counterparts that can provide only implicit, indirect and partial indications of dynamics.

The various events that comprise a comprehensive animation are presented via a set of episodes whose structures and sequencing mirror the behaviors that occur in the referent situation. Consequently, when an animation targets subject matter that is high in dynamic complexity, the animated depiction is correspondingly complex. This means that any one episode of such an animation may simultaneously display a rich assortment of relationships amongst a multitude of different entities and events that are distributed widely across the screen. As such an episode progressively unfolds over its duration, additional relationships will be revealed that are distributed across time rather than space. More broadly, the animation in its entirety may consist of a succession of several such episodes that are related to each other at

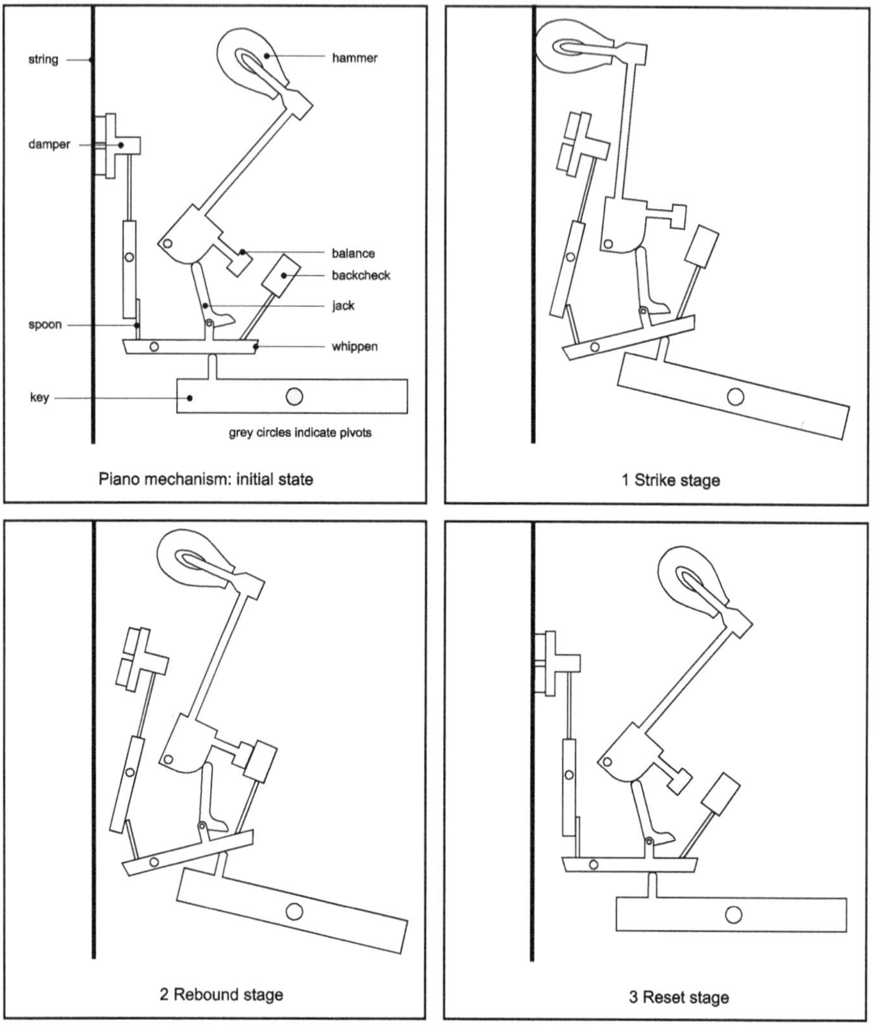

Fig. 1.1 The piano mechanism

a superordinate level. Further, some of the depicted entities and events may be compound assemblages whose component parts need to be considered either separately or in a coordinated fashion. These various contributions to complexity can be illustrated by referring to a comprehensive animation of a traditional upright piano mechanism that has been used for several empirical studies on learning from dynamic visualizations (e.g., Boucheix & Lowe, 2010; Lowe & Boucheix, 2011). Although the piano mechanism's component entities are depicted in a somewhat abstract and simplified manner (cf. Fig. 1.1), the dynamics portrayed in the animation are nevertheless a highly accurate representation of how a real piano operates.

The piano mechanism operates in three sequential stages and these are portrayed in the animation through three successive episodes:

1. the *Strike* episode that corresponds to the pianist depressing a key which causes the hammer to swing rapidly towards the piano string and strike it to produce a musical note;
2. the *Rebound* episode that follows immediately – while the pianist is still holding down the key – in which the hammer rebounds from the string and is stopped by the balance-backcheck collision;
3. The *Reset* episode that occurs when the pianist releases the key and allows the mechanism to return to its original configuration in readiness for a further key press.

Within each of these episodes, the mechanism's various constituent entities participate in events that propagate two tightly coordinated causal chains whose joint functions are fundamental to the piano's proper operation. The first of these causal chains is concerned with the hammer striking the string to set it into vibration, thus producing a musical note. The second is concerned with the damper that stops the string from vibrating when required. Tight coordination of these two causal chains is essential so that the damper is removed from the string before the hammer strikes and is replaced when the note is to finish sounding. This overarching functional relationship between hammer and damper operations is brought about by a specialized set of lower-level spatial and temporal relationships amongst the entities that are distributed along the path of each of the two causal chains.

In order for the piano mechanism to operate properly, the events involving its constituent entities that bring about the reciprocal actions of the hammer and damper must occur either simultaneously or in a rapid cascade. These multiple events happening at the same time or in quick succession occur in locations that are widely dispersed across the display area. There are also big differences between the entities that are engaged in these events and between the characteristics of the events themselves. For example, some of the entities are large (e.g., the hammer) while some are small (e.g., the jack); some are made up of various sub-entities (e.g., the hammer) while others are essentially a single piece (e.g., the jack); some parts move extensively (e.g., the head of the hammer), while other parts move relatively little (e.g., the jack); some entities perform events that have a single function (e.g., the key) while other entities perform several, depending on the operational episode in which they are engaged (e.g., the jack). As we shall see in the next section, such characteristics can have a profound effect on how learners process the available information.

1.3 Challenges for Learner Processing

The piano mechanism example illustrates how a comprehensive animation of complex subject matter confronts the learner with a rich and continuously varying flow of dynamic information. These features may impose substantial processing costs on learners that must be weighed against the potential benefits animations could bring. A major contributor to such costs is the limited capacity of the human information processing system, a constraint that can mean learners are simply ill-equipped to deal with the type of information presentation that complex comprehensive animations deliver (Lowe, 1999). We noted above that a fundamental challenge for learners comes from the fact that the presented information is distributed across multiple locations in the animated display with the diverse activities happening at these different sites occurring either at the same time or in very rapid succession. This is well illustrated in the piano mechanism animation where depression of the key at the bottom right of the display results almost immediately in all the mechanism's other constituent entities performing their own different movements in concert at widely dispersed locations. Although this multiplicity of simultaneous or near simultaneous movements is an accurate representation of how a piano mechanism works, it is poorly suited to how human information processing works. The situation is exacerbated for learners who are novices with respect to the topic addressed in the animation. Because they lack the domain specific background knowledge that would enable them to process the animation's information in a meaningful top-down fashion, their processing is primarily driven by its perceptual attributes (cf. Hegarty & Kriz, 2008; Kriz & Hegarty, 2007).

During visual perception, our eyes first perform a rapid characterization of the display in broad terms (e.g. shapes, etc.) then pick up more detailed information in a stepwise fashion by means of a series of foveal fixations that are made on relatively small individual areas (i.e., first holistic then analytical processing). Each of these fixations is separated from the next by a saccade in which the eye jumps between successive locations. Because the area covered by any single fixation is quite modest, the information we extract during our viewing of an animation is necessarily piecemeal and cumulative. In other words, we sample the displayed information serially rather than processing it all at once in a parallel fashion. This can sometimes be very challenging, especially when many different things are happening simultaneously across an animated display, which is typically the case when a conventionally designed animation depicts complex subject matter (cf. Ploetzner & Breyer, 2017, this volume). Extensive simultaneity forces viewers to use their limited visual attention selectively and allocate it as best they can amongst various aspects of the display. Fixating and shifting attention within a display take a small but finite time. The period between fixations is too long to allow a learner to visually interrogate the whole areas of an animated display before changes take place in the portrayal (keeping in mind that animations must present many frames per second in order to maintain the illusion of continuous dynamic change). For example, if learners focus their attention on the head of the piano's hammer, they cannot at the same time pick up detailed information about what the jack is doing. Eye tracking studies in which learners are required to study the piano mechanism animation have shown that in total, they fixate on only a small fraction of the presented information (Boucheix & Lowe, 2010; Lowe & Boucheix, 2011). Comprehensive animations such as this that have extensive simultaneity clearly present far more information than a learner can fully process in the time available. Further, learners likely lack strategies for exploring the animation in an efficient manner.

The fixations that learners make when viewing a complex, unfamiliar animation are not only restricted to a subset of the available information, but are also unevenly distributed (Boucheix, Lowe, & Soirat, 2006). This would not be of concern if learners devoted most of their attention to information that is most relevant to building a high quality mental model. However, what actually happens tends to be just the opposite. Because of limitations on human information processing capacity, the myriad aspects of a complex animated display compete for the learner's attention. When the information presented in an animation is unfamiliar to the viewer so that its processing is largely perceptual, conspicuous aspects receive far more visual attention than others that are relatively inconspicuous. Learners not only extract less information from the animation than they should, but the information they do extract is likely to be sub-optimal as raw material for mental model construction. With the piano animation example, research has revealed that learners typically neglected crucial information about the dynamics of the inconspicuous jack and how its actions are intimately related to the operation of other parts of the mechanism (Lowe & Boucheix, 2011). Most attention was instead devoted to far more conspicuous entities such as the hammer. This failure to extract information about the jack, an item that is high in thematic relevance but low in perceptual salience, severely compromised the quality of the mental model learners were ultimately able to build as a result of viewing the animation.

Because the defining feature that distinguishes animated from static graphics is their change over time, the information animations present is necessarily transitory. Given the rich array that complex animations present to learners, this transience can severely restrict the extent to which detailed analysis of the presented information is possible. However, it is only by analyzing this information very carefully that some of the more subtle – but nevertheless important – relationships that are the foundation for understanding can be detected. If learners try to distribute their attention relatively evenly across the animation, the information that they are able to extract is likely to be relatively superficial. If they instead try to undertake a more detailed analysis of one particular aspect of the animation, the penalty will be that other temporally coincident aspects situated elsewhere in the display are inevitably neglected. Both scenarios seriously prejudice learners' extraction of the type of raw material that is essential for constructing a high quality mental model. In the first case, superficial processing means that visually subtle but operationally crucial information can be missed. With the piano example, if details about the sophisticated shaping and construction of entities such as the jack and the hammer are not extracted, learners' capacity to represent the functional relations within the mechanism will be compromised. In the second case, neglecting much of the animation's information in order to characterize just one aspect in detail will effectively discard raw material and limit the scope of the mental model that it is possible for the learner to build. With the piano example, if extra attention was devoted to components in the hammer's causal chain at the expense of those components in the damper's causal chain, the mental model constructed as a result would be seriously deficient.

A further challenge for learner processing comes from the inevitable overwriting in a learner's short term memory of information presented earlier in the animation's time course (Lowe, 1999). This is a result of short term memory being the way station through which all incoming information must pass before being incorporated into long term memory. Short term memory's very limited capacity means that as fresh information is internalized by the learner, its previous contents are necessarily replaced with what is being newly acquired. Because animation is a transient form of representation, information presented prior to what is currently being portrayed is no longer externally available to the learner for purposes such as the comparison processes required to establish broad-scale temporal relationships. For example, if learners wish to compare the initial Strike episode of the piano's operation with its final Reset episode, they must contend with the fact that the intervening Rebound episode will have overwritten information about the Strike episode. This contrasts markedly with the situation for static graphics that depict dynamic subject matter. As a persistent form of external representation, static graphics give learners the opportunity to make extended, repeated and detailed comparisons of information that in reality would be widely separated in time. The lack of such opportunities with animated graphics – due to information overwriting – highlights the disconnect that exists between how animations present their subject matter and how human learners process information. However, the processing benefits associated with the persistence of static graphics come at the cost of sacrificing dynamic continuity and the consequent requirement for learners to make complex inferences about fine-grained temporal relationships.

Poor sequencing in the extraction of information from a complex animation is another tendency exhibited by learners who are novices regarding the depicted topic (Lowe, 2008). Again, this appears to be because their extraction of information is largely driven by its perceptual characteristics rather than by its relevance to the task of building a satisfactory mental model (cf. Wagner & Schnotz, 2017, this volume). In effect, the order in which learners extract different subsets of the available information depends on the relative perceptibility of those subsets. Information that stands out most from the rest of the display will be extracted first while other aspects will be extracted later in order of perceptibility (Fischer, Lowe, & Schwan, 2008). However, such sequencing is unlikely to be optimal in terms of mental model construction. It would be more efficient and effective if extraction occurred in a sequence that maximized opportunities for building meaningful relationships between subsets of information (as opposed to sequencing based merely on relative conspicuity). Effective mental modeling of complex content typically requires the construction of hierarchical structures in which knowledge is represented and interconnected at a variety of levels (Ploetzner & Lowe, 2014, 2017, this volume). A learner's development of such an internal representation is likely to be facilitated if the informational building blocks from which it is constructed are acquired in an order that fosters formation of the required multi-level relationships. This is unlikely if perceptual salience is the main criteria for sequencing acquisition of information.

As noted above, when processing of an animation is dominated by its perceptual characteristics, the relationships a learner extracts during viewing will likely be perceptually based. However, a high quality mental model cannot be built on the basis of such low level descriptive connections. Instead, linkages between entities must be represented in terms of the contributions they make with respect to the specific referent subject matter that is portrayed in the animation. For example, the mental model of a piano mechanism must ultimately represent the functional relationships that determine how the piano does its job. Such relationships are fundamental to the chains of interactions between components that connect the pianist's playing actions to the musical result produced by the piano mechanism. If learners are not able to progress from a descriptive to an explanatory characterization of the depicted relationships, they will be unable to construct a high quality mental model.

The hierarchical relational structure that is characteristic of complex subject matter is well exemplified by the piano mechanism. Local functional relationships between immediately adjacent parts (such as the key and the whippen) contribute to the overarching functions in the piano's operation (specifically, the coordinated actions of the hammer and damper). For a mental model to be of high quality, it needs to incorporate such hierarchical relationships and that in turn would require learners to extract multilevel information while viewing the animation. However, this is an unreasonable expectation under the circumstances that prevail in a comprehensive animation.

It is clear from the preceding discussion that comprehensive animations can pose a variety of processing challenges to learners. These challenges help to explain why learners so often fail to reap the anticipated educational benefits from conventionally designed animations. We contend that the shortcomings of comprehensive animations as tools for learning have their origins in the basis of their design and how their design features interact with characteristics of the human information processing system. Of particular concern is that the prevailing design approach is not informed by an understanding of how learners actually process animations. The next section offers a theoretical account of such processing that could inform an alternative and potentially more effective approach to animation design that draws on research-based principles rather than intuition and opinion.

1.4 Animation Processing and Learning

The Animation Processing Model (APM; Lowe & Boucheix, 2008a) addresses learning from conventional comprehensive animations, particularly those that present complex, unfamiliar subject matter. It characterizes such learning in terms of a

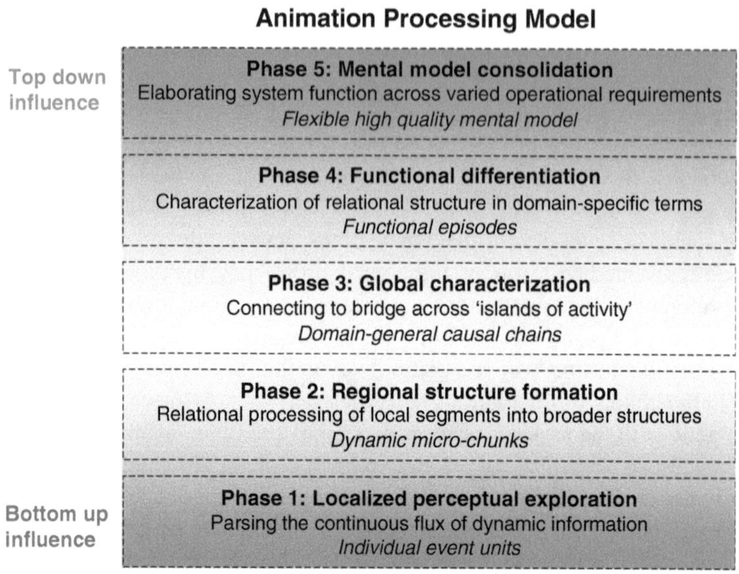

Fig. 1.2 The Animation Processing Model summary diagram

set of five interrelated processing phases that need to be carried out in order to construct a mental model from such a presentation (cf. Fig. 1.2). It assumes that the activities occurring during this processing are iterative and cumulative but not necessarily strictly sequential.

For the resulting mental model to be of high quality (i.e., for effective learning to have occurred), all these phases must be executed satisfactorily. The interacting bottom-up and top-down processes posited by the APM can be broadly characterized as being of two types: decomposition and composition. In simple terms, decomposition is analytical processing by which the learner breaks down the information presented in the animation whereas composition is synthetic processing by which the learner progressively assembles the products of such decomposition into the higher order knowledge structures that constitute a mental model. The main concern in the present chapter is with the challenges posed by decomposition because they appear to be at the heart of fundamental problems that learners can have in processing animations effectively. However, some aspects of composition will also be discussed because tackling decomposition alone will not ensure the quality of a learner's mental model.

APM Phase 1 processing is concerned with parsing the presented information and decomposing it into a form that the human information processing system can handle within its limited capacity. To perform this decomposition, learners must extract suitable *event units* from the animation as they study it (where an event unit is an entity plus its associated behavior). If learning from an animation is to be successful, the event units that a learner extracts must be relevant to the central theme of the learning task at hand. This is because they are the raw material (i.e., the basic building blocks) from which a mental model is constructed. If learners fail to extract these crucial high thematic relevance event units but instead mainly internalize less relevant information, the requisite raw material for high quality mental model construction will be lacking and learning will be compromised. As will now be explained, this compromised scenario is all too likely for comprehensive animations of the type being considered here (i.e., those presenting complex, unfamiliar subject matter).

According to the APM, a learner's initial decomposition of an animation is very much a bottom-up activity and so is based largely on the perceptual attributes of the display. This means that in selectively attending to the displayed information, learners tend to notice entities with visuospatial and temporal characteristics (i.e., appearance and behavior) that give them a high perceptual salience. As a consequence, there is a corresponding neglect of aspects with lower perceptual salience. In the context of learning from animated graphics, the behavior of entities is of particular note because the human perceptual system privileges dynamic information that contrasts with its surroundings. Entities having a high level of dynamic contrast are likely to capture the viewer's attention, irrespective of their thematic relevance. This results in information extraction being skewed towards aspects of the display that are most noticeable but not necessarily most relevant, a situation that provides a poor foundation for further compositional processing.

APM Phase 2 processing is a primary compositional activity founded on the event units extracted in Phase 1 as described above. It involves the grouping together of two or more adjacent event units into highly localized clusters that are perceived as being related to each other in some way (such as according to Gestalt principles). These regional groupings consisting of just a few event units are referred to as *dynamic micro-chunks*, a term which indicates their temporal character, their limited scope, and their cohesive nature. The relationships that bind event units into dynamic micro-chunks tend to be bottom-up rather than top-down unless the learner can recruit pre-existing background knowledge about the depicted topic. Once the learner has generated sufficient dynamic micro-chunks, it is possible for Phase 3 processing to begin in which these local chunks are linked together into larger assemblies that bridge more widely spaced regions. Although bottom-up relationships could still be involved in these linkages, Phase 3 would typically be expected to introduce some measure of top-down processing. One important outcome of this progressive linking activity is the formation of chains of interactions that connect causes to effects. However, at this stage, these causal chains may be essentially domain general in nature and not necessarily interpreted in terms of the overall functionality of the specific referent subject matter depicted by the animation. Nevertheless, when carried out effectively, Phase 3 processing ultimately results in a relatively exhaustive characterization of the network of relationships that can be discerned amongst the animation's component entities. As the various causal chains are identified, Phase Four processing may ensue in which domain-specific background knowledge – if available – can be invoked to set this network of relationships in the context of the particular functioning represented by the animation. Phase 5 processing extends this functionality to situations beyond those explicitly depicted in the animation and results in a powerful mental model that has the flexibility to be applied across a variety of scenarios.

The present chapter focuses mainly on APM Phases 1 and 2, and to a lesser extent on some aspects of Phase 3 processing. These are the phases in which bottom-up and domain-general processing generally play the more dominant role in how a learner deals with an animation. They are also crucial in laying the foundation for Phases 4 and 5 and so can have a profound effect on the quality of the mental model that a learner is able to construct from the study of an animation. Phase 1 is essentially about *decomposition* of the information flow that learners must deal with in a comprehensive animation. In contrast, the other two phases are more concerned with *composition* activities in which the event units extracted as a result of decomposition are progressively condensed into more and more inclusive relational structures. However, we have noted that successful processing of a comprehensive animation can be effectively derailed in Phase 1 because learners who are novices in the depicted domain are ill-equipped to extract the required high relevance event units. Even if they have some limited success in internalizing such event units, they face a further hurdle in combining them via Phase 2 and 3 processing into the higher order structures needed to build a high quality mental model.

In the next section, we examine several key types of intervention that researchers have investigated as possibilities for improving learning from animation.

1.5 Helping Learners to Cope: Conventional Approaches

With few exceptions (e.g., Boucheix, Lowe, & Bugaiska, 2015; Boucheix, Lowe, Putri, & Groff, 2013; Lowe & Boucheix 2011; Lowe & Mason, 2017, this volume; Mason, Lowe, & Tornatora, 2013; Ploetzner & Lowe, 2014, 2017, this volume), the interventions researchers have investigated thus far as possibilities for improving learning from animations have typically not been based on a detailed model of how learners process animations. Nevertheless, the general aim of interventions such as visual cueing, user control and segmentation has been to support more effective learner processing. These three examples will now be examined to illustrate the limitations of interventions that are applied to comprehensive animations without questioning how such animations were designed in the first place.

As noted above, there have been attempts to correct misdirection of attention due to salience-relevance mismatches by adding the type of color cues found to be effective in static graphics (cf. Van Gog, 2014). However, such cueing has been of little or no benefit in an animated context because the influence of the animation's *dynamic* contrast on attention allocation appears to be substantially more powerful than that of the *visuospatial* (e.g., color) contrast (Lowe & Boucheix, 2011). Modifying traditional color cues by making them dynamic (in order to better compete for the learners' attention) produces some additional benefit regarding learner extraction of thematically relevant information that would otherwise have a low perceptual salience (see De Koning & Jarodzka, 2017, this volume) but has little effect on ultimate mental model quality (Boucheix & Lowe, 2010). Even with such added support, it seems that the outcomes of APM Phase 1 processing are unsatisfactory in terms of providing the raw material learners need to build high quality mental models. Proper decomposition of comprehensive animations depicting complex unfamiliar subject matter remains a largely intractable problem for learners.

For some time, both educational practitioners and researchers have considered user control as likely to benefit learning from animations. User control can offer learners considerable flexibility in how animations may be interrogated during their study. Characteristics such as the animation's speed, direction and continuity can be altered, traditionally by means of video-like controls situated below the display area (but increasingly via the more direct forms of interaction that are available on tablet computers). Potential advantages posited for user control of animation range from reducing learners' cognitive load to permitting more intensive, recursive processing of the available information (Mayer & Chandler, 2001; Schwan & Riempp, 2004). However, empirical studies suggest that adding user control to a comprehensive animation sometimes fails to confer the anticipated benefits. User control not only requires learners to redirect some of their finite processing resources to control activities and hence away from the central learning task, it also carries the implicit assumption that the target learners will be able to interrogate the animation productively. Research suggests that the extra requirement of exercising user control can have a negative influence on learning from animation (Bétrancourt & Réalini, 2005; Boucheix, 2008). Further, the interrogations undertaken by learners who are novices with respect to the depicted subject matter tend to be deficient with respect to where they look in an animation, when they look, and what they look at (Lowe, 2008).

An intervention that is somewhat more constraining than adding open-ended user control is for the original animation to be broken into segments before being presented to learners. This typically involves slicing the time course of the animation into shorter pieces that are often described as 'meaningful' units. Each of these segments might be one of the individual episodes that together constitute the whole animation. By providing pauses between successive segments and a 'Continue' button, segmentation reduces the amount of information that has to be dealt with on a continuing basis. Segmentation is supposed to benefit learning from animation in two ways (Spanjers, Van Gog, & Van Merrienboer, 2010; Spanjers et al. 2011; Wong, Leahy, Marcus, & Sweller, 2012). Firstly, reducing the amount of information to be dealt at one go by segmenting the animation and inserting inter-segment pauses allows learners to manage processing of essential information in a way that helps avoid cognitive overload by giving them more processing time. Secondly, the subdivision of the animation into temporally separated 'meaningful' units gives learners cues about the underlying structure of the referent subject matter. However, merely cutting an animation into short sections as described above does not alter the *within*-segment complexity of the representation. This can be illustrated by the piano animation example. Although it would be reasonable to segment the piano's operation on the basis of its three episodes (Strike, Rebound, and Reset), this does nothing to ameliorate the processing challenges that learners encounter during the presentation of an individual stage. Each of the segments still confronts learners with a comprehensive representation of the subject matter dynamics, with all the attendant demands that were identified earlier. A further issue is that reports of research into the effect of segmentation on learning from animation tend to lack a precise definition of how the meaningfulness of a unit of segmentation is determined. In some cases, it almost seems to be assumed that simply chopping off small pieces of an animation will make them more meaningful because they are more manageable in working memory. Typically, the breaking of the animation into segments appears to be done on the basis of experimenter opinion rather than according to a more formal and principled analytical approach. With complex animations that are not mere strings of successive events, segmenting the animation in this way runs the risk of destroying the continuity of events and so compromising internalization of the overarching temporal relationships that are integral to a coherent mental model of the referent subject matter. This is especially problematic if the animation contains multiple events that occur in parallel or have extensive temporal overlap. For example, choosing an animation segment according to the boundaries of one specific event out of a set of staggered overlapping events inevitably cuts across other events in that set and so disrupts their continuity. Such disruption is inconsistent with the notion of segmenting animations into meaningful units.

None of the interventions discussed above involve changes in the assumptions underlying the animation's original design. In all cases, the animations remain comprehensive in how they depict the subject matter's dynamics. In these types of intervention, the user or the designer merely introduces some modifications to the pre-existing presentation regime but does not directly challenge its core design premises. Our contention is that the dominant current approach to animation design is at odds with some key characteristics of human information processing. The corollary is that no amount of superficial manipulation of comprehensive animations with respect to how they present the referent subject matter will produce worthwhile improvements in learning, especially when the depicted topic is complex and unfamiliar. Instead, a root-and-branch change is needed. The next section presents an alternative approach to designing animated graphics that is intended to circumvent some major disadvantages of conventional comprehensive animations with respect to learner processing. We have termed this a *Composition Approach* to designing animations to signal a fundamental change in thinking about the relationship between an animated graphic and the learners for whom it is intended. The Composition Approach is intended to re-orient the viewpoint of designers, from their traditional attitudes to the attributes of the *external* representation (the animation) to a deeper consideration of the *internal* representation (mental model) that the learner needs to compose when processing an animation.

1.6 A Composition Approach to Animation Design

We have noted that APM Phase 1 processing tends to be a major stumbling block for learners. Even when learners are supported by dynamic cues that attempt to strongly direct attention to the most relevant aspects of the display, the quality of the mental models they compose is relatively low. The interventions discussed in the previous section essentially side step a major issue – that learners are simply ill-equipped to carry out the proper decomposition of comprehensive animations. This invites the question: if decomposition is such an intractable impediment for learners, would it be possible to design educationally effective animations that helps relieve learners of this burden?

If the five phases of the APM are considered collectively, all but the first phase are concerned with the progressive composition of event units into the increasingly interrelated and hierarchically organized knowledge structures that ultimately result in a mental model. Phase 1 processing can therefore be regarded as essentially a preliminary enabling activity whose purpose is to provide the raw material for this later and more central composition activity. The Composition Approach (Lowe & Boucheix, 2012a) accepts the reality of learners' failures to extract appropriate event units from a comprehensive animation and aims to reduce that processing burden. This reduction allows some of the learner's scarce internal processing resources to be released for use in activities that are more directly concerned with effective learning. As its name indicates, the main focus of the Composition Approach is on helping learners to compose a high quality mental model. Instead of expecting learners to decompose a conventional comprehensive animation, it supplies them with parcels of dynamic information that are consistent with the results of an 'ideal' decomposition. The overall concept is that learners are furnished with a 'kit of parts', each part of which is a small animated assembly comprising two or possibly more interconnected event units. We term these assemblies (i.e., parts making up the 'kit') *relation sets* to indicate both that they are held together by *intra* event unit relations and that they are capable of being linked to other relation sets via *inter* event unit relations. Supplying a series of relation sets to learners allows them to build simple, individual events progressively into the complex, integrated network of temporal chains and functional relationships that a high quality mental model requires. This is consistent with simple-to-complex sequencing recommendations made in the broader instructional design literature (e.g., Van Patten, Chao, & Reigeluth, 1986). In contrast with static depictions of dynamic phenomena, relation sets retain temporal continuity of the subject matter and so reduce the need for learners to make complex inferences about behavior. This is a particular advantage for learners with low prior knowledge of the subject matter who are ill equipped to make such inferences.

Figure 1.3 shows three successive snapshots from such a relation set that covers one aspect of the piano mechanism's operation (the actual relation set is a continuous animation). Two types of event units comprise this relation set – one representing the activity of the piano key and the other representing the consequent activity

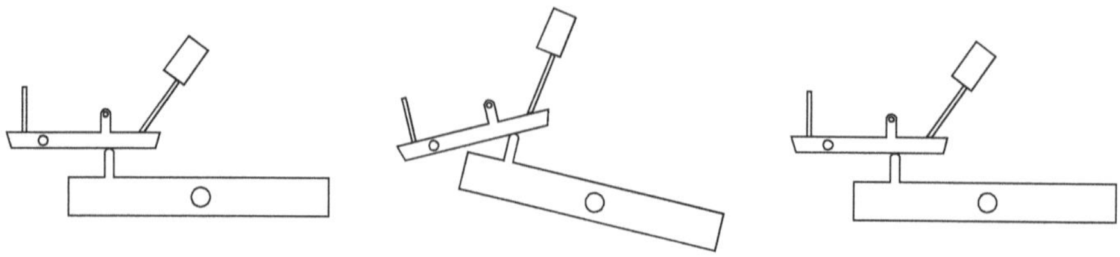

Fig. 1.3 Snapshots from an example relation set (key-whippen). When the right end of the piano key is depressed, its left end pushes on the whippen so that it tilts. These actions then reverse

of the whippen that is in contact with the key's riser. This key-whippen relation set shows the start of a causal chain (the 'cause') that ultimately ends in the hammer striking the piano string to sound a musical note (the 'effect'). An example Composition Approach animation for the piano mechanism would encompass the complete causal chain by first presenting the key-whippen relation set and then, by replacement, presenting successive pairs of event units (whippen-jack, jack-hammer, hammer-string) as relation sets that progressively work along the causal chain's path. Once this succession of relation sets comprising the strike stage of the piano mechanism had finished, the rebound then reset phases would be presented in a similar manner using appropriate relation sets. An analogous progressive approach would then be used to present the relation sets involved in the damper's operation during the strike, rebound and reset stages of the mechanism's functioning. The net effect of this approach is to take behaviors that in reality occur simultaneously or in very rapid succession and spread those actions out over a much wider time span. As a result, the amount of simultaneity or near simultaneity is greatly reduced so that fuller attention can be devoted to each aspect of the subject matter. The slavish adherence to behavioral realism that characterizes comprehensive animation designs is relinquished and instead the presentation regime is manipulated so that it is better matched with the capacities of the human information processing system. There is a considerable reduction both in the amount of information that the learner must process per unit time and in the spatial dispersal of event units that compete for the learner's attention.

The composition approach takes account of various aspects of the APM with the overarching goal of helping learners obtain the raw material they need to compose a high quality mental model (i.e., one that is appropriate, accurate and sufficiently comprehensive). At a broad level, its incremental presentation of the referent subject matter is consistent with the iterative, cumulative character of learners' animation processing. More specifically, relation sets are designed to accord with the way learners actually extract and internalize information from animations. The aim is to make the properties of relation sets consistent with those of the dynamic micro chunks identified by the APM as central to Phase 2 processing. Not only are stringent limits placed on the size of relation sets (in the piano example, just two event units), but they must also be made up of adjacent entities that directly interact in ways that contribute to the target system's functionality. The limited number of event units presented at any one time dramatically reduces competition for the learner's attention compared with what would be the case in a comprehensive animation. This means that low salience, high relevance aspects (such as the behavior of the piano's jack) are much less likely to be neglected. Further, the requirement that entities be adjacent makes it easier for foveal vision to extract detailed information about their characteristics and interactions than would be the case if they were widely separated. Each individual relation set therefore not only provides information about its component event units (i.e., entities and their associated behaviors), but also shows the dynamic relationship between those event units (e.g., how changes in the inclination of the piano key affect the orientation of the whippen).

However, the establishment of these *intra*-set relations is only part of the story. According to the APM, effective mental model building requires learners to progressively condense such local dynamic structures into larger and larger hierarchically-structured assemblages. The further mental linkages needed for this aspect of composition are fostered by providing opportunities for individual relation sets to be interconnected via *inter*-set relationships. In the piano mechanism example this is done by having successively presented relation sets share a common event unit. For example, the key-whippen relation set is followed by a whippen-jack relation set (the animation cross-fades from one to the next). The presence of the common whippen event unit is intended to help the learner compose a more inclusive key-whippen-jack dynamic structure. This shared event unit approach is continued along the entire length of each causal chain. The net result is that learners have the opportunity to extract information that is more accurate and comprehensive than would otherwise be the case.

Relation sets are central to the Composition Approach. They must be carefully tailored to ensure that the mental model constructed from them is appropriate for targeted learning task. For this reason, the attributes of relation sets (such as their size and constitution) cannot be based on designer intuitions or the opinions of subject matter experts. Rather, relation sets are devised to target explicit learning outcomes on the basis of a detailed, systematic analysis of the dynamic subject matter plus a careful consideration of the proclivities and constraints of the human information processing system. Various sources of knowledge are recruited in this process. It begins with a time-based characterization of all the event units that are to be represented in the portrayal. Figure 1.4 shows such an event unit analysis for the piano mechanism example. Each row specifies one of the entities that make up the depicted system and the various behaviors it exhibits over the time course of the system's operation are shown from left to right. The result is a detailed mapping of all the constituent event units that specifies their relative durations and temporal distributions. This mapping can reveal potential sources of processing problems that learners would encounter if the subject matter was presented as a comprehensive animation. Detecting such issues involves examining the event unit analysis to identify circumstances where there are likely to be significant challenges to learner processing. For example, it can be seen from the left-hand section of Fig. 1.4 that during the strike stage, many and varied simultaneous and cascading event units would compete for the learner's attention. A similar situation regarding the overlapping of events is also revealed for the final reset stage (essentially, the strike stage in reverse).

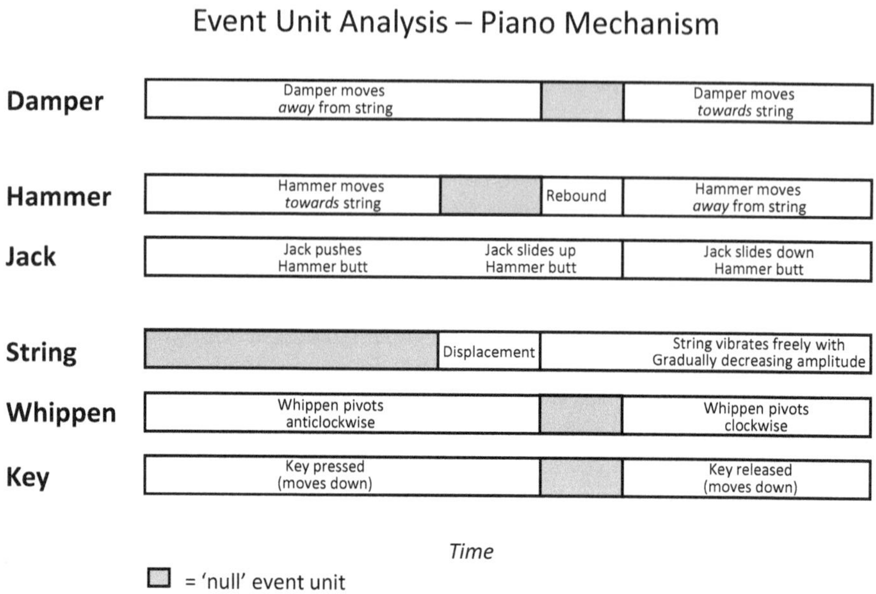

Fig. 1.4 Partial event unit analysis for operation of the piano mechanism (from when the key is pressed until its release). Note (i) the extensive simultaneity, and (ii) multiple actions of jack

By inspecting the nature of these competing event units (e.g., the size and shape of the entities and their relative movements), it can be deduced that domain novices would likely neglect the jack due to the far greater perceptibility of the entities that surround it. As already mentioned, this is indeed what occurs when learners are given a comprehensive animation of the piano mechanism. An event unit analysis combined with other sources of knowledge about the subject matter and human information processing are used as the basis for designing relation sets that avoid these problems by presenting information in a highly constrained and systematic fashion. This reduces learners' decomposition burden and allows them to optimize allocation of their limited information processing resources to the central task of composing a high quality mental model.

1.7 Relation Sets and Sequencing

This chapter identified learners' failure to extract the required event units as a fundamental problem that can occur in their processing of comprehensive animations. Deficiencies in this basic raw material would undoubtedly prejudice mental model construction. However, a high quality mental model is far more than a collection of the necessary event units. These building blocks must be assembled into hierarchical structures to properly represent the rich multilevel relationships that are central to the referent's functioning. For this reason, the *sequencing* of relation sets is an integral part of the composition approach to designing educational animations. Relation sets need to be sequenced in a way that facilitates learners' formation of appropriate superordinate connections within the emerging representational structure. In the piano example of the Composition Approach given above, the order for presentation of relation sets followed the progress of each of the mechanism's two causal chains (starting with depression of the piano key as the cause, and ending with either the hammer's or the damper's interaction with the string as the effects). However, this particular form of sequencing is only one amongst a number of possibilities. For example, it would be feasible to reverse this order and 'work backwards' from the effects to their causes. Another possibility would be to start with the whippen because it is responsible for the coordinating the interrelated actions of the hammer and damper. In that case, relation sets from both the hammer and the damper causal chains could conceivably be progressively presented in concert to indicate the higher order relationship involved in their coordinated actions. Empirical studies are needed to compare the effectiveness of such alternatives.

Given that centrality of sequencing in the Composition Approach, it is important to consider what might be the effect of providing learners with a full complement of relation sets but not presenting them in an appropriate order. This could be done by using a random presentation sequence that lacked any meaningful superordinate form of organization (such as a causal chain). The fact that the presentation consisted of small relation sets (rather than a full comprehensive animation) would expected to substantially reduce the amount of information the learner would have to deal with at any one time. According to a simple cognitive load based explanation (Ayres & Pass, 2007), such a reduction should result in better learning than would occur from a corresponding comprehensive animation. However, this is not what would be predicted from the perspective of the APM because of the importance it attaches to the role of establishing inter-event unit relationships in constructing mental models.

1.8 Composition Versus Comprehensive Design

Empirical evidence concerning the effectiveness of the Composition Approach comes from a recent study by Lowe and Boucheix (2016), part of which is reported below. Participants (university psychology students) were randomly assigned to study one of three animated versions depicting the piano mechanism's operation (n = 20 per condition). The Comprehensive version presented a conventionally designed behaviorally realistic portrayal of the mechanism in operation. Two other versions both used relation sets comprised of pairs of event units but differed in how those relation sets were sequenced and constituted. In the Contiguous animation, the relation sets were sequenced according to the progress of actions through the key-hammer and key-damper causal chains and were made up of immediately adjacent (contacting) event units (cf. Fig. 1.5). This version therefore corresponded to the

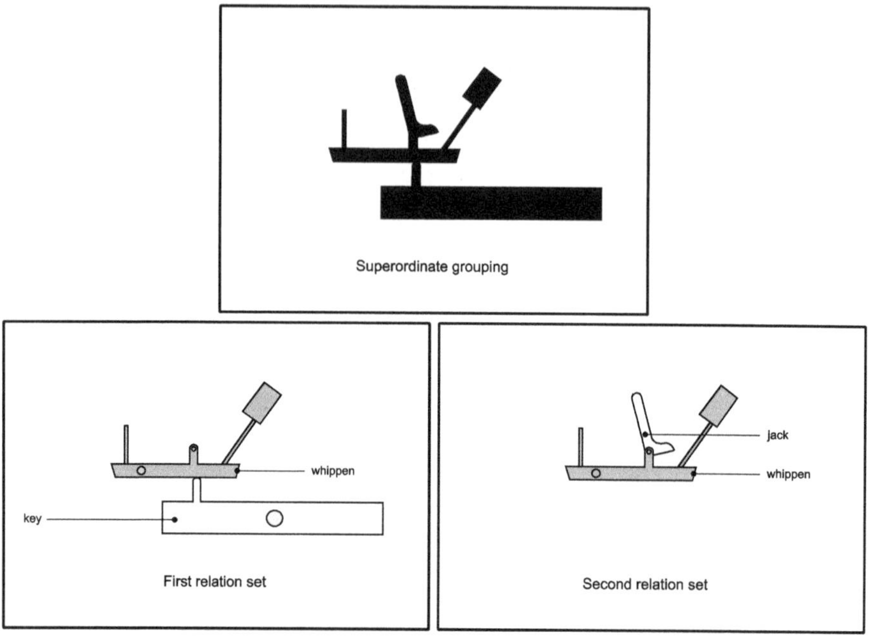

Fig. 1.5 Example Contiguous relation sets. Note (i) contact relationship between event units in each relation set (key-whippen; whippen-jack), (ii) adjacency of successive relation sets, (iii) sharing of event unit (whippen) between the two relation sets. Sequencing and constitution of relation sets facilitates composition of the hierarchical structures necessary for building high quality mental model

Composition Approach as detailed above. In the Non-Contiguous animation, the relation sets were sequenced in a quasi-random order with the proviso that the event units comprising the relation sets were not in contact (cf. Fig. 1.6). In both the Contiguous and Non-contiguous versions, there were cross-fades between successive relation sets. Before studying their allocated animated versions, participants in all three conditions were shown a static picture of the whole piano mechanism for 30 s.

In both the Contiguous and Non-Contiguous versions, event simultaneity was present but reduced to a practical minimum by having just two event units displayed together in each relation set. However, the Contiguous animation should have several potential processing advantages over its Non-Contiguous counterpart (cf. Figs. 1.5 and 1.6). First, the two sites of activity are closer together in the Contiguous version so fewer fixations and shorter transitions between fixations would be needed in order to characterize what was happening. More efficient interrogation of the presented information would therefore be possible. This difference is exemplified by the key-whippen relation set (Contiguous) compared with the key-hammer relation set (Non-Contiguous).

Second, the Contiguous version shows a direct relationship between each pair of event units that comprised a relation set due to their physical contact (e.g., the tail

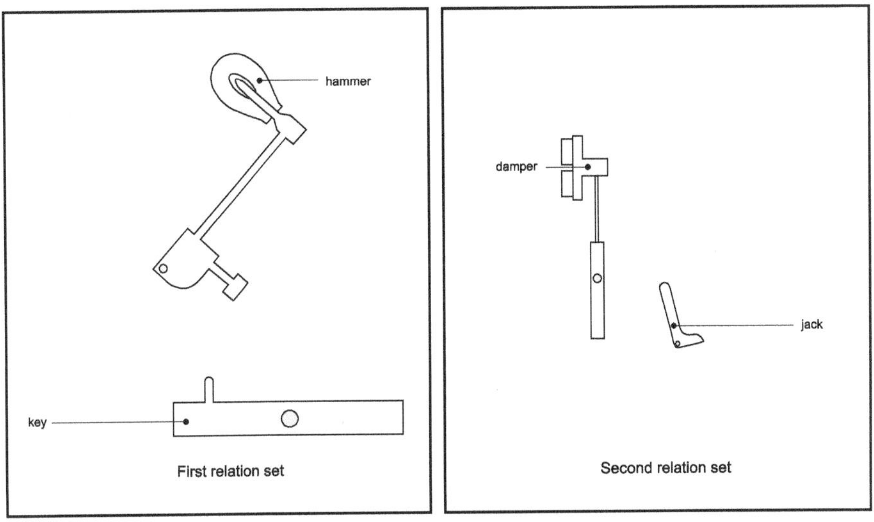

Fig. 1.6 Example Non-contiguous relation sets. Note (i) no contact relationships between event units in each relation set, (ii) separation of relation sets, (iii) no sharing of event units between two relation sets. Prejudicial to building higher order knowledge structures

of the key pushes on the whippen). Although relationships certainly exist between the event units comprising a Non-Contiguous relation set, they are indirect (e.g., there is clearly a relationship between the downwards movement of the key head and the sideways movement of the hammer but there is no direct contact between them). The relationships between event units in the Contiguous version could be understood in terms of domain-general interactions that are common in the everyday world and so did not rely on the learner having specialized background knowledge about piano functionality. This would not be the case with the non-contiguous version where there was no perceptual basis for making sense of the observed pattern of movements.

Third, the sequencing of the Contiguous version in way that specifically facilitates the making of inter-relation set connections would foster the formation of higher level information structures from lower level precursors. For example, in the Contiguous version, the key-whippen relation set could be linked to the whippen-jack relation set via their common event unit and so form a higher level key-whippen-jack structure with extended temporal continuity. Establishing such superordinate relationships would be considerably more difficult for a Non-Contiguous sequence in which the key-hammer relation set was followed by the damper-jack relation set.

There were three dependent measures. In the cross-task, participants were presented with a static picture of the original piano mechanism and required to manipulate a cross marked on each of the mechanism's entities in order to indicate the direction and extent of that entity's movement. For the second task, participants produced a written text that explained as fully as possible how the piano mechanism works. Outputs from this second task were used as a measure of mental model quality. The third measure was a transfer test involving a static picture of a novel historical piano mechanism that had major structural differences from the mechanism depicted in the original piano animation (despite being able to perform the same musical functions). This was also a cross-task where participants' manipulation of crosses marked indicated how they predicted each of the marked entities would move during operation of the novel mechanism.

For the original piano mechanism, the overall mean cross-task scores and standard deviations were 50.2 (16.9), 43.4 (12.3) and 45.4 (13.1) for the Contiguous, Non-Contiguous and Comprehensive conditions respectively. An ANOVA indicated that these scores were not significantly different ($F(2,57) = 1,22$, $p = 0.30$), which was as expected because the cross-task targeted only the highly localized behaviors of individual entities (i.e., single isolated movements). However, an ANOVA showed the mean mental model scores (percentages) of 43.2 (11.2), 27.5 (18.2) and 30.7 (15.9) for the Contiguous, Non-Contiguous and Comprehensive versions respectively did differ significantly ($F(2,57) = 5,81$, $p = 0.005$, $\eta p^2 = 0.17$). Post hoc comparisons revealed that participants in the Contiguous condition obtained significantly higher scores than those in the Non-Contiguous and Comprehensive conditions ($F(1,57) = 10.40$, $p = 0.002$, $d = 1.04$; $F(1,57) = 6.61$, $p = 0.012$, $d = .91$) but there was no significant difference between scores for those latter two conditions ($F(1, 57) = 0.42$, $p = 0.51$).

These results indicate not only that the Composition Approach to design produces animations that are far more effective than conventional comprehensive animations, but also that the constitution and progressive sequencing of relation sets play a crucial role in the benefits achieved. Simple regression analysis indicated a high correlation between mental model quality scores for the original piano animation and scores from the transfer cross-task for the novel piano mechanism ($R = 0.38$, $R^2 = 0.14$, $F(1,58) = 9,60$, $p = 0.003$). Being able to predict how entities in a novel piano mechanism are likely to move is a very stringent test of mental model quality because it shows that the participant's internal representation of the original mechanism contains coherent, hierarchically structured information that encompasses the overarching functional principles of the device. Without such information, it would be difficult to generate a further mental model for the novel mechanism that was capable of being run in order to make the required predictions.

This study showed that the design of an animation according to the Composition Approach helped learners to generate a mental model of far higher quality than was developed from a conventional comprehensive animation. These superior results are attributed to the advantages gained by adopting a principled approach to design that is founded on a research-based account of the processes involved in learning from animation. Implementation of the Composition Approach requires considerably more analysis of the referent subject matter's dynamics and how they are likely to affect learner processing than is typically the case for conventionally-designed educational animations.

1.9 Conclusion

The Composition Approach is a major departure from prevailing animation design orthodoxies in its rejection of the privileged status that behavioral realism has in conventional comprehensive animations. This departure is justified on the grounds that designing educational animations on the basis of intuition and experience results in dynamic visualizations that can be challenging for learners to process and prejudicial to building high quality mental models. The mismatches between design features of these animations and the realities of learner processing are particularly problematic when the subject matter is complex and unfamiliar to the target audience. Instead of remaining faithful to the dynamics of the referent subject matter, the Composition Approach deliberately manipulates how such information is presented to learners. This manipulation is informed by the Animation Processing Model and is particularly focused upon decreasing the need for learners to decompose an animation themselves. The staged presentation of carefully tailored relation sets greatly reduces the excessive simultaneity that occurs when comprehensive animations present myriad events as they would actually happen. This use of relation sets spreads information through time that would otherwise be too temporally concentrated for learners to process effectively (cf. Sanchez & Wiley, 2017, this volume). However, reduction in processing demands is only one side of the Composition Approach coin. The other is the systematic sequencing of those relation sets in order to facilitate progressive development of the hierarchical relationships that are required for a high quality mental model.

In the study referred to above, the mental model scores for those in the Composition Approach condition were nearly half as large again as scores for the comprehensive animation group. Although this result is encouraging, these Composition Approach scores were nevertheless less than 50% of the possible maximum and so are certainly not indicative of a particularly high quality mental model. However, it should be noted that this initial incarnation of the Composition Approach addresses only some stages of the APM (those that are mainly concerned with domain-general bottom-up processes). It does not tackle the more top-down aspects of the APM (Phases 4 and 5) where domains-specific background knowledge is so important. Future research could investigate whether an expansion of the approach to include these aspects would result in further improvements in mental model quality. Another aspect that deserves research attention is the extent to which the results obtained for the piano mechanism would generalize to other types of subject matter. In principle, it appears that relation sets can be devised equally well for other types of content ranging from physical to biological systems (Lowe & Boucheix, 2012b, 2013). It remains to be seen if implementation of a Composition Approach using such relation sets produces similar improvements in mental model quality. It is also possible that as a result of having worked with animations designed according to the Composition Approach, learners may be better able to handle conventional comprehensive animations. This is because experience in building mental models from relation sets could prepare them for carrying out decomposition of comprehensive animations more effectively than they otherwise would. Rather than merely following along with the overall action of a complex animation in an holistic sense (cf. Paik & Schraw, 2013), they may adopt a more analytical approach to the information being presented.

The gains in mental model quality that result from using a Composition Approach are substantially greater than those generally achieved when interventions such as visual cueing are applied to comprehensive animations (De Koning, Tabbers, Rikers, & Paas, 2007; Van Gog, 2014). Keeping in mind that the main use for visual cues in such animations is to assist learners in decomposing the continuous flux of dynamic information they encounter, the use of the Composition Approach opens up new and perhaps more productive possibilities for interventions such as cueing. For example, instead of using visual cues to direct attention to low salience, high relevance aspects of a comprehensive animation, they could be used to help learners appreciate higher level functional relationships (cf., Lowe & Boucheix, 2008b). In this way, the application of visual cues could assist with the primary activity of composition rather than with the peripheral and very much second order issue of mere decomposition. For example, cues could be used to foster continuity between different but related causal chains, such as those involved in the actions of the hammer and the damper in the piano mechanism. Continuity is vital for conveying the high level relationships that link otherwise disparate individual events and episodes together into the coherent overarching knowledge structures we need acquire in order to make sense of this behavior. The possibility of using visual cues and other interventions within an animation designed according to the composition approach warrants empirical investigation, particularly with regard to its potential for further increasing mental model quality by addressing the types of processing involved in APM Phases 4 and 5.

References

Ayres, P., Marcus, N., Chan, C., & Qian, N. (2009). Learning hand manipulative tasks: When instructional animations are superior to equivalent static representations. *Computers in Human Behavior, 25*, 348–353.

Ayres, P., & Pass, F. (2007). Making instructional animations more effective: A cognitive load approach. *Applied Cognitive Psychology, 21*, 695–700.

Bétrancourt, M., & Réalini, N. (2005, June). *Faut il vraiment laisser le déroulement d'une animation sous le contrôle de l'apprenant?* Paper presented at the 11th Journées d'Etude sur le Traitement Cognitif des Systèmes d'Information Complexes (JETCSIC), Nice.

Boucheix, J.-M. (2008). Young learners' control of technical animations. In R. K. Lowe & W. Schnotz (Eds.), *Learning with animation: Research implications for design* (pp. 208–234). New York: Cambridge University Press.

Boucheix, J.-M., & Lowe, R. K. (2010). An eye tracking comparison of external pointing cues and internal continuous cues in learning with complex animations. *Learning and Instruction, 20*, 123–135.

Boucheix, J.-M., Lowe, R. K., & Bugaiska, A. (2015). Age differences in learning from instructional animations. *Applied Cognitive Psychology, 29*, 524–535.

Boucheix, J.-M., Lowe, R. K., Putri, D. K., & Groff, J. (2013). Cueing animations: Dynamic signaling aids information extraction and comprehension. *Learning and Instruction, 25*, 71–84.

Boucheix, J.-M., Lowe, R. K., & Soirat, A. (2006, August). *Online processing of a complex technical animation: Eye tracking investigation during verbal description*. Paper presented at the comprehension of text and graphics conference, University of Nottingham.

De Koning, B. B., & Jarodzka, H. (2017). Attention guidance strategies for supporting learning from dynamic visualizations. In R. Lowe & R. Ploetzner (Eds.), *Learning from dynamic visualization – Innovations in research and application*. Berlin: Springer (this volume).

De Koning, B. B., Tabbers, H. K., Rikers, R. M. J. P., & Paas, F. (2007). Attention cueing as a means to enhance learning from an animation. *Applied Cognitive Psychology, 21*, 731–746.

De Koning, B. B., Tabbers, H. K., Rikers, R. M. J. P., & Paas, F. (2009). Towards a framework for attention cueing in instructional animations: Guidelines for research and design. *Educational Psychology Review, 21*, 113–140.

Fischer, S., Lowe, R. K., & Schwan, S. (2008). Effects of presentation speed of a dynamic visualization on the understanding of a mechanical system. *Applied Cognitive Psychology, 22*, 1126–1141.

Hegarty, M., & Kriz, S. (2008). Effects of knowledge and spatial ability on learning from animation. In R. K. Lowe & W. Schnotz (Eds.), *Learning with animation: Research implications for design* (pp. 304–356). New York: Cambridge University Press.

Jenkinson, J. (2017). The role of craft-based knowledge in the design of dynamic visualizations. In R. Lowe & R. Ploetzner (Eds.), *Learning from dynamic visualization – Innovations in research and application*. Berlin: Springer (this volume).

Kriz, S., & Hegarty, M. (2007). Top-down and bottom-up influences on learning from animations. *International Journal of Human Computer Studies, 65*, 911–930.

Lowe, R. K. (1999). Extracting information from an animation during complex visual learning. *European Journal of Psychology of Education, 14*, 225–244.

Lowe, R. K. (2008). Learning from animation: Where to look, when to look. In R. K. Lowe & W. Schnotz (Eds.), *Learning with animation: Research implications for design* (pp. 49–68). New York: Cambridge University Press.

Lowe, R. K., & Boucheix, J.-M. (2008a). Learning from animated diagrams: How are mental models built? In G. Stapleton, J. Howse, & J. Lee (Eds.), *Diagrammatic representation and inference* (pp. 266–281). Berlin: Springer.

Lowe, R. K., & Boucheix, J.-M. (2008b). Supporting relational processing in complex animated diagrams. In G. Stapleton, J. Howse, & J. Lee (Eds.), *Diagrammatic representation and inference* (pp. 391–394). Berlin: Springer.

Lowe, R. K., & Boucheix, J.-M. (2011). Cueing complex animation: Does direction of attention foster learning processes? *Learning and Instruction, 21*, 650–663.

Lowe, R. K., & Boucheix, J.-M. (2012a). Dynamic diagrams: A composition alternative. In P. Cox, B. Plimmer, & P. Rogers (Eds.), *Diagrammatic representation and inference* (pp. 233–240). Berlin: Springer.

Lowe, R. K., & Boucheix, J.-M. (2012b). Addressing challenges of biological animations. In E. de Vries & K. Scheiter (Eds.), *Proceedings of the meeting of the EARLI special interest group on comprehension of text and graphics* (pp. 217–129). Grenoble: University of Grenoble.

Lowe, R. K., & Boucheix, J.-M. (2013, August). *Principled animation design: A key to improving learning*. Paper presented at the 15th Biennial EARLI conference for Research on Learning and Instruction, Munich.

Lowe, R. K., & Boucheix, J.-M. (2016). Principled animation design improves comprehension of complex dynamics. *Learning and Instruction, 45*, 72–84.

Lowe, R., Boucheix, J. M., & Fillisch, B. (2017). Demonstration tasks for assessment. In R. Lowe & R. Ploetzner (Eds.), *Learning from dynamic visualization – Innovations in research and application*. Berlin: Springer (this volume).

Lowe, R., & Mason, L. (2017). Self-generated drawing: A help or hindrance to learning from animation? In R. Lowe & R. Ploetzner (Eds.), *Learning from dynamic visualization – Innovations in research and application*. Berlin: Springer (this volume).

Lowe, R. K., & Schnotz, W. (2014). Animation principles in multimedia learning. In R. E. Mayer (Ed.), *The Cambridge handbook of multimedia learning* (2nd ed., pp. 513–546). New York: Cambridge University Press.
Mason, L., Lowe, R. K., & Tornatora, M. C. (2013). Self-generated drawings for supporting comprehension of a complex animation. *Contemporary Educational Psychology, 38*, 211–224.
Mayer, R. E., & Anderson, R. B. (1991). Animations need narrations: An experimental test of a dual coding hypothesis. *Journal of Educational Psychology, 83*, 484–490.
Mayer, R. E., & Anderson, R. B. (1992). The instructive animation: Helping students build connections between words and pictures in multimedia learning. *Journal of Educational Psychology, 84*, 444–452.
Mayer, R. E., & Chandler, P. (2001). When learning is just a click away: Does simple user interaction foster deeper understanding of multimedia messages? *Journal of Educational Psychology, 93*, 390–397.
Mayer, R. E., & Moreno, R. (2002). Animation as an aid to multimedia learning. *Educational Psychology Review, 14*, 87–99.
Paik, E. S., & Schraw, G. (2013). Learning with animation and illusions of understanding. *Journal of Educational Psychology, 105*, 449–457.
Ploetzner, R., & Breyer, B. (2017). Strategies for learning from animation with and without narration. In R. Lowe & R. Ploetzner (Eds.), *Learning from dynamic visualization – Innovations in research and application*. Berlin: Springer (this volume).
Ploetzner, R., & Lowe, R. K. (2012). A systematic characterization of expository animations. *Computers in Human Behavior, 28*, 781–794.
Ploetzner, R., & Lowe, R. K. (2014). Simultaneously presented animations facilitate the learning of higher-order relationships. *Computers in Human Behavior, 34*, 12–22.
Ploetzner, R., & Lowe, R. (2017). Looking across instead of back and forth – How the simultaneous presentation of multiple animation episodes facilitates learning. In R. Lowe & R. Ploetzner (Eds.), *Learning from dynamic visualization – Innovations in research and application*. Berlin: Springer (this volume).
Sanchez, C. A., & Wiley, J. (2017). Dynamic visuospatial ability and learning from dynamic visualizations. In R. Lowe & R. Ploetzner (Eds.), *Learning from dynamic visualization – Innovations in research and application*. Berlin: Springer (this volume).
Schnotz, W., & Lowe, R. K. (2008). A unified view of learning from animated and static graphics. In R. K. Lowe & W. Schnotz (Eds.), *Learning with animation: Research implications for design* (pp. 304–356). New York: Cambridge University Press.
Schwan, S., & Riempp, R. (2004). The cognitive benefits of interactive videos: Learning to tie nautical knots. *Learning and Instruction, 14*, 293–305.
Spanjers, I. A. E., Van Gog, T., & Van Merrienboer, J. J. G. (2010). A theoretical analysis of how segmentation of dynamic visualizations optimizes students' learning. *Educational Psychology Review, 22*, 411–423.
Spanjers, I. A. E., Wouters, P., van Gog, T., & van Merriënboer, J. J. G. (2011). An expertise reversal effect of segmentation in learning from animated worked out examples. *Computers in Human Behavior, 27*, 46–52.
Tversky, B., Heiser, J., Mackenzie, R., Lozano, S., & Morrison, J. (2008). Enriching animations. In R. K. Lowe & W. Schnotz (Eds.), *Learning with animation: Research implications for design* (pp. 263–285). New York: Cambridge University Press.
Van Gog, T. (2014). The signaling (or cueing) principle in multimedia learning. In R. E. Mayer (Ed.), *The Cambridge handbook of multimedia learning* (2nd ed., pp. 263–278). New York: Cambridge University Press.
Van Patten, J., Chao, C. I., & Reigeluth, C. M. (1986). A review of strategies for sequencing and synthesizing instruction. *Review of Educational Research, 56*, 437–471.
Wagner, I., & Schnotz, W. (2017). Learning from static and dynamic visualizations: What kind of questions should we ask? In R. Lowe & R. Ploetzner (Eds.), *Learning from dynamic visualization – Innovations in research and application*. Berlin: Springer (this volume).
Wong, A., Leahy, W., Marcus, N., & Sweller, J. (2012). Cognitive load theory, the transient information effect and e-learning. *Learning and Instruction, 22*, 449–457.

Chapter 2
Learning from Animations: From 2D to 3D?

Stephan Schwan and Frank Papenmeier

2.1 Introduction

Although traditional animations usually present their content in a two-dimensional manner, there is a growing body of dynamic visualizations that make use of three-dimensional depictions. For the learning of science in particular, many topics demand comprehension of events unfolding in space, ranging from operations of sophisticated machines to chemical reactions of large organic molecules (cf. Jenkinson, 2017, this volume; McGill, 2017, this volume). Such topics often also require comprehension of complex three-dimensional objects based on their inspection from different sides, be they anatomical structures (cf. Berney & Bétrancourt, 2017, this volume) or archaeological artefacts, among others. This has raised questions as to whether the introduction of three-dimensional space in animations fosters learning and knowledge acquisition and how 3D animations should be designed in order to achieve their goals (Dalgarno & Lee, 2010).

The present chapter gives an overview of the role of three-dimensional animations in learning. It starts by characterizing three-dimensional animations as part of a larger transformation of animation production from conventional, drawing-based animations to animations that are generated by applying certain computational methods to numerical data sets. In the main part of this chapter, the scope of three-dimensional animations is described and classified, with three characteristics of three-dimensional animations and their implications for memory and learning being discussed in finer detail. First, the introduction of three-dimensional digital

S. Schwan (✉)
Leibniz-Institut für Wissensmedien, Tubingen, Germany
e-mail: s.schwan@iwm-tuebingen.de

F. Papenmeier
University of Tübingen, Tubingen, Germany
e-mail: frank.papenmeier@uni-tuebingen.de

R. Lowe, R. Ploetzner (eds.), *Learning from Dynamic Visualization*,
DOI 10.1007/978-3-319-56204-9_2

depictions of objects and scenes opens up a broader range of animation possibilities than are available with traditional two-dimensional animations. These possibilities include both the dynamics of content (e.g., a machine in motion) and the dynamics of visual presentation (e.g., continuous movement of a virtual camera). Second, conceptual differences between three-dimensional monoscopic and stereoscopic presentations are described. In particular, monoscopic presentations (also called 2.5D, pseudo-3D, or synoptic) project three-dimensional content onto a two-dimensional plane (e.g., a computer screen). Because both of the viewer's eyes receive the same information, a truly three-dimensional impression is not possible. In contrast, in stereoscopic (3D) presentations, a viewer's left and right eye receive two slightly different views, creating the impression of true three-dimensional depth. Third, three-dimensional presentations also have a profound impact on users' modes of interaction: Besides being able to control the temporal aspects of an animation by actions such as starting, stopping, or rewinding, learners may also determine the animation's spatial characteristics by controlling position and movement of the virtual camera. The chapter ends by considering the implications of these innovations for animation research and practice.

2.2 From Two-Dimensional to Three-Dimensional Animations

Production of animations has undergone major transformations during the last two decades. Traditionally, animated cartoons were laboriously created on a frame-by-frame basis, similar to the painting of sequences of thousands of individual pictures in classical Disney cartoon films (Johnston & Thomas, 1981). With the advent of software tools such as Adobe Flash (first release in 1997), this process was substantially simplified and mechanized. Now, individual graphic objects could be digitally defined and animated. For example, by specifying the start and end point of an object's movement, the software could automatically determine and render the object's intermediate positions along a path. The appearance of objects in terms of size or color could also be easily transformed. By means of these possibilities, complex digital animations could be built out of simple graphic elements and commands. Yet, these software tools were based on the metaphor of visually presenting information on a flat plane, like a canvas or a sheet of paper. Accordingly, most animations built with these tools were two-dimensional in nature. In contrast, the current generation of software tools for animation design, like Unity or Blender, supplements previous ones by employing the metaphor of space that contains voluminous movable objects that extend in three-dimensions. Additionally, these tools not only allow the specification of three-dimensional objects and spatial layouts but also include so-called 'physics engines' that model the kinematics of objects according to predetermined physical laws.

With respect to space, recent animation tools allow digital objects to be defined as volumes (instead of two-dimensional shapes) on the basis of three-dimensional coordinates – width, height, and depth (cf. Jenkinson, 2017, this volume; McGill, 2017, this volume). Therefore, the projection screen now defines a window to a scene extending into depth. This move towards three-dimensionality has been accompanied by innovations in hardware technology that allow for a stereoscopic presentation, further enhancing viewers' impressions that they are perceiving actual spatial depth. With respect to dynamics, the digital definition of objects as sets of numerical values and coordinates allows for continuous transformations both of the objects themselves, (e.g., as morphs from one shape to another; Soemer & Schwan, 2012), of the movement of object parts, and of the whole objects' trajectories in three-dimensional space, according to predetermined mechanical or biological principles. With respect to camera position and perspective, recent animation tools allow for an easy definition of virtual cameras including their position, lens, and rotation, specifying both the distance and angle from which an animated scene is presented to the viewer. In addition, cameras may be set into motion, allowing for complex camera movements during the course of an animation. The net result is that animations made using these new facilities converge with films with their complex repertoire of established design principles (Bordwell & Thompson, 1979).

2.3 Types of Three-Dimensional Expository Animations

The observed trend towards using three-dimensional animations instead of two-dimensional ones has led to a broadened range of different animation types. In their meta-analysis of research on learning with animations, Höffler and Leutner (2007) had not yet referred to the distinction between two-dimensional and three-dimensional animations. Similarly, Ploetzner and Lowe (2012) have provided a detailed and comprehensive classification of animations used in research until then, where again the question of animations' spatial structure played only a minor role. This reflects the fact that to date, the animated material used in empirical research has not made much use of three-dimensional opportunities. But in order to accommodate probable future developments, further differentiations of the taxonomy's spatial dimension seem to be necessary. These differentiations relate to both the three-dimensional structure and the three-dimensional presentation of animations. While the three-dimensional structure concerns the addition of a third dimension to animations as well as the placement and motion of the virtual camera towards the three-dimensional scene, the three-dimensional presentation concerns the distinction between monoscopic (2.5D) and stereoscopic (3D) presentations of space.

2.4 Three-Dimensional Structure and Dynamic Camera Viewpoints

Most often, conventional expository animations have been used to show dynamically unfolding events from a fixed, stationary point of view. Examples range from piano mechanics (Boucheix, Lowe, Putri, & Groff, 2013; Lowe & Boucheix, 2017, this volume) and pendulum clocks (Fischer, Lowe, & Schwan, 2008), to organic systems (de Koning, Tabbers, Rikers, & Paas, 2007) and intercellular processes (Huk, Steinke, & Floto, 2010) to biological movement patterns (Imhof, Scheiter, Edelmann, & Gerjets, 2012; Lowe, Schnotz, & Rasch, 2011). Following Tversky, Morrison, and Bétrancourt (2002), one can ask for which content and under which conditions the addition of a third dimension to animations conforms to the principles of congruence and apprehension. According to these principles, the structure and content of an external representation should correspond to the desired structure and content of the internal representation (congruence principle) and be readily and accurately perceived and comprehended (apprehension principle).

Compared to two-dimensional depictions, adding a third dimension heightens the complexity of an illustration because spatial relations between the various elements have to be coded on three instead of two axes. Three-dimensional depictions also typically introduce occlusions and foreshortening (i.e. optical distortions of objects extending along the depth axis). In some circumstances, occlusions may help to understand otherwise unavailable spatial relationships. On the other hand, both occlusions and foreshortenings carry the danger of making understanding of the presentation more difficult for the learner. Thus if, for example, the elements of a mechanical system can be neatly arranged in two dimensions without loss of information, a two-dimensional depiction of the mechanical system in motion should be preferred over a three-dimensional one for ease of comprehension. In contrast, if the three-dimensional arrangement of the relevant elements in operation carries important information that cannot by easily depicted in two-dimensional graphics, and if this knowledge of this spatial structure should form an important element of the learner's mental model, a three-dimensional visualization should be preferred over a two-dimensional one for reasons of congruence.

When using three-dimensional animations, care must be taken to identify a viewpoint from which the relevant elements in operation can best be seen, avoiding occlusions and extreme foreshortenings. This often leads to an oblique viewpoint, showing the animated events not from a frontal perspective, but instead at an angle between about 30° to 60° degrees (Fischer, Lowe, & Schwan, 2008; Huk et al., 2010). This corresponds to the notion of canonical views as introduced by Palmer, Rosch, and Chase (1981; see also Blanz, Tarr, & Bülthoff, 1999). Compared to other viewing perspectives, canonical ones maximize the number of an object's visible surfaces and the visibility of its characteristic parts. Therefore, objects presented from canonical views are more accurately and easily identified than from other, non-canonical views.

To date, three-dimensional animations of dynamically unfolding events are a rare exception in the empirical literature on learning with animations. Following Tversky et al.'s (2002) principle of apprehension, at least for conventional animations with a fixed viewpoint, researchers seemingly have tended to avoid the additional complexity of a three-dimensional depiction in favor of an easier to grasp two-dimensional variant. Accordingly, few studies have systematically compared such two-dimensional and three-dimensional animations of similar content, finding at best mixed evidence for an advantage of the latter one (Huk, 2006; Huk et al., 2010). Still, under what circumstances – that is, for what content, what learning tasks, and what kind of learners – three-dimensional animations with a fixed viewpoint may better support the learners than two-dimensional ones is an open question that has to be addressed in future studies. For example, studies with static material indicate that three-dimensional depictions may be particularly suited for shape identification and discrimination but not for identification of relative positions in space (St. John, Cowen, Smallman, & Oonk, 2001). Also, adequate interpretation of three-dimensional depictions seems to require a high level of spatial ability (Huk, 2006; Huk et al., 2010; Khooshabeh & Hegarty, 2010).

It can be argued that if a dynamic event is simple enough or 'flat enough' to be intelligible from one stationary viewpoint, making the depiction three-dimensional will add little or nothing to its comprehensibility. In contrast, if a dynamic event is more complex and the interplay of its elements takes place not only in a flat plane, but extends into space, three-dimensionality may substantially enhance intelligibility.

Comprehensibility of an event taking place in space may also be facilitated by introducing a dynamic change of viewpoint. Accordingly, going three-dimensional has introduced a second important class of animations that show objects or scenes from changing viewpoints instead. Further, changing viewpoints are not only used for depicting dynamic events but also for depicting spatially extended static objects or scenes. Here, the impression of dynamics is not due to a moving or changing object but instead due to the observer's viewpoint (brought about by the virtual camera) moving through three-dimensional space. With regard to expository animations, such a moving viewpoint may serve a number of different purposes. Accordingly, several types of expository camera movements can be distinguished, including movement for completeness, for establishing connections, for regulating the focus of attention, and for decorative purposes.

Camera Movement for Completeness In many educational contexts, learners have to develop an appropriate mental representation of complex, three-dimensional objects, be they molecules, anatomical structures, or reconstructions of archaeological artifacts or buildings. In all these cases, inspecting the target object from one side alone may not be sufficient to fully understand its elements and their spatial relations because from a given viewpoint, relevant parts may be located on a hidden side or be occluded by other elements. Also, relative to a given viewpoint, visibility of surface planes extending into depth may suffer from foreshortening. To avoid these problems and allow the learner to make a comprehensive inspection of the object, an animation may present a 360° circular movement of the camera around an object.

Basic research has demonstrated that mental representation of objects and scenes is largely viewpoint dependent (Diwadkar & McNamara, 1997; Tarr, 1995); that is, viewers do not normally tend to develop an abstract, viewpoint independent representation but instead store a set of individual views. On later occasions, in order to identify the object or scene from a novel view, viewers start with the stored view that most closely matches the novel one and try to align both views by mental rotation. The more discrepant the two views are, the longer this process takes and the more error prone it becomes. (Diwadkar & McNamara, 1997). Therefore, the more different viewpoints of an object or scene to-be-learned that are presented to a learner, the more flexible his or her resulting mental representation will be. This finding holds not only for static objects and scenes, but also for events that dynamically unfold in space. Here again, presenting the event from different viewpoints facilitates identification from novel perspectives, indicating a more flexible mental representation (Garsoffky, Schwan, & Hesse, 2002).

Because a particularly dense variety of views result from a continuous movement around an object or a scene, providing an animation that offers such movement conforms to Tversky et al.'s (2002) congruence principle. Also, because continuous change of viewpoint around an object or a scene is in accordance with everyday experience, it can be assumed to conform to Tversky et al.'s (2002) apprehension principle as well. However, as research from the field of anatomy learning has demonstrated, these assumptions hold only for learners with sufficient spatial ability (Garg, Norman, Spero, & Maheswari, 1999; Nguyen, Nelson, & Wilson, 2012). Both Garg et al. (1999) and Nguyen et al. (2012) found that learners with low levels of spatial ability benefited from the presentation of a small set of key views (similar to canonical views) as opposed to a large, comprehensive set of views interconnected via continuous and uniform camera movements. It seems that low spatial ability learners find difficulties in combining the dense set of views into an integrated mental representation, possibly due to the transience of viewpoint-specific information that imposes high processing demands on working memory.

In order to reduce processing demands while providing learners with animations encompassing a larger sample of views, thus balancing completeness of presentation with required processing resources, several design options come to mind. One solution could be to substantially slow down the speed of camera movement, while another, discussed by Garg, Norman, Eva, Spero, and Sharan (2002), could be to provide a small set of key views and let the camera "wiggle" around these views within a range of about 10° to provide some additional three-dimensional information. A third option, which will be discussed further below, is to give the learners the opportunity to interactively control speed and trajectory of viewpoint position.

Finally, while for some topics circular movements around an object or scene tend to provide learners with a more complete impression of the content, other topics such as astronomy, geography, or archaeology require continuous movement of camera along a linear path (for an example from astronomy see Eriksson, Linder, Airey, & Redfors, 2014). To date, little is known whether the provision of continuous camera movements following a given trajectory instead of a set of distinct, but overlapping views indeed leads to a better understanding of the respective content.

Camera Movement for Viewpoint Optimization Many instances of dynamically unfolding events can be decomposed into a sequence of individual steps. Think, for example, of the assembling of a machine along a production line or of the process of digestion along the gastrointestinal tract. Learning about and understanding such events requires building a mental model based on the comprehension of the individual steps and linking them according to underlying principles of causality (Lowe & Boucheix, 2008; Narayanan & Hegarty, 2002). While for some events or processes a single viewpoint may suffice for all steps to be intelligible to a viewer, other events may require shifts of viewpoints during the presentation's time course in order to present each step from an optimal perspective. This may be achieved by moving the virtual camera along a predefined path, stopping at certain moments at particular points that provide viewers with a privileged sight of the current step of the event.

Empirical evidence for this design strategy comes from studies that demonstrate the processing advantages of canonical views both for individual objects (Palmer, Rosch, & Chase, 1981; Blanz, Tarr, & Bülthoff, 1999) and for ongoing events (Garsoffky, Schwan, & Huff, 2009). Compared to other views, canonical views provide an optimal perspective on an object or scene, as manifested by viewers' preferences and also by memory advantages. In case of events, views perpendicular to an event's main axis of change or movement have been shown to be beneficial for processing and therefore to qualify as canonical views (Garsoffky et al., 2009). Because the main axis of movement may shift during the course of an event canonical views should shift accordingly. In conventional films, switching from one canonical view to another is typically achieved by abrupt viewpoint changes in form of film cuts. This is partly due to the fact that for real world film recordings continuous camera movements are difficult to create. In contrast, numerical definition of objects and events via digitalization allows for creating animations in which even complex predefined camera movements are easily implemented. Therefore, although both film cuts and continuous camera movements have become equally viable options for building animations, several empirical comparisons have provided evidence in favor of camera movements. For example, some learning topics require observers to simultaneously pay attention to several moving objects, like molecules in a chemical reaction, or players' moves on a playing field. Here, basic research has demonstrated that continuous movement of observers' viewpoint does hardly impede the attentional tracking of several moving objects (Meyerhoff, Huff, Papenmeier, Jahn, & Schwan, 2011), while film cuts do (Huff, Jahn, & Schwan, 2009). Also, a study of Garsoffky, Huff, and Schwan (2007) showed that memory for a complex dynamic event (an animated scene from a basketball game) was significantly higher for continuous compared to abrupt in-between changes of viewpoint induced by film cuts.

While camera movements for completeness typically deal with static objects or scenes, camera movements for viewpoint optimization can include both camera motion *and* motion of objects or object parts. Therefore, learners have to disentangle both types of movement in order to comprehend the mechanism or event to be learned. Findings from Liu et al. (2005) indicate that during perception, viewers are successful in separating even extreme movements of whole scenes (due to camera pans or rotations) from relative movements of objects within that scene. But, on the other hand, in these studies, tracking multiple objects is so demanding that only little scene related information is processed and elaborated (Jahn, Papenmeier, Meyerhoff, & Huff, 2012), casting doubts on the appropriateness of such types of animations for learning. Also, while changing viewpoints during an event sequence may provide an optimized view for each step of the event, fostering comprehension of individual event steps, it also implies that different steps are seen from different viewpoints, possibly making it more difficult for the learner to appropriately link these steps causally in his or her mental model. Therefore, in terms of the Animation Processing Model proposed by Lowe and Boucheix (2011, 2017, this volume), viewpoint optimization by camera movement may facilitate parsing of the event into discrete steps (Phase 1) and the local processing of these steps (Phase 2), but may prove detrimental for connecting those steps into a causal chain (Phase 3). However, to our knowledge, to date no empirical research from the field of instructional design has addressed the topic of learning dynamic content from dynamically changing viewpoints.

Camera Movement for Regulating Focus of Attention Even if arranged on a flat plane perpendicular to the line of sight (as often in the case of conventional animations), complex animations often include multiple entities that require attention from the viewer. A growing body of literature has shown that due to the transience of animations, learners may tend to overlook some relevant elements or dynamics because they are distracted by other more perceptually salient parts of the animation (Lowe & Schnotz, 2014). In order to guide learners' attention through an animation that requires multiple attentional foci, several cueing options have been developed and empirically tested, including, for example, arrows, shading, or color coding (de Koning, Tabbers, Rikers, & Paas, 2009). Virtual cameras allow for another, yet empirically largely unexplored cueing alternative, namely, change of camera distance from medium long shots (showing the whole scene) to close-ups (showing one particular detail of the scene), either by means of a camera track or by zooming-in. In cinematography, use of camera distance for guiding viewers' attention has a long tradition, and so-called analytical editing of scenes, by which an event is decomposed into various single shots that are shown from different distances, can be considered one of the keystones of Hollywood cinema (Bordwell & Thompson, 1979).

Compared to arrows or color coding, reduction of camera distance could operate more unobtrusively. Also, it not only guides learners' attention to a relevant part of the animation, but also presents this part in an enlarged manner, showing more details and simultaneously keeping other possible distracting elements out of the frame. On the other hand, attention guidance via reduction of camera distance is less precise because arrows or color coding more clearly indicate which of the pictorial elements are intended to be looked at. Also, in a close-up, only a restricted section of the whole event is displayed, implying that some important contextual information may be missing that would otherwise be represented in memory (Papenmeier, Huff, & Schwan, 2012). Relating these considerations to the Animation Processing Model proposed by Lowe and Boucheix (2011, 2017, this volume), regulating focus of attention by camera movement may again facilitate parsing of the event into discrete steps due to the regular variations of distance from far to close and vice versa (Phase 1), and may also facilitate local processing of these steps because of its closer framing and its pictorial enlargement (Phase 2). On the other hand, due to the loss of "the whole picture" of the event, it may prove detrimental for connecting these steps into a causal chain (Phase 3). But once again, to our knowledge, little empirical research from the field of instructional design exists on the topic of guiding attention by variations in virtual camera distance. First empirical evidence on the cueing functions of zoom-ins comes from a study that was recently conducted by Glaser, Lengyel, Toulouse, and Schwan (in press). Taking three-dimensional reconstructions of ancient Roman buildings as the to-be-learned subject matter, these authors found that compared to static views and zoom-outs, learners in the zoom-in condition looked at the central part of the scene significantly longer, indicating that zoom-ins may indeed serve an attention focusing purpose.

Camera Movement for Decorative Purposes Finally, using camera movements to transform static depiction of scenes to animations with dynamically changing visual information is often used as a strategy to catch and hold viewers' attention in informal learning contexts. For example, museums and exhibitions today make heavy use of screens and displays for expository purposes. However, in the museum context, such displays have to compete with other exhibits for visitors' attention (Schwan, Lewalter, & Grajal, 2014). Building on evidence that dynamic visual stimuli attract more attention than static ones (Mital, Smith, Hill, & Henderson, 2011), many displays in museums come in the form of visualizations which are animated by complex camera movements, for example, as "fly-throughs" of reconstructed excavation sites in archaeological exhibits. Similar arguments also apply to science documentaries on TV or on the Internet. Here again, filmmakers tend to avoid static digital pictures in favor of dynamic ones in order to hold viewers' attention and prevent them from zapping to other competing channels. But besides their attention catching and holding purposes, the camera movements often seem to be only partly motivated by further, more learning-related intentions, similar to the ones discussed above. Therefore, they bear a strong resemblance to the use of decorative pictures and seductive details in multimedia learning material (Magner, Schwonke, Aleven, Popescu, & Renkl, 2014; Rey 2012). Yet, the implications of such decorative uses of camera movements in animations still await further empirical investigation.

2.5 Three-Dimensional Presentation: Adding Stereoscopic Cues

In the field of instructional design, the term "3D" is used in a broad sense to characterize representations that, in contrast to "2D", include a third axis of depth, thereby giving objects volume and defining the spatial layout of a given scene in three dimensions (cf. Jenkinson, 2017, this volume; McGill, 2017, this volume). However, pictorial representations (in contrast to haptic models, for example) are not truly three-dimensional but instead evoke only an impression of three-dimensionality on the basis of projection on a flat surface. To achieve this impression, they make use of a number of different pictorial cues. Perceptual psychology informs us that one large group of static pictorial cues operates monoscopically, requiring just one eye for the impression of depth in space (Vishwanath & Hibbard, 2013). These static depth cues include occlusion, size constancy, converging lines, and texture gradients. Motion parallax, which is the computation of relative distances due to observer movement, constitutes a further rather effective monoscopic depth cue. Besides these monoscopic cues, recent technological advancements have opened up the possibility for the addition of stereoscopic depth cues. In these cases, the term "3D" does not mark the difference to 2D regarding an animations three-dimensional structure (e.g., Huk et al., 2010), but instead the difference between stereoscopic and monoscopic viewing (Carrier, Rab, Rosen, Vasquez, & Cheever, 2012; Khooshabeh & Hegarty, 2010). In order to avoid confusions, we propose to use the term "2.5D" for three-dimensional monoscopic presentations, while restricting the term "3D" for three-dimensional stereoscopic presentations.

While monoscopic presentations do not require advanced technology but can be viewed on ordinary screens (e.g., Berney & Bétrancourt, 2017, this volume; Jenkinson, 2017, this volume; McGill, 2017, this volume), stereoscopic viewing requires special equipment. Several different technologies have been developed for stereoscopic viewing (Mendiburu, 2009). Currently, most applications operate by a combination of a specific display technology together with the use of corresponding glasses. Typically, the screen displays two separate, slightly different pictures to each of the eyes, either simultaneously or in brief succession. Viewers mentally fuse the two pictures into a single percept that appears to be truly three-dimensional, with the strength of the 3D impression depending on inter-ocular distance between cameras and the distance between projection screen and viewer. Differences relate to the way these two pictures are separated, either by combining differently colored pictures and corresponding filtering glasses, alternating pictures and the respective shuttered glasses ("active glasses"), or using polarized light, again together with the respective filtering glasses ("passive glasses"). Also, so-called autostereoscopic displays have been developed that do not require additional glasses, but use prismatic screens projecting two slightly different pictures to the viewer's eyes instead. The various 3D technologies all have their advantages and disadvantages. Stereoscopic pictures viewed with active glasses have a brighter tone but viewing suffers from flickering pictures and the heavy weight of the glasses. In contrast, passive glasses are more lightweight and do not show flicker but viewing suffers from darker pictures. Finally, for autostereoscopic screens no glasses are needed, but they have a very limited resolution and the 3D impression is strongly dependent on the particular viewing position in front of the screen.

Preparing expository animations for 3D presentation requires careful consideration of several detrimental effects resulting from the perceptual specifics of stereoscopic projection (Meesters, Ijsselsteijn, & Seuntiens, 2004; Mendiburu, 2009). These include cardboard effects (objects appear unnaturally flat), puppet theatre effects (objects appear miniaturized), image ghosting (objects appear to have a second shadow contour), and keystone effects (distortions of vertical parallaxes). But even when designed appropriately, 3D should not be considered more "natural" than other presentation techniques because it is presented on a flat surface and therefore still requires a dissociation of convergence and accommodation Together, these characteristics may contribute to feelings of visual fatigue, visual discomfort, eyestrain, and headaches, which has been reported for a substantial proportion of viewers (Lambooij, Fortuin, Heynderickx, & Ijsselstein, 2009; Ukai & Howarth, 2008).

Hence, from an instructional perspective, the question arises under which circumstances the introduction of 3D instead of 2.5D for purposes of learning and knowledge acquisition is justified, given the necessity of a complex technology (displays, glasses), the additional costs of the appropriate design of stereoscopic material, together with the dangers of visual fatigue or discomfort, and the fact that about 5–10% of the population suffer from stereo blindness (i.e. the inability to perceive stereoscopic projections as three-dimensional; Lambooij et al., 2009). Because stereoscopic presentation has been introduced only quite recently, empirical evidence is sparse and mixed at best.

In general, both 2.5D and 3D provide a third dimension that may be beneficial for building appropriate mental representations, particularly when extension in space is relevant for comprehension. But while going from 2D to 2.5D may add some important information, going from monoscopic 2.5D to stereoscopic 3D is a smaller step because monoscopic presentations already include a rich array of spatial cues. Accordingly, recent findings indicate that learners benefit from the addition of stereopsis only under specific circumstances. More particularly, in basic memory research, several studies have found a stereo advantage for recognition of static objects, especially if these objects are presented from novel views (Bennett & Vuong, 2006; Burke, 2005). This was the case even for displays with strong monocular depth cues (shading; Lee & Saunders, 2011). But on the other hand, for recognition of a large set of photos of natural scenes, Valsecchi and Gegenfurtner (2012) found a stereo advantage only for a small subset of pictures. This positive effect of stereoscopic presentations was even more restricted in cases of animated learning material. In a series of studies, Papenmeier and Schwan (2016) investigated the role of stereoscopy for memorizing complex molecule-like structures. They found that viewers did not benefit from stereoscopic presentation while learning the stimulus material. In contrast, however, if the depictions of molecules were presented stereoscopically in a subsequent memory test, learners outperformed participants who had to solve the memory test with monoscopic test items. This indicates that stereoscopic information is not included in the memory representation that is built during the learning phase but that the benefit of stereoscopic information is restricted to phases of reactivating object memory for purposes of recognition.

The finding that memory and learning benefit only to a small degree from stereoscopic over monoscopic three-dimensional dynamic presentations is also corroborated by studies with material from various other fields. For example, in a path analysis of possible memory effects of stereoscopic versus monoscopic movie screenings, substantial effects on emotions and immersion but neither direct nor indirect effects on memory for the films' content were found (Carrier et al., 2012). Similarly, using dental anatomy as a learning topic, Khooshabeh and Hegarty (2010) could not find an advantage of stereoscopic animations for tasks of visualizing a cross section of molar teeth. For learning abdominal anatomy, Luursema, Verwey, Kommers, and Annema (2008) found that for novices provision of stereoscopic animations facilitated localization but not identification (naming) of the various anatomical parts. In accordance with these findings, two recent reviews of the effectiveness of stereoscopic displays in medicine come to similar conclusions (McIntire, Havig, & Geiselman, 2014; Van Beurden, Ijsselstein & Juola, 2012). In medical practice, stereopsis has been shown to improve diagnosis (e.g., 3D ultrasound visualizations) and decrease the time needed for minimally invasive surgery (MIS) procedures and, more generally, for tasks involving the manipulation of objects. In contrast, its uses for training and learning are less clear. Analyzing the results of 11 experiments for medical training and learning, McIntire et al. (2014) found that four experiments showed an advantage of stereopsis, four experiments found mixed results, while the remaining three experiments showed no difference between 2.5D and 3D learning material.

Overall, these findings suggest that the suitability of stereopsis for purposes of learning and knowledge acquisition is limited. Not only does a substantial part of the population suffer from stereo blindness and many users of stereoscopic glasses report having experienced eyestrain and headaches, but also the learning gains seem to be small and restricted to certain types of learning content that has a strong spatial component but lacks strong monocular depth cues (McIntire et al., 2014). Accordingly, comparing the suitability of stereopsis in chemistry education, Trindate, Fiolhais, and Almeida (2002) found benefits of stereoscopic presentations only for comprehension of crystalline structures, but not for phase transitions or orbital structures, indicating that possible advantages of 3D presentations are strongly topic dependent.

2.6 Adding Interactivity to Three-Dimensional Visualizations

A conventional animation often allows learners to control its temporal parameters in terms of starting/stopping, varying presentation speed from slow to fast motion, and also changing presentation direction from forward to backward and vice versa (Schwan & Riempp, 2004). While some conventional animations give learners rudimentary control over its spatial characteristics by letting them switch between two different two-dimensional views (Meyer, Rasch, & Schnotz, 2010), the underlying numerical description of digital animations now substantially broadens possibilities for controlling the spatial parameters of three-dimensional animations by the users. But whereas control of temporal parameters can easily be done with predefined, fixed animations, user dependent variation of spatial parameters requires online computing of the animation and can therefore currently only be done on computer devices with sufficient processing power.

In general, user control provides the opportunity for an animation's characteristics to be adapted to a learners' individual cognitive needs (Schwan & Riempp, 2004). For example, giving learners the option to control the pace of multimedia learning material has been shown to facilitate learning and understanding (pacing principle; Hasler, Kersten, & Sweller, 2007; Mayer & Chandler, 2001; Wouters, Tabbers, & Paas, 2007). In the case of spatial characteristics, options for control encompass all parameters discussed in the previous sections, including continuous camera movements regulating distance through zoom-ins and zoom-outs as well as selection of appropriate, canonical viewpoints. This gives learners the freedom to freely explore a complex object or scene or even a dynamically unfolding event from different perspectives. Typically, learners spontaneously use these options, not only regarding an animation's temporal characteristics (Schwan & Riempp, 2004) but also regarding its spatial characteristics. In particular, changing the angle of view and zooming in/out have been found to be prominent types of interactivity that are heavily used in 3D environments (Yuan, Calic, & Kondoz, 2012).

However, on the other hand, having control over the virtual camera places some additional burden on the learners because they have to appropriately plan and execute changes in camera position. In comparison to predefined system controlled trajectories of the camera, this may lead both to more extraneous cognitive load and also to the danger of choosing suboptimal camera positions (Keehner, Hegarty, Cohen, Khooshabej, & Montello; 2008). Therefore, the benefits of freely exploring a three-dimensional animation in a self-guided manner may be outweighed by its cognitive costs. This may be the reason why most empirical studies that have directly compared system-controlled (non-interactive) and user-controlled (interactive) three-dimensional animations have either found no differences between the two conditions or even advantages of the system-controlled versions (Keehner et al., 2008; Khooshabeh & Hegarty, 2010; Nguyen et al., 2012; Papenmeier & Schwan, 2016).

Whether learners indeed benefit from interactively controlling the spatial parameters of a three-dimensional animation depends on a number of factors. First, successful control of three-dimensional animations seems to require an above average level of spatial abilities (Garg et al., 2002; Huk, 2006). Learners with low spatial abilities may experience high cognitive demands because interactive control requires additional planning and monitoring of content-related activities over and above the cognitive demands that result from building an appropriate spatial mental representation. Second, learners need to have appropriate strategies for controlling the spatial parameters of the visualization. In particular, they should be able to use an animation's control options to identify and focus on canonical viewpoints that provide the most informative perspectives on a given object or scene (Garg et al., 2002; Keehner et al., 2008). While in the Keehner et al. (2008) study about one half of the learners were able to spontaneously identify these key views, a substantial portion of learners failed to do so, indicating that they lacked the necessary strategies. But it should be kept in mind that in most studies, participants were not familiar with interactive, three-dimensional animations. Instead, it was the first time they had such interactive 3D systems and they had been given only a brief introduction into the system. Therefore, further research should investigate whether training or routinely practicing such tasks for an extended period of time would enable users to develop appropriate strategies for dealing with this type of visualizations. Additionally, almost all of the studies have investigated the role of interactivity for animations of the "complete view of static objects" type (mostly with anatomical topics). An even more demanding type of animations presents dynamic events in which canonical views change during its course. As discussed above, canonical viewpoints may shift during the course of event, requiring a time-dependent planning of the moves of the virtual camera, most probably overwhelming even learners with high levels of spatial abilities. Under these conditions, system-controlled three-dimensional animations would be expected to better facilitate learning than user-controlled types.

Additional measures may also help learners to control the spatial parameters of three-dimensional animations in better ways. In particular, the cognitive costs of executing position changes and movements of the virtual camera may be reduced by the use of devices that allow for a natural interaction with 6 degrees of freedom, like 3D mice or Wii controllers instead of keyboards or 2D mice (Yuan et al., 2012). Reducing the cognitive costs of planning is probably more difficult to achieve. Also, a better spatial orientation of the viewers can be achieved by including a visible coordinate system that updates according to the users interactions with the animation (Stull, Hegarty, & Mayer, 2009). Further, current technology also allows for systems of graded interactivity where learners can choose between different levels of interactivity, depending on their prior knowledge and their cognitive prerequisites. Instead of offering novices the whole range of possible interactions, such systems could, for example, restrict viewpoint positions to a set of meaningful ones and let users switch between them.

2.7 Conclusions and Outlook

Three-dimensional animations can be seen to embody the fundamental transition from sketching to computing that has taken place in recent years. This transition, which is still underway, has profound implications for the development of digital learning material. Being based on numerical descriptions, learning content can be visualized in many different ways – from simple two-dimensional wireframes to detailed stereoscopic renderings. Learning content can also be computationally transformed and its appearance can be flexibly controlled and modified by the learners. This may even go beyond pure graphic visualizations, opening up possibilities for haptic interactions with 3D prints (Preece, Williams, Lam, & Weller, 2013).

Within this broad range of options, going 3D does not simply add a third dimension to conventional animations, but instead complements them by animations that show static objects or scenes from changing viewpoints. Here, the impression of dynamics is not due to a moving or changing object or scene, but instead due to a moving viewpoint of the observer. Certainly, both principles can be combined, resulting in animations with changing objects or events from changing viewpoints. Also, the notion of interactivity is broader in the context of three-dimensional animations. While traditional animations focus on allowing learners to control the pace of an animation, interactive three-dimensional depictions often allow learners to control their relative viewing position as well; that is, they may interactively approach or retreat, zoom in and out, rotate around an event, or pursue even more complicated trajectories.

From a psychological perspective, these opportunities have implications for learning and understanding. In general, in comparison to two-dimensional representations, animated three-dimensional representations are both more detailed and more complex, with implications for three relevant learning issues. First, three-dimensional animations allow for a precise definition of viewpoint trajectories that may guide the viewers' attention to relevant parts of objects or events; that is, instead of providing learners with a fixed perspective, viewpoints can be flexibly adapted in terms of viewing angle and distance during the course of an animation. Additionally, camera movement may serve a range of different purposes, including completeness of view, optimizing viewpoints, guiding attention, or simply making the presentation more appealing.

Hence, questions of pedagogically, perceptually, and cognitively guided selection of appropriate viewpoints arise (Garsoffky, Schwan, & Huff, 2009). While extension into depth, changing distances, and moving viewpoints are relatively new approaches in the design of instructional animations, they have a long tradition in other fields, particularly in cinematography. Furthermore, filmic design principles have received some attention from empirical research on cognition and perception of film in recent years (Smith, Levin, & Cutting, 2012; Schwan, 2013). Therefore, while the boundaries between animation and film get more and more blurred (McClean, 2007), research findings from cognitive film studies may provide some guidance for animation design as well.

Second, 3D provides a third dimension that may be beneficial for building up appropriate mental representations, particularly when extension in space is relevant for comprehension. However, the term "3D" should be differentiated into monoscopical three-dimensional presentations ("2.5D") and stereoscopic 3D-presentations. But whereas going from 2D to 3D opens up the field for a much broader range of animations because not only events or moving objects but also the continuous changes of viewpoint brought into effect by movements of the virtual camera come into play, introducing stereoscopic 3D does not add much to the instructional options of animation beyond providing an additional depth cue.

Accordingly, recent results show that learners benefit from addition of stereopsis only under specific circumstances, indicating that stereoscopic information supports the construction of mental representations only in the absence of other depth cues such as depth from motion (Papenmeier & Schwan, 2016).

Third, 3D also adds more degrees of freedom for learner control and can be combined with touch, gesture, or head-motion based interfaces instead of mouse or keyboard. In accordance with assumptions of embodied cognition, coupling complex 3D presentations with the possibility for haptic manipulation and haptic feedback has been shown to enhance learning and deepen understanding (Bivall, Ainsworth, & Tibell, 2011). But while more natural, increases in 3D interactivity may also have its costs in terms of increased requirements for appropriately planning and monitoring content-related activities.

Taken together, from a conceptual perspective, existing taxonomies have to be complemented and differentiated with regard to these new forms of animation. Taking the taxonomy proposed by Ploetzner and Lowe (2012) as a starting point, the spatial characteristics of animations should include not only 2D, but also 2.5D (three-dimensional monoscopic) and 3D (three-dimensional stereoscopic) presentations. The taxonomy should also be complemented with a distinction between event dynamics and viewpoint dynamics, offering many new opportunities for future research on the role of animations for learning and knowledge acquisition.

References

Bennett, D. J., & Vuong, Q. C. (2006). A stereo advantage in generalizing over changes in viewpoint on object recognition tasks. *Perception & Psychophysics, 68*, 1082–1093.

Berney, S., & Bétrancourt, M. (2017). Learning three-dimensional anatomical structures with animation: Effects of orientation references and learners' spatial ability. In R. Lowe & R. Ploetzner (Eds.), *Learning from dynamic visualization — Innovations in research and application.* Berlin: Springer (this volume).

Bivall, P., Ainsworth, S., & Tibell, L. A. E. (2011). Do haptic representations help complex molecular learning? *Science Education, 95*, 700–719.

Blanz, V., Tarr, M. J., & Bülthoff, H. H. (1999). What object attributes determine canonical views? *Perception, 28*, 575–599.

Bordwell, D., & Thompson, K. (1979). *Film art: An introduction.* New York: McGraw Hill.

Boucheix, J.-M., Lowe, R. K., Putri, D. K., & Groff, J. (2013). Cueing animations: Dynamic signaling aids information extraction and comprehension. *Learning and Instruction, 25*, 71–84.

Burke, D. (2005). Combining disparate views of objects: Viewpoint costs are reduced by stereopsis. *Visual Cognition, 12*, 705–719.

Carrier, L. M., Rab, S. S., Rosen, L. D., Vasquez, L., & Cheever, N. A. (2012). Pathways for learning from 3D technology. *International Journal of Environmental & Science Education, 7*, 53–69.

Dalgarno, B., & Lee, M. J. W. (2010). What are the learning affordances of 3-D virtual environments? *British Journal of Educational Technology, 41*, 10–32.

De Koning, B. B., Tabbers, H. K., Rikers, R. M. J. P., & Paas, F. (2007). Attention guidance in learning from complex animation: Seeing is understanding? *Learning and Instruction, 20*, 111–122.

De Koning, B. B., Tabbers, H. K., Rikers, R. M. J. P., & Paas, F. (2009). Towards a framework for attention cueing in instructional animations: Guidelines for research and design. *Educational Psychology Review, 21*, 113–140.
Diwadkar, V. A., & McNamara, T. P. (1997). Viewpoint dependence in scene recognition. *Psychological Science, 8*, 302–307.
Eriksson, U., Linder, C., Airey, J., & Redfors, A. (2014). Who needs 3D when the universe is flat? *Science Education, 98*, 412–442.
Fischer, S., Lowe, R. K., & Schwan, S. (2008). Effects of presentation speed of a dynamic visualization on the understanding of a mechanical system. *Applied Cognitive Psychology, 22*, 1126–1141.
Garg, A. X., Norman, G. R., Eva, K. W., Spero, L., & Sharan, S. (2002). Is there any real virtue of virtual reality? The minor role of multiple orientations in learning anatomy from computers. *Academic Medicine, 77*, S97–S99.
Garg, A., Norman, G. R., Spero, L., & Maheswari, P. (1999). Do virtual computer models hinder anatomy learning? *Academic Medicine, 74*, S87–S89.
Garsoffky, B., Huff, S., & Schwan, S. (2007). Changing viewpoints during dynamic events. *Perception, 36*, 366–374.
Garsoffky, B., Schwan, S., & Hesse, F. W. (2002). Viewpoint dependency in the recognition of dynamic scenes. *Journal of Experimental Psychology: Learning, Memory, and Cognition, 28*, 1035–1050.
Garsoffky, B., Schwan, S., & Huff, M. (2009). Canonical views of dynamic scenes. *Journal of Experimental Psychology: Human Perception and Performance, 35*, 17–27.
Glaser, M., Lengyel, D., Toulouse, C., & Schwan, S. (in press). Designing computer-based learning contents: Influence of digital zoom on attention. *Education Technology Research and Development*.
Hasler, B. S., Kersten, B., & Sweller, J. (2007). Learner control, cognitive load and instructional animation. *Applied Cognitive Psychology, 21*, 713–729.
Höffler, T. N., & Leutner, D. (2007). Instructional animation versus static pictures: A meta-analysis. *Learning and Instruction, 17*, 722–738.
Huff, M., Jahn, G., & Schwan, S. (2009). Tracking multiple objects across abrupt viewpoint changes. *Visual Cognition, 17*, 297–306.
Huk, T. (2006). Who benefits from learning with 3D models? The case of spatial ability. *Journal of Computer Assisted Learning, 22*, 392–404.
Huk, T., Steinke, M., & Floto, C. (2010). The educational value of visual cues and 3D-representational format in a computer animation under restricted and realistic conditions. *Instructional Science, 38*, 455–469.
Imhof, B., Scheiter, K., Edelmann, J., & Gerjets, P. (2012). How temporal and spatial aspects of presenting visualizations affect learning about locomotion patterns. *Learning and Instruction, 22*, 193–205.
Jahn, G., Papenmeier, F., Meyerhoff, H. S., & Huff, M. (2012). Spatial reference in multiple object tracking. *Experimental Psychology, 59*, 163–173.
Jenkinson, J. (2017). The role of craft-based knowledge in the design of dynamic visualizations. In R. Lowe & R. Ploetzner (Eds.), *Learning from dynamic visualization — Innovations in research and application*. Berlin: Springer (this volume).
Johnston, O., & Thomas, F. (1981). *Disney animation: The illusion of life*. New York: Walt Disney Productions.
Kheener, M., Hegarty, M., Cohen, C., Khooshabeh, P., & Montello, D. R. (2008). Spatial reasoning with external visualizations: What matters is what you see, not whether you interact. *Cognitive Science, 32*, 1099–1132.
Khooshabeh, P., & Hegarty, M. (2010). Inferring cross-section: When internal visualizations are more important than properties of external visualizations. *Human-Computer Interaction, 25*, 119–147.

Lambooji, M., Fortuin, M., Heynderickx, I., & Ijsselsteijn, W. (2009). Visual discomfort and visual fatigue of stereoscopic displays: A review. *Journal of Imaging Science and Technology, 53*, 1–14.
Lee, Y. L., & Saunders, J. A. (2011). Stereo improves 3D shape discrimination even when rich monocular shape cues are available. *Journal of Vision, 11*, 1–12.
Liu, G., Austen, E. L., Booth, K. S., Fisher, B. D., Argue, R., Rempel, M. I., et al. (2005). Multiple-object tracking is based on scene, not retinal, coordinates. *Journal of Experimental Psychology: Human Perception and Performance, 31*, 235–247.
Lowe, R., & Boucheix, J. M. (2008). Learning from animated diagrams: How are mental models built? In G. Stapleton, J. Howse, & J. Lee (Eds.), *Diagrammatic representation and inference* (pp. 266–281). Berlin: Springer.
Lowe, R., & Boucheix, J.-M. (2017). A composition approach to design of educational animations. In R. Lowe & R. Ploetzner (Eds.), *Learning from dynamic visualization — Innovations in research and application*. Berlin: Springer (this volume).
Lowe, R., & Schnotz, W. (2014). Animation principles in multimedia learning. In R. E. Mayer (Ed.), *The Cambridge handbook of multimedia learning* (pp. 513–546). Cambridge, MA: Cambridge University Press.
Lowe, R., Schnotz, W., & Rasch, T. (2011). Aligning affordances of graphics with learning task requirements. *Applied Cognitive Psychology, 25*, 452–459.
Luursema, J. M., Verwey, W. B., Kommers, P. A. M., & Annema, J. H. (2008). The role of stereopsis in virtual anatomical learning. *Interacting with Computers, 20*, 455–460.
Magner, U. I. E., Schwonke, R., Aleven, V., Popescu, O., & Renkl, A. (2014). Triggering situational interest by decorative illustrations both fosters and hinders learning in computer-based learning environments. *Learning and Instruction, 29*, 141–152.
Mayer, R. E., & Chandler, P. (2001). When learning is just a click away: Does simple user interaction foster deeper understanding of multimedia messages? *Journal of Educational Psychology, 93*, 390–397.
McClean, S. T. (2007). *Digital storytelling*. Cambridge, MA: MIT Press.
McGill, G. (2017). Designing instructional science visualizations in the trenches: Where research meets production reality. In R. Lowe & R. Ploetzner (Eds.), *Learning from dynamic visualization — Innovations in research and application*. Berlin: Springer (this volume).
McIntire, J. P., Havig, P. R., & Geiselman, E. E. (2014). Stereoscopic 3D displays and human performance: A comprehensive review. *Displays, 35*, 18–28.
Meesters, L. M. J., Ijsselsteijn, W. A., & Seuntiens, P. J. H. (2004). A survey of perceptual evaluations and requirments of three-dimensional TV. *IEEE Transactions on Circuits and Systems for Video Technology, 14*, 381–391.
Mendiburu, B. (2009). *3D movie making*. London: Routledge.
Meyer, K., Rasch, T., & Schnotz, W. (2010). Effects of animation's speed of presentation on perceptual processing and learning. *Learning and Instruction, 20*, 136–145.
Meyerhoff, H. S., Huff, M., Papenmeier, F., Jahn, G., & Schwan, S. (2011). Continuous visual cues trigger automatic spatial target updating in dynamic scenes. *Cognition, 121*, 73–82.
Mital, P. K., Smith, T. J., Hill, R. L., & Henderson, J. M. (2011). Clustering of gaze during dynamic scene viewing is predicted by motion. *Cognitive Computation, 3*, 5–24.
Narayanan, N. H., & Hegarty, M. (2002). Multimedia design for communication of dynamic information. *International Journal of Human-Computer Studies, 57*, 279–315.
Nguyen, N., Nelson, A. J., & Wilson, T. D. (2012). Computer visualizations: Factors that influence spatial anatomy comprehension. *Anatomical Sciences Education, 5*, 98–108.
Palmer, S., Rosch, E., & Chase, P. (1981). Canonical perspective and the perception of objects. In J. Long & A. Baddeley (Eds.), *Attention and performance IX* (pp. 135–151). Hillsdale: Erlbaum.
Papenmeier, F., Huff, M., & Schwan, S. (2012). Representation of dynamic spatial configurations in visual short-term memory. *Attention, Perception, & Psychophysics, 74*, 397–415.
Papenmeier, F., & Schwan, S. (2016). If you watch it move, you'll recognize it in 3D: Transfer of depth cues between encoding and retrieval. *Acta Psychologica, 164*, 90–95.

Ploetzner, R., & Lowe, R. (2012). A systematic characterization of expository animations. *Computers in Human Behavior, 28*, 781–794.
Preece, D., Williams, S. B., Lam, R., & Weller, R. (2013). "Let's get physical": Advantages of a physical model over 3D computer models and textbooks in learning imaging anatomy. *Anatomical Sciences Education, 6*, 216–224.
Rey, G. D. (2012). A review of research and a meta-analysis of the seductive detail effect. *Educational Research Review, 7*, 216–237.
Schwan, S. (2013). The art of simplifying events. In A. P. Shimamura (Ed.), *Psychocinematics: Exploring cognition at the movies* (pp. 214–226). New York: Oxford University Press.
Schwan, S., Lewalter, D., & Grajal, A. (2014). Understanding and engagement in places of science experience: Science museums, science centers, zoos and aquariums. *Educational Psychologist, 49*, 70–85.
Schwan, S., & Riempp, R. (2004). The cognitive benefits of interactive videos: Learning to tie nautical knots. *Learning and Instruction, 14*, 293–305.
Smith, T. J., Levin, D., & Cutting, J. E. (2012). A window on reality: Perceiving edited moving images. *Current Directions in Psychological Science, 21, 107–113.*
Soemer, A., & Schwan, S. (2012). Visual mnemonics for language learning: Static pictures vs. animated morphs. *Journal of Educational Psychology, 104*, 565–579.
St. John, M., Cowen, M. B., Smallman, H. S., & Oonk, H. M. (2001). The use of 2D and 3D displays for shape-understanding versus relative-position tasks. *Human Factors, 43*, 79–98.
Stull, A. T., Hegarty, M., & Mayer, R. E. (2009). Getting a handle on learning anatomy with interactive three-dimensional graphics. *Journal of Educational Psychology, 101*, 803–816.
Tarr, M. J. (1995). Rotating objects to recognize them: A case study on the role of viewpoint dependency in the recognition of three-dimensional objects. *Psychonomic Bulletin & Review, 2*, 55–82.
Trindade, J., Fiolhais, C., & Almeida, L. (2002). Science learning in virtual environments: a descriptive study. *British Journal of Educational Technology, 33*, 471–488.
Tversky, B., Morrison, J. B., & Bétrancourt, M. (2002). Animation: Can it facilitate? *International Journal of Human-Computer Studies, 57*, 247–262.
Ukai, K., & Howarth, P. A. (2008). Visual fatigue caused by viewing stereoscopic motion images: Background, theories, and observations. *Displays, 29*, 106–116.
Valsecchi, M., & Gegenfurtner, K. R. (2012). On the contribution of binocular disparity to the long-term memory for natural scenes. *PlosOne, 7(11)*, e49947.
Van Beurden, M. H. P. H., IJsselsteijn, W. A., & Juola, J. F. (2012). Effectiveness of stereoscopic displays in medicine: A review. *3D Research, 3*, 1–13.
Vishwanath, D., & Hibbard, P. B. (2013). Seeing in 3-D with just one eye: Stereopsis without binocular vision. *Psychological Science, 24*, 1673–1685.
Wouters, P., Tabbers, H. K., & Paas, F. (2007). Interactivity in video-based models. *Educational Psychology Review, 19*, 327–342.
Yuan, H., Calic, J., & Kondoz, A. (2012). Analysis of user requirements in interactive 3D video systems. *Advances in Human-Computer Interaction, 2012*, 1–11. ID 343197.

Chapter 3
Looking Across Instead of Back and Forth: How the Simultaneous Presentation of Multiple Animation Episodes Facilitates Learning

Rolf Ploetzner and Richard Lowe

3.1 Give It a Try

Assume you are interested in sailing and you want to understand how a yacht can sail. In order to study the subject matter, you obtain a textbook on sailing (e.g., Bark, 2009; Overschmidt & Gliewe, 2009). It describes and depicts some of the physical principles that apply to sailing. Using a bird's eye view, the visualizations present schematic and idealized depictions of the main courses that a yacht can sail in relation to the wind direction. Each of the visualizations shows a compass, the wind direction, the yacht's hull and sail, and various forces that act on the yacht. Figure 3.1 presents the course termed "broad reach" in which the yacht sails off the wind, but not directly downwind. The course termed "close hauled" is displayed in Fig. 3.2. Here the yacht sails as close as possible towards the wind direction.

In broad reach, the yacht's hull is oriented very differently with respect to the wind direction than in close hauled. Is the orientation of the yacht's sail – with respect to both the wind direction and the yacht's hull – also different between broad reach and close hauled? And what about the forces? Which of the forces differ in magnitude and/or direction in broad reach as opposed to close hauled? Would you be able to formulate a higher-order relationship that captures the observable differences?

How did you attempt to provide answers to the questions raised above? Maybe you started by looking at Fig. 3.1 in an effort to first grasp its overall visuospatial arrangement, secondly to distinguish units within the overall arrangement, and

R. Ploetzner (✉)
University of Education at Freiburg, Freiburg, Germany
e-mail: rolf.ploetzner@ph-freiburg.de

R. Lowe
Curtin University, Perth, Australia

University of Burgundy, Dijon, France
e-mail: r.k.lowe@curtin.edu.au

R. Lowe, R. Ploetzner (eds.), *Learning from Dynamic Visualization*,
DOI 10.1007/978-3-319-56204-9_3

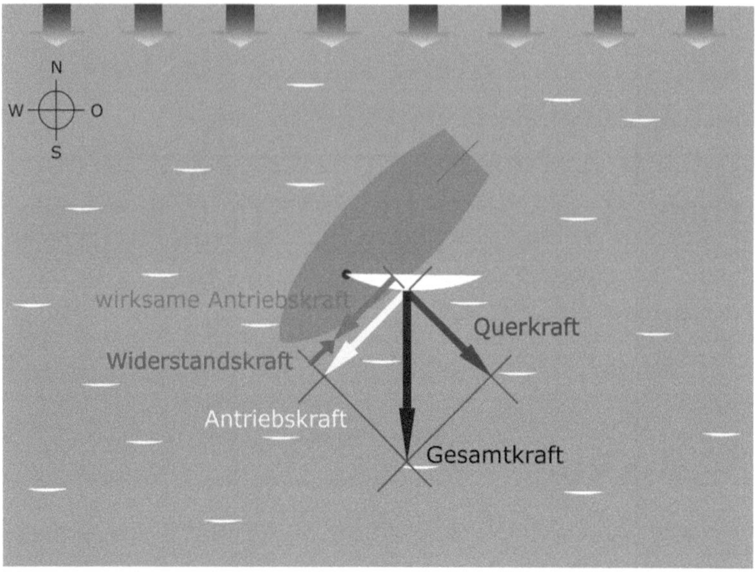

Fig. 3.1 The forces that act on a yacht in broad reach (Gesamtkraft [resultant force], Antriebskraft [driving force], Widerstandskraft [resistance force], wirksame Antriebskraft [effective driving force], Querkraft [drifting force])

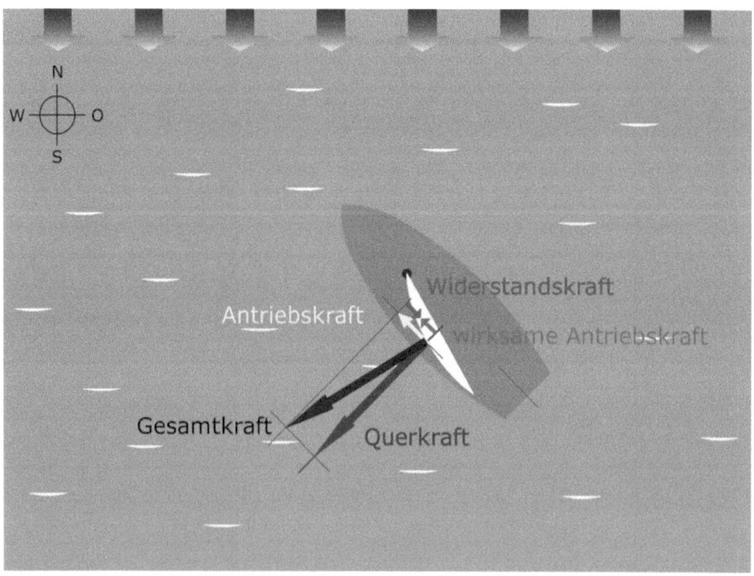

Fig. 3.2 The forces that act on a yacht in close hauled

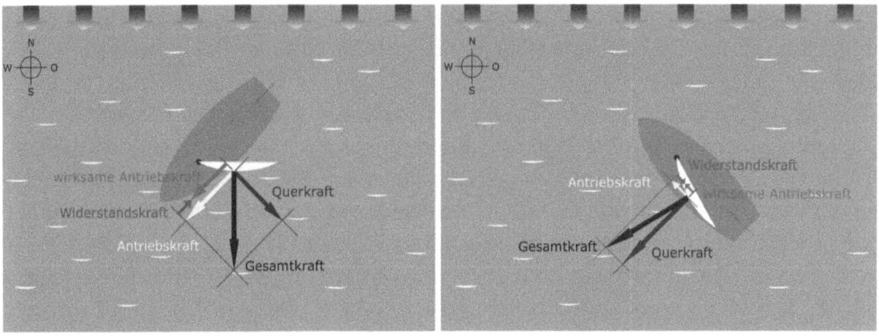

Fig. 3.3 Broad reach and close hauled displayed next to each other, 'optimized' for compare and contrast activities

thirdly to analyze relationships between the identified units. Next, perhaps you looked up Fig. 3.2 and once more invoked these same general processes. During your analysis of Fig. 3.2, you may have tried to recall what was presented in Fig. 3.1. Perhaps you were able to recall some of the information presented in Fig. 3.1, but probably not all of it. You therefore looked up Fig. 3.1 again to refresh your memory and/or to deepen your analysis. You may have also looked back and forth between the two figures several times and might have questioned yourself as to why the figures are set out so poorly. In fact, this type of layout could be considered poor design in respect to static learning material because it violates important multimedia design principles regarding spatial contiguity (cf. Mayer & Fiorella, 2014) and split-attention (cf. Ayres & Sweller, 2014). Perhaps in desperation you even considered folding the pages in such a way that the two figures would be located next to each other because that would make it much easier to compare and contrast the two figures by just looking across them. Please don't damage your book! We have re-arranged the figures for you in Fig. 3.3.

3.2 Presenting Multiple Animation Episodes Sequentially or Simultaneously

What you just experienced has some parallels with learning from many behaviorally realistic animations that present multiple episodes one after the other, i.e. sequentially. The sailing animation used in the present study consists of not only two, but four episodes. Each episode depicts one of four courses that a yacht can sail in relation to the wind direction: running (i.e., directly downwind), broad reach, close hauled, and tacking (i.e., sailing against the wind). Any one of these episodes depicts a specific set of local relationships between entities such as the wind direction, the yacht's hull and sail, and the forces that act on the yacht. However, there are also higher-order relationships connecting the individual episodes at a more general level. These relationships link the wind direction to the sail's orientation, the sail's orientation to the directions and magnitudes of the different forces, and the directions and magnitudes of the forces to the yacht's speed. For instance, the relationship that links the wind direction to the sail's orientation could be expressed as "the closer the yachts sails to the wind direction, the closer the sail is oriented towards the hull." In order to develop a comprehensive and hierarchically structured mental model of the animated subject matter, the learner needs to internally represent not only local relationships from within individual episodes, but also higher-order between-episode relationships that encompass information from the animation as a whole. In this situation, constructing a satisfactory mental model requires learners to compare and contrast relevant entities, as well as their local relationships, across individual episodes in order to identify and internalize higher-order relationships.

When episodes are presented sequentially, there can be considerable temporal separation between two instances of event units that need to be compared or contrasted in order to establish inter-episode relations. For example, a learner may wish to compare and contrast material found in the first episode on running with corresponding material in the third episode on close hauled. This would require the learner to extract relevant information from the first episode, store it in memory until the corresponding material in the third episode appears, and then carry out compare and contrast operations between the internal and external representations. In the meantime, the learner is likely to have been engaged in intra-episode processing of the second episode on broad reach in order to make sense of it in its own right. Such intervening processing will most probably result in the overwriting of material stored from the first episode (cf. Lowe, 1999) and thereby severely impede the compare and contrast processes necessary for establishing higher-order relationships.

If the animation was user-controllable, the learner might address this problem by re-inspecting the first episode to identify, extract, and memorize the required material and then skip to the appropriate sequence in the third episode in order to establish a relationship. However, the interrogation involved would not only be inefficient in terms of processing, but also prone to error because of the reliance on memory reliability.

Presenting component episodes of an animation simultaneously, in contrast, offers affordances considerably more suited to identifying and extracting high-level cross-episode relationships than does sequential presentation. However, it requires a major spatiotemporal manipulation of the information that involves a substantial departure from what could actually occur in real life. In reality, it is of course impossible for the same yacht to simultaneously sail on four different courses. Nevertheless, if an animation were to "play tricks with space and time," as suggested by Tversky, Heiser, Machenzie, Lozano, and Morrison (2008), previously unavailable affordances would be offered to the learner. In particular, when different episodes are displayed simultaneously on the same screen, comparisons and contrasts of corresponding material can be made directly and efficiently via scans across the display. The simultaneous presentation of episodes essentially eliminates the need to memorize relevant information over the time taken to present intervening episodes in a sequential animation. It changes the nature of the processing task that learners are required to perform (cf. Zhang & Norman, 1994): instead of requiring them to relate an internal representation to an external representation, simultaneous presentation allows for repeated perceptual switching between two or more external representations. It is also relatively easy for the learner to shift between within-episode and between-episode interrogation without appreciable processing overheads. With respect to the animation as a whole, simultaneous presentation allows learners to move relatively seamlessly between different levels of relationship which should help in building a more coherent, hierarchically structured mental model of the subject matter.

3.3 Different Types of Animation and Inductive Processes

Many learning tasks require students to induce higher-order relationships from learning material such as expository animations. According to Holland, Holyoak, Nisbett, and Thagard (1986), induction encompasses "... all inferential processes that expand knowledge in the face of uncertainty" (p. 1). A major goal of induction is "… to learn about the variability of the environment" (Holland et al., 1986, p. 22). In order to learn about the variability – and constancy – of the environment, students need to recognize both differences and regularities across varying situations. Regularities can rely on shared features of entities as well as on shared relations between entities (cf. Klauer & Leutner, 2012; Klauer & Phye, 2008). According to Klauer and Leutner (2012), compare and contrast processes are the 'silver bullets' for identifying such regularities.

Educational research has investigated learning from a diverse range of expository animations (for a recent review see Ploetzner & Lowe, 2012) and it is evident that there are many different approaches to the presentation of information in animations. Even the animation episodes used in the present study could be displayed in many different ways (cf. Fig. 3.4): each episode could be presented with or without explanatory text, only a single episode or multiple episodes could be shown, and multiple episodes could be presented either sequentially or simultaneously – to mention only a few of the many possibilities. How then does a student need to process each type of presentation in order to learn successfully?

One important type of expository presentation consists of animations that are accompanied by verbal explanations. The explanations might be provided to the learner in either written or spoken form. According to the modality principle (cf. Ginns, 2005; Low & Sweller, 2014; Mayer, 2009), students learn more successfully from animations with narration than from animations with on-screen text. For instance, Kombartzky, Ploetzner, Schlag, and Metz (2010), as well as Ploetzner and Schlag (2013), investigated how learning from the four sailing episodes can be supported by a cognitive learning strategy when the episodes are presented one after the other and are accompanied by spoken explanations (see also Ploetzner & Breyer, 2017, this volume). Commonly the explanations spell out relevant entities, the features of these entities, and the relations between them. That is, in a narrated anima-

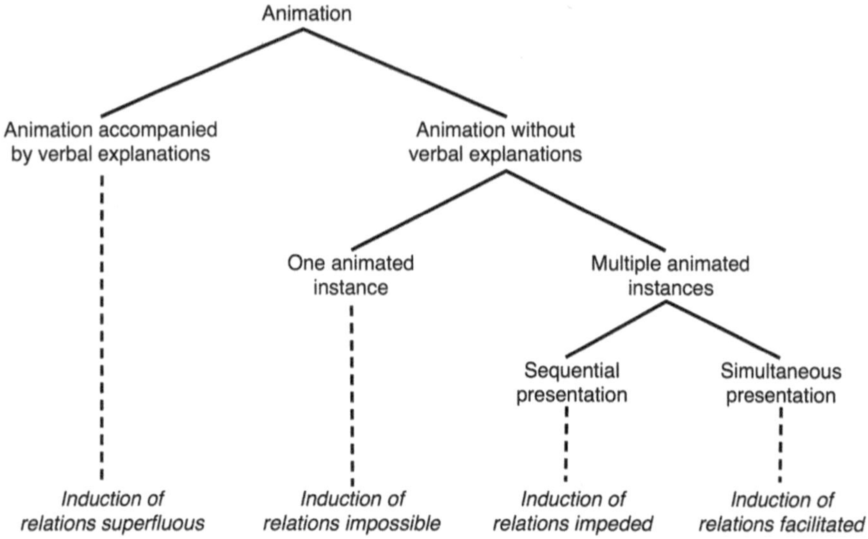

Fig. 3.4 Different types of animation and how they are related to inductive processes

tion, the learner might not be required to induce the regularities in the animated subject matter by her- or himself. According to the Cognitive Theory of Multimedia Learning (CTML; Mayer, 2009, 2014), the learner needs to mentally construct a pictorial and a verbal model by processing the corresponding external representations and then integrating both mental representations into one coherent mental model of the animated subject matter. Therefore, in order to understand the relevant relationships, the learner is required to significantly engage in inter-representation processing.

A second important type of expository animation is made up of animations that are not accompanied by verbal explanations. In this case, the learner is required to recognize the regularities in the animated subject matter without assistance from a narration. According to the Animation Processing Model (APM; Lowe & Boucheix, 2008, 2011, 2017, this volume; Lowe & Schnotz, 2014), learning from animations without verbal explanations progresses as a cumulative activity in which bottom-up and top-down processes interact in order to construct an increasingly comprehensive mental model of the animated subject matter. If the learner, however, lacks relevant domain-specific prior knowledge, then her or his activities will be mostly limited to bottom-up processes. That is, the learner starts with breaking down the continuous flux of information in the animation into individual event units (Phase 1 of the APM) and then successively combines them into broader regional structures (Phase 2 of the APM). Subsequently, the learner could proceed to link regional structures by establishing higher-order relationships that may cover the animation's entire spatial and temporal scope (Phase 3 of the APM). If the learner does not possess the required domain-specific prior knowledge, and does not receive complementary information such as verbal explanations, it is rather unlikely that she or he will be able to progress to Phases 4 and 5 of the Animation Processing Model (cf. Lowe, 2004). Therefore, in order to recognize the relevant relationships, especially in Phase 3 of the APM, the learner needs to engage in considerable intra-representation processing.

Expository animations very often visualize just one instance of a dynamic process (cf. Ploetzner & Lowe, 2012). That is, the animation displays the process in just one specific situation. For instance, an animation shows just one course that a yacht could sail in relation to a specific wind direction. Because a single instance of a process does not reveal the process's variability, the learner can recognize neither differences nor regularities. The induction of higher-order relationships is impossible under these circumstances. However, if an expository animation visualizes multiple instances of a dynamic process, the learner can compare and contrast the different instances in order to identify differences between the instances as well as regularities across the instances. In this case, the design of the animation might either impede or facilitate the required compare and contrast activities.

The sequential presentation of multiple animation instances demands that the learner selects information from one instance, stores it in memory until the corresponding entities in another instance appear, and then compares and contrasts the internal and external representations. In the meantime, the learner may process additional instances and further burden her or his working memory. Thus, it is likely that the cognitive processes necessary for inducing higher-order relationships are impeded by the sequential presentation of multiple animation instances. Presenting multiple animation instances simultaneously, in contrast, offers affordances to the learner that are considerably better suited to supporting the required cognitive processes. For instance, the comparing and contrasting of corresponding entities can be done directly and efficiently via scans across the display. Furthermore, the need to memorize information is fundamentally reduced. The learner can also easily shift from within-instance analysis to between-instance analysis without noticeable processing overheads. Therefore, the compare and contrast processes necessary for inducing higher-order relationships should be facilitated by the simultaneous presentation of multiple animation instances. Figure 3.5 shows how the four sailing episodes are presented in one display.

3.4 Study

Ploetzner and Lowe (2014) conducted an experimental study in order to investigate how the sequential and simultaneous presentation of multiple animation episodes influences the learning of higher-order relationships.

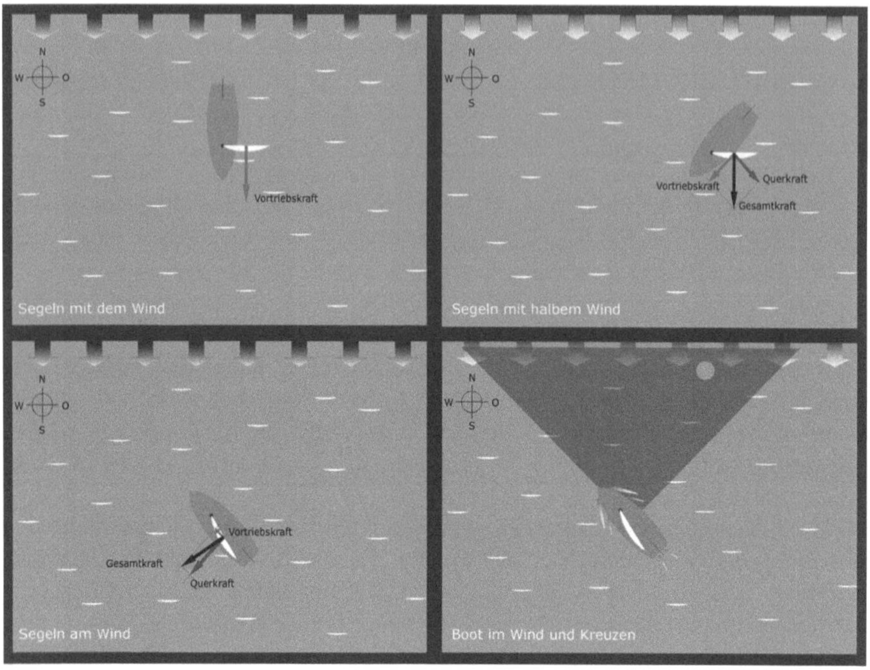

Fig. 3.5 Four animation episodes displayed simultaneously

3.4.1 Design and Hypotheses

Learning from two different versions of a sailing animation was investigated. Both versions consisted of four animation episodes. Each episode portrays a course that a yacht can sail in relation to the wind direction: running, broad reach, close hauled and tacking. While one group of students learned from a sequential presentation of the four episodes (sequential group), another group of students learned from a simultaneous presentation of the same episodes (simultaneous group). Pre- and posttests were administered before and after the learning sessions.

As delineated above, due to the specific requirements and affordances of each presentation, it was hypothesized that the students would process the sequential and simultaneous presentations in different ways. In particular, it was assumed that the simultaneous presentation would result in more visual transitions between the individual episodes than would the sequential presentation. That is, the affordances offered by the simultaneous presentation with respect to comparing and contrasting different episodes should be reflected in the frequency with which the students shifted their visual attention from one episode to another. Furthermore, it was hypothesized that the simultaneous presentation would lead to more bi-directional visual transitions between the different episodes than would the sequential presentation. While simultaneously presented episodes enable learners to directly shift their visual attention between episodes in either direction with equal ease, sequentially presented episodes are likely to favor a more linear processing of the episodes. Finally, it was assumed that the simultaneous presentation would result in more successful learning of higher-order relationships than would the sequential presentation. Because simultaneously presented episodes facilitate necessary compare and contrast processes, they should therefore support the identification of higher-order relationships of the animated subject matter.

3.4.2 Participants, Material, and Procedure

A total of 60 pre-service teacher students volunteered for the study. They received financial compensation for their participation in the study. None of the students who participated in the study had experience in sailing. The students were randomly assigned to the sequential group (25 female and 5 male students, mean age = 21.83 years, $SD = 2.15$) and the simultaneous group (26 female and 4 male students, mean age = 22.00 years, $SD = 3.43$). The eye movements of eight randomly selected students from each group were recorded while they studied the animation episodes.

The students whose eye movements were not recorded participated in groups of up to four individuals, whereas the students whose eye movements were recorded participated individually. Each student was individually seated in front of a computer. To begin, the students completed a pretest that consisted of six items assessing the students' prior knowledge of the principles that apply to sailing. Next, the students studied three printed pages that depicted and explained the graphic entities shown in the animations: a compass, arrows indicating the wind direction, an arrow representing the magnitude and direction of a force, a parallelogram and how it is used to resolve a force into its component forces, and a buoy that the yacht is to reach by tacking. Thereafter, printed instructions informed the students that they could study the animation for up to 9 min, make use of the media player to start, stop, forward and rewind the animation, and watch the animation as often as they wished within the limits of the learning time. During the subsequent learning phase, each student watched the animation by taking advantage of a standard media player on the computer. Lastly, the students completed a posttest. The test contained 14 multiple-choice items that were provided in a verbal format, as well as 10 open items that were presented in a mixed verbal-graphic format. Each item required the students to make use of higher-order relationships that bridge two or more sailing courses, as well as to apply these relationships to sailing courses different from those visualized in the animations.

Students' eye movements were recorded while they watched the animation. The recording device was a SensoMotoric Instruments (SMI) RED binocular remote eye tracker. It consisted of a 22 in. widescreen display with a resolution of 1680 px. (width) × 1050 px. (height), infrared light emitting diodes, and eye tracking cameras. For each student, a nine-point calibration procedure was conducted until horizontal and vertical accuracy was at least 1.0°. In the sequential group, average horizontal and vertical accuracies were $M = 0.44°$ ($SD = 0.17°$) and $M = 0.65°$ ($SD = 0.15°$) respectively. In the simultaneous group, average horizontal and vertical accuracies were $M = 0.58°$ ($SD = 0.12°$) and $M = 0.39°$ ($SD = 0.27°$). After calibration, eye movements were recorded at a sampling rate of 60 Hz. In addition, the displays on the computer screens were recorded using the SMI Video Analysis Package, a program that permits the capture and analysis of dynamic stimuli. With respect to the recorded eye movements, three kinds of gaze-based events were distinguished: fixations, saccades, and transitions. Fixations were defined as events in which the gaze remained for at least 80 ms within a maximum radius of 100 pixels (cf. Blignaut & Beelders, 2009). Saccades were defined as gaze movements from one fixation to another fixation. Transitions were defined as saccades that take place between areas of interests, i.e. they start in one area of interest and end in another area of interest (cf. Holmqvist et al. 2011).

The SMI analysis software BeGaze was used to determine how often the students performed saccades within and transitions between different areas of interest. The analysis was conducted at two different levels of detail. At the first level, four areas of interest were defined within the display. Each area of interest covered the complete spatial region and temporal extent of one of the four sailing courses: running, broad reach, close hauled, or tacking. That is, whenever a sailing course was visible in the screen capture, the corresponding area of interest was active. For all screen captures from the sequential animation, only one area of interest was active at a time. However, for captures from the simultaneous animation, all areas of interest were active at once. Figure 3.6 shows the four areas of interest for a capture from the simultaneous animation.

At the second, more detailed level, three areas of interest were defined. These areas referred to the sailing courses running, broad reach, and close hauled. Each area of interest covered the spatial region and temporal extent of the corresponding course in which the yacht's hull, the sail, and the different forces are shown. Because no forces were shown in the fourth course, tacking, this course was excluded from the analysis. Figure 3.7 shows an area of interest for a capture taken from the sequential animation. For both levels of detail, we determined the frequencies of saccades that took place within the different areas of interest as well as the frequencies of transitions that took place between the different areas of interest. Due to the limits of the eye tracker's resolution, it was not possible to conduct a satisfactory analysis at an even more fine-grained level for individual components such as the sail and the single forces.

3.4.3 Results

Table 3.1 shows the transition matrices for the areas of interest that covered the complete sailing courses. The sequential group ($M = 1340.5$, $SD = 198.6$) produced significantly more fixations overall than the simultaneous group ($M = 1169.3$, $SD = 76.0$; $t(14) = 2.28$, $p < 0.05$; $d = 1.13$). The sequential group ($M = 1003.5$, $SD =$

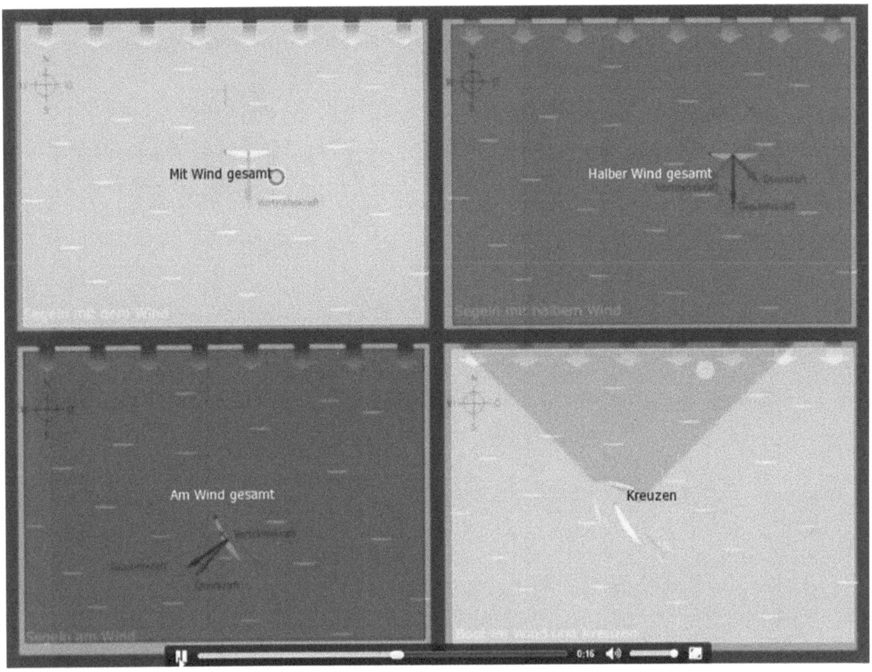

Fig. 3.6 Four areas of interest that cover the complete sailing courses in the simultaneous animation

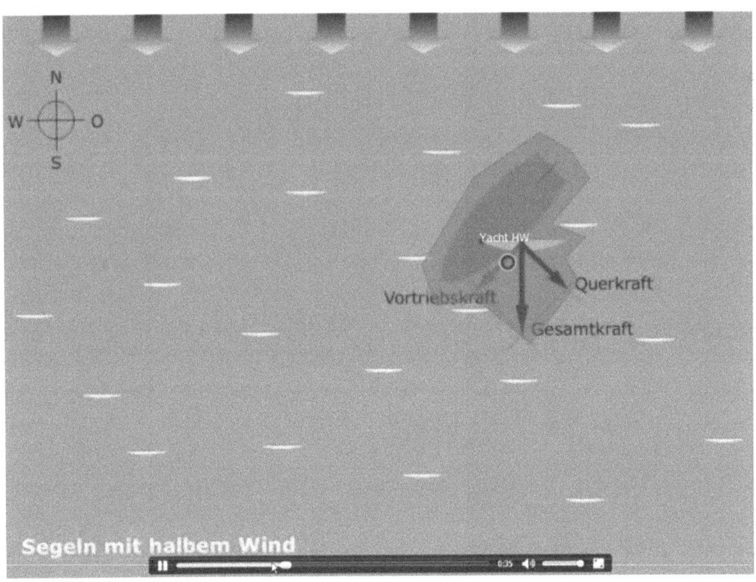

Fig. 3.7 An area of interest that covers specific details of a sailing course in the sequential animation

Table 3.1 The average frequencies (*M*) and standard deviations (*SD*) of the saccades within the areas of interest (values located on the principal diagonal) and the transitions between the areas of interest (values located above and below the principle diagonal)

		Running		Broad reach		Close hauled		Tacking	
Group	AOI	*M*	*SD*	*M*	*SD*	*M*	*SD*	*M*	*SD*
Sequential	Running	146.6	76.5	7.0	3.3	0.3	0.7	0.5	0.9
(n = 8)	Broad reach	2.3	2.6	260.5	71.5	10.0	3.0	0.1	0.4
	Close hauled	0.5	1.0	4.3	3.2	267.0	79.0	11.8	6.3
	Tacking	1.8	2.6	0.5	1.4	6.6	8.7	283.9	52.7
Simultaneous	Running	77.4	28.0	18.6	5.7	11.6	4.5	2.9	2.0
(n = 8)	Broad reach	16.4	6.0	150.9	41.0	13.0	5.0	12.4	6.0
	Close hauled	13.5	6.5	7.9	5.6	205.5	56.1	21.1	10.5
	Tacking	2.8	2.3	11.6	5.0	17.9	8.9	182.4	56.1

230.7) also showed significantly more saccades and transitions (i.e., saccades and transitions across the complete transition matrix) than the simultaneous group (M = 765.8, SD = 95.7; t-test for independent groups with inhomogeneous variances, $t(9.3) = 2.69$, $p < 0.05$; $d = 1.35$).

As predicted, the sequential group ($M = 45.5$, $SD = 22.6$) exhibited significantly fewer transitions between the different areas of interest (i.e., values located above and below the principal diagonal of the transition matrix) than the simultaneous group ($M = 149.6$, $SD = 25.5$; $t(14) = -8.64$, $p < 0.001$; $d = 4.32$). Figure 3.8 exemplifies transitions that occurred between different episodes during learning from the simultaneous animation. Conversely, the sequential group ($M = 958.0$, $SD = 215.2$) exhibited significantly more saccades within the different areas of interest (i.e., values located on the principal diagonal of the transition matrix) than the simultaneous group ($M = 616.1$, $SD = 99.46$; $t(14) = 4.08$, $p < 0.01$; $d = 2.04$). Furthermore, the sequential group made 85.7% more transitions in one direction (i.e., transitions located above the principal diagonal of the transition matrix; $M = 29.6$, $SD = 9.5$) than in the opposite direction (i.e., transitions located below the principal diagonal of the transition matrix; $M = 15.9$, $SD = 13.6$). In contrast, the simultaneous group made only 14.7% more transitions in one direction ($M = 79.6$, $SD = 13.2$) than in the opposite direction ($M = 70.0$, $SD = 13.3$). This difference between the two groups is significant (t-test for independent groups with inhomogeneous variances, $t(7)$ = 2.66, $p < 0.05$; $d = 1.33$).

The results found with respect to the areas of interest that covered the complete sailing courses were entirely consistent with the results yielded with respect to the areas of interest that only covered the yacht's hull, the sail, and the different forces. Table 3.2 presents the corresponding transition matrices. Again, the sequential group (M = 138.0, SD = 42.9) showed significantly more saccades and transitions than the simultaneous group ($M = 69.9$, $SD = 33.8$; $t(14) = 3.53$, $p < 0.01$; $d = 1.76$).

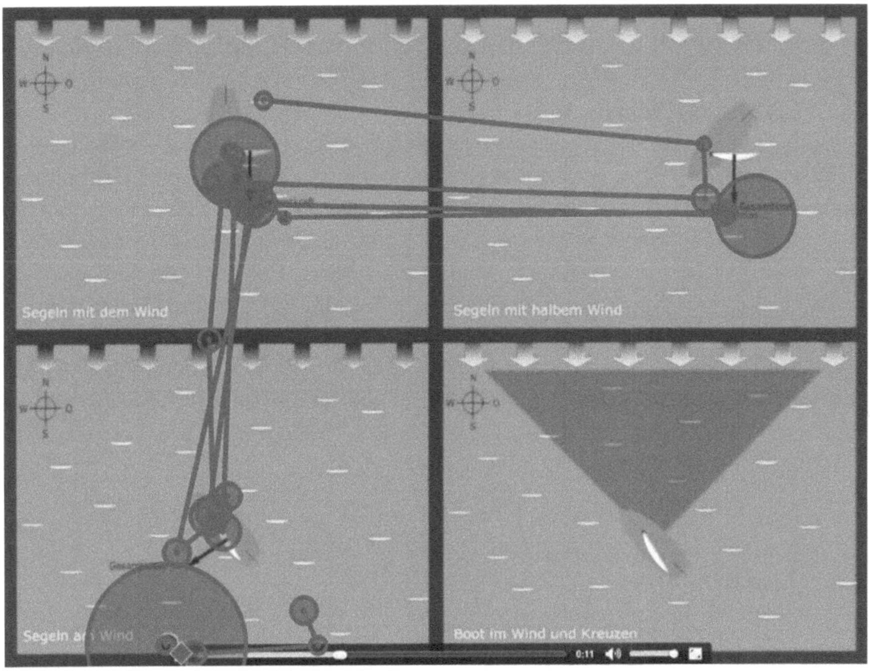

Fig. 3.8 A scan path that visualizes a student's eye movements of the past 10 s. *Larger circles* indicate longer fixation times

Table 3.2 The average frequencies (*M*) and standard deviations (*SD*) of the saccades within the areas of interest (values located on the principal diagonal) and the transitions between the areas of interest (values located above and below the principle diagonal)

		Running		Broad reach		Close hauled	
Group	AOI	*M*	*SD*	*M*	*SD*	*M*	*SD*
Sequential	Running	12.9	7.8	0	0	0	0
($n = 8$)	Broad reach	0	0	58.5	23.9	0.1	0.4
	Close hauled	0.1	0.4	0	0	66.4	22.4
Simultaneous	Running	9.8	15.6	2.0	2.1	2.0	2.4
($n = 8$)	Broad reach	1.6	2.1	18.4	11.7	1.8	1.7
	Close hauled	1.9	2.1	0.9	1.1	31.6	21.3

However, the sequential group ($M = 0.25$, $SD = 0.46$) exhibited significantly fewer transitions between the areas of interest than the simultaneous group ($M = 10.1$, $SD = 7.4$; t-test for independent groups with inhomogeneous variances, $t(7.1) = -3.78$, $p < 0.01$; $d = 1.88$). Conversely, the sequential group ($M = 137.8$, $SD = 43.2$) exhibited significantly more saccades within the different areas of interest than the simultaneous group ($M = 59.8$, $SD = 29.0$; $t(14) = 4.24$, $p < 0.01$; $d = 2.12$). Furthermore, the sequential group showed just one transition between the areas of interest in either direction ($M = 0.13$ and $SD = 0.35$ for transitions in one direction; $M = 0.13$

and $SD = 0.35$ for transitions in the opposite direction). In contrast, the simultaneous group showed several transitions between the areas of interest in both directions ($M = 5.8$ and $SD = 4.5$ for transitions in one direction; $M = 4.4$ and $SD = 3.0$ for transitions in the opposite direction).

Both groups of students possessed very little prior knowledge about the mechanisms underlying sailing (sequential group $M = 0.12$ (1.94%), $SD = 0.36$; simultaneous group $M = 0.07$ (1.11%), $SD = 0.25$). The difference between the sequential group and the simultaneous group in prior knowledge was not significant ($t(58) = 0.61$, *n.s.*). Furthermore, prior knowledge did not significantly correlate with the performance on the posttest ($r = 0.08$, *n.s.*). Therefore, prior knowledge was not considered any further. An analysis of the posttest items revealed an acceptable overall reliability (Cronbach's $\alpha = 0.76$). On the posttest, the simultaneous group performed significantly better than the sequential group (simultaneous group $M = 47.40$ (69.79%), $SD = 9.91$; sequential group $M = 42.10$ (61.91%), $SD = 8.62$; $t(58) = 2.20$, $p < 0.05$; Cohen's $d = 0.57$).

3.5 Discussion

The study reported in this chapter compared the educational effectiveness of the sequential and simultaneous presentation of animation episodes. It yielded three main results. First, the simultaneous presentation resulted in significantly more visual transitions between the episodes than the sequential presentation. Second, the simultaneous presentation lead to significantly more bi-directional visual transitions between the episodes than the sequential presentation. Third, the learning of higher-order relationships was significantly better from the simultaneously presented episodes than from the sequentially presented episodes.

The first two results were consistently found at two levels of detail: (1) the broad scale areas of interest that covered the complete sailing courses and (2) the finer grained areas of interest that covered only the yacht's hull, the sail, and the different forces. These results likely reflect the different affordances that simultaneous and sequential presentations of animated episodes offer to learners. While the simultaneous presentation invites learners to shift their visual attention back and forth between episodes, the sequential presentation suggests that the episodes be processed one after the other. Nevertheless, even with simultaneously presented episodes, uni-directional transitions occurred slightly more often than bi-directional transitions. This tendency may reflect the standard "reading order" from left to right and from top to bottom. The arrangement of the four sailing episodes is consistent with this possibility (cf. Fig. 3.5). Regarding the effects of different presentation formats on learning from pictorial representations, the present results have similarities with those from Lowe, Schnotz, and Rasch (2011). They found that variations in the spatiotemporal arrangement of a set of pictures portraying kangaroo locomotion affected the learners' performance on a sequencing task. The arrangement that provided learners with more affordances for comparing and contrasting important relationships between the kangaroo's body parts resulted in the best performance.

Although the first two results do not directly verify that learners compared the simultaneously presented episodes, the third result suggests that they went beyond a mere mechanistic shifting of their visual attention between episodes. In light of all three results, it appears plausible that learners actively compared and contrasted the episodes in an effort to identify higher-order relationships of the animated subject matter. However, it would be pedagogically unwise to conclude from these results that learners should be instructed to simply shift their visual attention as often as possible between simultaneously presented episodes. According to the APM, it is the comparisons and contrasts made between co-present episodes in order to establish meaningful relationships that are crucial here – the repeated shifts of visual attention are merely perceptual indicators of this deeper processing. Furthermore, the APM also suggests that successful learning from simultaneously as well as sequentially presented episodes requires both within-episode and between-episode interrogation. Without both, it would be difficult for the learner to construct the hierarchical knowledge structure that characterizes a well-developed mental model.

Furthermore, although the simultaneous presentation of animated episodes makes the display much more complex and provides considerably more information to the learners, it did not negatively affect learning as might be expected from the perspective of theoretical frameworks such as the Cognitive Load Theory (e.g., Ayres & Sweller, 2014; van Gog, Paas, & Sweller, 2010). In fact, quite the opposite occurred; it seems that learners are able to regulate their interrogation of an animation in order to avoid being overwhelmed. Perhaps a more sophisticated view of juvenile and adult learners is required in terms of their ability to adapt to complex information environments, for instance, on the basis of perceptual as well as cognitive techniques and strategies (cf. Ploetzner, 2016; Ploetzner, Lowe, & Schlag, 2013; Ploetzner & Schlag, 2013; Kombartzky, Ploetzner, Schlag, & Metz, 2010; see also Ploetzner & Breyer, 2017, this volume). This possibility fits well with the theoretical framework provided by the APM. From this perspective, an animation design that contains deliberate spatiotemporal manipulation of the referent subject matter does not necessarily prejudice learning simply because it results in a more 'difficult' display. Rather, the key issue is what affordances for task-appropriate processing are made available to learners as a result of that manipulation. In the case of converting sequentially occurring episodes into a simultaneous format, it appears that the benefits of being able to carry out the comparisons and contrasts necessary to establish higher-order relationships outweigh the possible costs associated with a more complex and information-rich display.

The induction of higher-order relationships relies on the identification of regularities in the learning material (cf. Holland et al., 1986; Klauer & Leutner, 2012). If the learning material is an expository animation that is not accompanied by verbal explanations, this implies that more than just one instance of a dynamic process needs to be visualized because a single instance of a process does not reveal the process's variability. As a consequence, it is impossible for the learner to identify regularities in the animated subject matter. If multiple instances of a dynamic process are to be displayed to the learner, they can be presented sequentially, simultaneously, or as a combination of both. For instance, the four sailing courses could be presented to the learner as a sequence of several simultaneous presentations with each presentation showing two animation episodes next to each other: initially running and broad reach, then broad reach and close hauled (see Fig. 3.3), and finally close hauled and tacking. Because the results of the study reported in this chapter suggest that the simultaneous presentation of animation episodes affords learners to engage in compare and contrast processes and – as a consequence of this engagement – facilitates the induction of higher-order relationships, a combination of sequential and simultaneous presentations might be especially favorable if a larger number of animation episodes are to be displayed.

In multimedia learning environments, the induction of higher-order relationships does not commonly rely on learning from animation but rather on learning from simulation. While animations merely imitate dynamic processes by presenting fixed sequences of pre-manufactured images, simulations computationally model dynamic processes by making use of formal modeling techniques (cf. Plass & Schwartz, 2014; Ploetzner & Lowe, 2012). Simulations are frequently employed in discovery learning (de Jong & Lazonder, 2014), whereby the learner repeatedly modifies the values of parameters that the simulation offers to the learner via the user interface. By applying the underlying model to the chosen parameter values, the simulation generates symbolic or pictorial representations that describe or visualize the consequences of these modifications. Thereafter, the learner interrogates the generated representations in an attempt to discover regularities. Thus, in contrast to learning from animation, when learning from simulation it is not the designer, but rather the learner who takes responsibility for deciding which instances of a dynamic process are visualized and in which order. Educational research, however, has convincingly demonstrated that discovery learning from simulation poses manifold challenges to learners and therefore requires extensive guidance (cf. Clark, Yates, Early, & Moulton, 2011; de Jong & Lazonder, 2014; de Jong & van Joolingen, 1998). This is especially true if the learners lack the methodological skills for appropriately organizing the discovery process.

When learners do not possess the methodological skills needed to learn effectively from simulation, then learning from multiple animation episodes might be a more promising alternative, particularly if related episodes are presented simultaneously. As in learning from simulation, learning from multiple animation episodes still requires the learner to systematically interrogate the different episodes in order to identify regularities within, as well as across, episodes. In contrast to learning from simulation, however, learners are no longer required to produce sufficiently informative episodes by themselves. Instead, the animation designer supports the learner by providing episodes that cover variability and constancy in the animated subject matter such that the induction of the relevant relationships becomes not only feasible but also realistic.

Acknowledgements This research was supported by the German Academic Exchange Service (DAAD) and the Australian Technology Network (ATN) within the "Joint Research Cooperation Scheme". We thank Onno Bahns and Marie Kösters for supporting the data collection and analysis.

References

Ayres, P., & Sweller, J. (2014). The split-attention principle in multimedia learning. In R. E. Mayer (Ed.), *The Cambridge handbook of multimedia learning* (2nd ed., pp. 206–226). New York: Cambridge University Press.

Bark, A. (2009). *Sportküstenschifferschein* [License for coastal sport skippers] (9. Aufl.). Bielefeld: Delius Klasing Verlag.

Blignaut, P., & Beelders, T. (2009). The effect of fixational eye movements on fixation identification with a dispersion-based fixation detection algorithm. *Journal of Eye Movement Research, 2*, 1–14.

Clark, R. E., Yates, K., Early, S., & Moulton, K. (2011). An analysis of the failure of electronic media and discovery-based learning: Evidence for the performance benefits of guided learning methods. In K. H. Silber & W. R. Foshay (Eds.), *Handbook of improving performance in the workplace* (pp. 263–297). San Francisco, CA: Pfeiffer.

de Jong, T., & Lazonder, A. W. (2014). The guided discovery principle in multimedia learning. In R. E. Mayer (Ed.), *The Cambridge handbook of multimedia learning* (2nd ed., pp. 371–390). New York: Cambridge University Press.

de Jong, T., & van Joolingen, W. R. (1998). Scientific discovery learning with computer simulations of conceptual domains. *Review of Educational Research, 68*, 179–201.

Ginns, P. (2005). Meta-analysis of the modality effect. *Learning and Instruction, 15*, 313–331.

Holland, J. H., Holyoak, K. J., Nisbett, R. E., & Thagard, P. R. (1986). *Induction – Processes of inference, learning, and discovery*. Cambridge, MA: MIT Press.

Holmqvist, K., Nyström, M., Andersson, R., Dewhurst, R., Jarodzka, H., & van de Weijer, J. (2011). *Eye tracking – A comprehensive guide to methods and measures*. New York: Oxford University Press.

Klauer, K. J., & Leutner, D. (2012). *Lehren und Lernen – Einführung in die Instruktionspsychologie* (2. Aufl.) [Teaching and learning – Introduction to the psychology of instruction]. Weinheim: Beltz Verlag.

Klauer, K. J., & Phye, G. D. (2008). Inductive reasoning: A training approach. *Review of Educational Research, 78*, 85–123.

Kombartzky, U., Ploetzner, R., Schlag, S., & Metz, B. (2010). Developing and evaluating a strategy for learning from animation. *Learning and Instruction, 20*, 424–433.

Low, R., & Sweller, J. (2014). The modality principle in multimedia learning. In R. E. Mayer (Ed.), *The Cambridge handbook of multimedia learning* (2nd ed., pp. 227–246). New York: Cambridge University Press.

Lowe, R. K. (1999). Extracting information from an animation during complex visual learning. *European Journal of Psychology of Education, 14*, 225–244.

Lowe, R. K. (2004). Interrogation of a dynamic visualization during learning. *Learning and Instruction, 14*, 257–274.

Lowe, R. K., & Boucheix, J.-M. (2008). Learning from animated diagrams: How are mental models built? In G. Stapleton, J. Howse, & J. Lee (Eds.), *Diagrammatic representations and inference – Proceedings of the fifth international conference, diagrams 2008* (pp. 266–281). Berlin: Springer.

Lowe, R. K., & Boucheix, J.-M. (2011). Cueing complex animations: Does direction of attention foster learning processes? *Learning and Instruction, 21*, 650–663.

Lowe, R., & Boucheix, J.-M. (2017). A composition approach to design of educational animations. In R. Lowe, & R. Ploetzner (Eds.), *Learning from dynamic visualization — Innovations in research and application*. Berlin: Springer (this volume).
Lowe, R. K., & Schnotz, W. (2014). Animation principles in multimedia learning. In R. E. Mayer (Ed.), *The Cambridge handbook of multimedia learning* (2nd ed., pp. 513–546). New York: Cambridge University Press.
Lowe, R., Schnotz, W., & Rasch, T. (2011). Aligning affordances of graphics with learning task requirements. *Applied Cognitive Psychology, 25*, 452–459.
Mayer, R. E. (2009). *Multimedia learning* (2nd ed.). New York: Cambridge University Press.
Mayer, R. E. (2014). Cognitive theory of multimedia learning. In R. E. Mayer (Ed.), *The Cambridge handbook of multimedia learning* (2nd ed., pp. 43–71). New York: Cambridge University Press.
Mayer, R. E., & Fiorella, L. (2014). Principles for reducing extraneous processing in multimedia learning: Coherence, signaling, redundancy, spatial contiguity, and temporal contiguity principles. In R. E. Mayer (Ed.), *The Cambridge handbook of multimedia learning* (2nd ed., pp. 279–315). New York: Cambridge University Press.
Overschmidt, H., & Gliewe, R. (2009). *Das Bodenseeschifferpatent A + D* [Yachtmaster A + D for Lake Constance]. Bielefeld: Delius Klasing Verlag.
Plass, J. L., & Schwartz, R. N. (2014). Multimedia learning with simulations and microworlds. In R. E. Mayer (Ed.), *The Cambridge handbook of multimedia learning* (2nd ed., pp. 729–761). New York: Cambridge University Press.
Ploetzner, R. (2016). *Understanding explanatory animation — An introduction to cognitive design and user support*. Freiburg: eXplanatory!media.
Ploetzner, R., & Breyer, B. (2017). Strategies for learning from animation with and without narration. In R. Lowe, & R. Ploetzner (Eds.), *Learning from dynamic visualization — Innovations in research and application*. Berlin: Springer (this volume).
Ploetzner, R., & Lowe, R. K. (2012). A systematic characterisation of expository animations. *Computers in Human Behavior, 28*, 781–794.
Ploetzner, R., & Lowe, R. K. (2014). Simultaneously presented animations facilitate the learning of higher-order relationships. *Computers in Human Behavior, 34*, 12–22.
Ploetzner, R., Lowe, R., & Schlag, S. (2013). A systematic characterization of cognitive techniques for learning from textual and pictorial representations. *Journal of Education and Learning, 2*, 78–95.
Ploetzner, R., & Schlag, S. (2013). Strategic learning from expository animations: Short- and mid-term effects. *Computers & Education, 69*, 159–168.
Tversky, B., Heiser, J., Mackenzie, R., Lozano, S., & Morrison, J. (2008). Enriching animations. In R. K. Lowe & W. Schnotz (Eds.), *Learning with animation: Research implications for design* (pp. 263–285). New York: Cambridge University Press.
van Gog, T., Paas, F. G. W. C., & Sweller, J. (2010). Cognitive load theory: Advances in research on worked examples, animations, and cognitive load measurement. *Educational Psychology Review, 22*, 375–378.
Zhang, J., & Norman, D. A. (1994). Representations in distributed cognitive tasks. *Cognitive Science, 18*, 87–122.

Chapter 4
Learning from Static and Dynamic Visualizations: What Kind of Questions Should We Ask?

Inga Wagner and Wolfgang Schnotz

4.1 Introduction

Dynamic visualizations have become widely used in learning environments. They can be either animations or videos.[1] What makes them different from static graphics is that they portray continuous temporal change of a subject matter by triggering perception of continuous change (cf. Lowe & Schnotz, 2014, p. 515). Besides being credited with motivating learners, dynamic visualizations are generally assumed to be inherently well suited to conveying information about dynamic processes. However, research regarding their learning effectiveness has produced mixed results (Lowe & Schnotz, 2008; Tversky, Heiser, Mackenzie, Lozano, & Morrison, 2008; Tversky, Morrison, & Bétrancourt, 2002). It has become clear that there is no inherent relationship between the form of representation used to display information and learning effectiveness. On the one hand, information about static content can be conveyed not only by static graphics but also by dynamic visualizations. For example, a static object can be rotated or the assembly of an object from its parts can be demonstrated by a dynamic visualization. On the other hand, information about dynamic content can be conveyed by either a dynamic visualization or a series of static graphics. A meta-analysis by Höffler and Leutner (2007) found a larger effect size for representational, non-decorative dynamic visualizations than for corresponding static graphics, if the learning aim was to acquire procedural motor knowledge such as how to assemble a machine gun (Spangenberg, 1973). All in all, however, research to date has not clearly indicated a superiority of dynamic visualizations over static graphics. In this chapter, we deal only with learning about

[1] The frequently made distinction between (computer-generated) animations and (taken) videos refers to the technique of production, which is not relevant to their psychological (perceptual and cognitive) processing. We use the term "dynamic visualization" to cover both.

I. Wagner (✉) • W. Schnotz
University of Koblenz-Landau, Mainz, Germany
e-mail: wagneri@uni-landau.de; schnotz@uni-landau.de

R. Lowe, R. Ploetzner (eds.), *Learning from Dynamic Visualization*,
DOI 10.1007/978-3-319-56204-9_4

dynamic content. We will argue that whether learning from dynamic visualizations is superior to learning from static graphics is not an appropriate question in this context. There are more fruitful questions such as the following:

(a) What kinds of mental representation are to be constructed, by which learners, and for what purposes?
(b) What kind of information should be emphasized? That is, what relative emphases should be given to spatial versus temporal characteristics of the dynamic content?

We first make the point that learning about dynamics can have different forms and serve different purposes. If the learning task is to recognize specific movement patterns, then learning about dynamics might take place primarily at the perceptual level and lead to an internal perceptual representation. However, if the learning task is to draw inferences from the depicted content, then learning about dynamics might take place primarily at the cognitive level and lead to an internal cognitive representation, a so-called mental model. Next, two empirical studies are described that compared learning from dynamic visualizations with learning from static graphics at both perceptual and cognitive levels. Based on the results of these studies, we argue that it may be more fruitful to analyze graphic displays in terms of the extent to which they emphasize temporal or spatial characteristics of the depicted subject matter, assuming that dynamic and static graphics should be regarded more as being on a continuum than as a strict dichotomy.

4.2 Perceptual and Cognitive Representation of Dynamics

Knowledge about dynamic content can have different forms and serve different purposes. For example, physicians use their patients' movement patterns to diagnose specific diseases while biologists use movement patterns of animals for recognition and categorization of the animals or of their behavior. In these cases, the individual needs a fine-grained perceptual representation of movement patterns. The focus is on answering "What is it?" questions. A rather different kind of knowledge about dynamic content is needed when a student tries to understand how technical devices such as bicycle pumps or biological organs such as nerve cells work. In this case, the individual needs a cognitive representation that captures why a specific sequence of events occurs. The individual is expected to answer questions such as "Why and how does it work?". We assume that answering what-questions and answering why-and-how-questions require different kinds of mental representations. In the first case, learners need to process information in a way that results in a perceptual representation of the dynamics. In the second case, they are required to engage in cognitive processing that leads to a more abstract kind of mental representation – a mental model of the dynamics at hand (see also Lowe & Boucheix, 2017, this volume). In the next sections, we consider perceptual and then cognitive learning about dynamics.

4.2.1 Perceptual Learning About Dynamics

Human perception is based on pre-attentive and data-driven bottom-up processes guided by highly automated visual routines (Neisser, 1976; Ullman, 1984). We live in a richly dynamic environment that entails temporally stable as well as temporally changing components. It is therefore unsurprising that the perceptual system is well equipped to extract spatial-temporal invariants from dynamic situations. The concept of invariants does not imply that there is no movement or no change in time at all. Rather, it means that *patterns* of movement or change are invariant and, thus, predictable. The gaits of humans, dogs, and horses, for example, are characterized by specific recurring spatial-temporal patterns, although the configuration of feet and limbs changes continuously. As demonstrated by Marey (1874), the human perceptual system is well equipped to detect such invariants and to encode them in terms of dynamic perceptual schemata. For example, if a person dressed in black but wearing lights at the shoulders, elbows, hands, hips, knees and feet is standing still in front of a wall covered by black velvet, it is impossible to recognize that this is a human being. Only a seemingly arbitrary distribution of light spots can be perceived. If the person starts walking, however, the observer immediately recognizes the presence of a human figure. The invariant pattern in the movement of the lights is sufficient to activate the appropriate dynamic perceptual schema that allows the display to be recognized as a person walking. Similarly, we can acquire schemata of the various movement patterns of animals. This allows us to distinguish different types of horse gaits such as walking, trotting, and galloping, for example. These schemata represent the typical (i.e. the most frequently encountered) spatial-temporal patterns.

Accordingly, perceptual learning about dynamics involves the construction of dynamic perceptual schemata that capture invariants within re-occurring movement patterns. In order to be sufficiently distinctive with regard to the details of movement, such perceptual schemata need to be sufficiently fine-grained with respect to temporal information. Because the temporal granularity of dynamic visualizations is usually higher than that employed in a series of static graphics, one can hypothesize that learning about dynamics at the perceptual level should be more effective from dynamic visualizations than from static graphics. Based on these considerations, one can formulate the following hypothesis:

(H1) If individuals are expected to attain knowledge about dynamics at the perceptual level in order to categorize different kinds of movement patterns, learning from dynamic visualizations should be more effective than learning from a set of static graphics.

In addition to its higher effectiveness, perceptual learning about dynamics from dynamic visualizations should also require less mental effort from learners than learning from a corresponding set of static graphics. If a complete perceptual mental representation of a process is to be built up with the help of static graphics, learners have to simulate the process internally by themselves and to fill the "gaps" between the depicted graphics. Such gap filling is not necessary when learning occurs with the help of a dynamic visualization. Thus, learning from static graphics requires learners to perform internal processes that consume their mental capacities so that perceptual learning from static graphics might be more effortful than perceptual learning from dynamic visualizations. This leads to the following hypothesis:

(H2) Perceptual learning about dynamics from dynamic visualizations should require less mental effort than learning from a set of static graphics.

In order to test these hypotheses, a learning experiment about the acquisition of perceptual representations of visible movements was performed (cf. Wagner, 2013).

4.2.2 Study I: Perceptual Representations of Visible Movement Patterns

Method 58 university students of social and educational sciences (age: $M = 23.1$ years; $SD = 3.9$ years; sex: 81.0% female) participated in the study. Participants were randomly distributed to two different experimental conditions. One group ($n = 30$) learned about the dynamic patterns of four different horse gaits (gallop, trot, walk, and running walk) with the help of four dynamic visualizations. The other group ($n = 28$) learned about the same horse gaits with the help of eight static graphics for each gait. The learning task for the participants was to encode the different movement patterns in order to be able to recognize and distinguish between the different horse gaits after the learning phase. Participants were instructed as follows: "Please carefully study the horse gaits that will be shown to you. After the learning phase you will be given a test in which dynamic visualizations and static graphics of the horse gaits will be presented to you. After each item, you will be asked to decide which of the horse gaits you have just seen." In the static graphics condition, all graphics in the allocated set were presented simultaneously on the screen. In each condition, presentation time for a single gait was 10 s. The presentation order of horse gaits was rotated between participants to avoid sequential effects in learning. Figure 4.1 shows a snapshot from the dynamic learning material depicting a trotting horse. Figure 4.2 shows a set of static graphics presenting a horse's walk.

After the learning phase, participants took a computerized test that required them to recognize the different horse gaits that they had just seen. This test consisted of 16 dynamic pictorial items and 16 static pictorial items that showed the different horse gaits. Participants had to decide which horse gait was shown by each item and respond via a pencil and paper answer sheet. Both dynamic and static item formats were integrated into the posttest so that there was no bias towards either of the two learning conditions due to how the test material was presented. The difficulty of the dynamic test items was varied by using different presentation times, ranging from 10 s (easiest) to 2.5 s (most difficult). The difficulty of the static items was varied by using different numbers of static graphics to represent one gait, ranging from 8 static graphics (easiest) to 2 static graphics (most difficult). The order of dynamic and static items was also varied to avoid sequential effects between the items. In

Fig. 4.1 Example of a snapshot from the presentation of horse gaits in the dynamic condition (horse trotting)

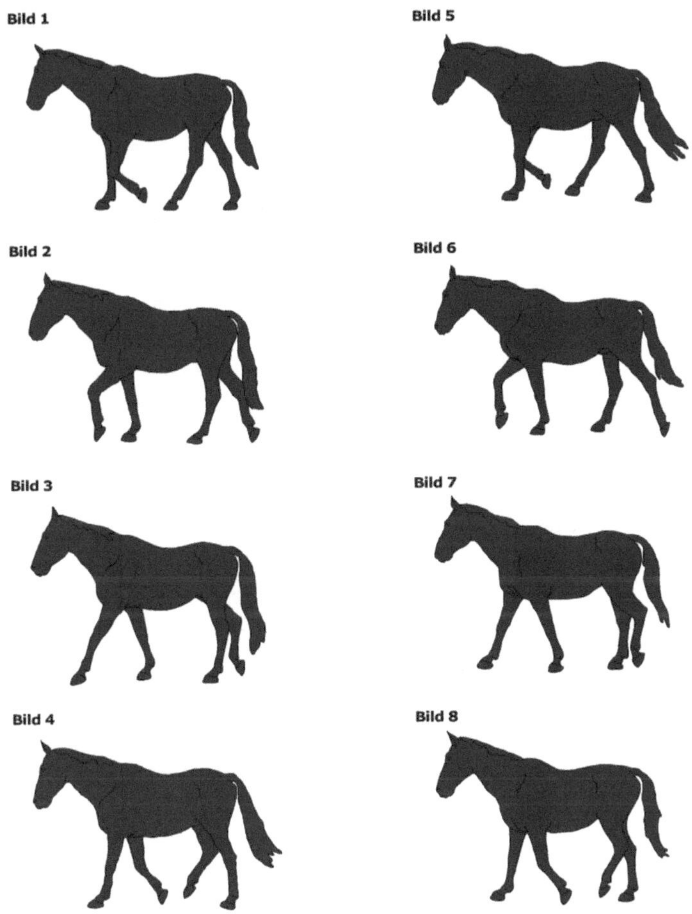

Fig. 4.2 Example of the presentation of horse gaits in the static graphic condition (horse walking)

Table 4.1 Means and standard deviations of the dependent variables

Criterion	Dynamic visualization ($n = 30$)			Static graphics ($n = 28$)		
All items (max. score = 32)	*M* 25.38	M_{rel} 0.79	*SD* 6.05	*M* 24.36	M_{rel} 0.76	*SD* 6.49
Dynamic items (max. score = 16)	*M* 13.50	M_{rel} 0.84	*SD* 3.27	*M* 12.57	M_{rel} 0.79	*SD* 3.71
Static items (max. score = 16)	*M* 11.97	M_{rel} 0.75	*SD* 3.15	*M* 11.79	M_{rel} 0.74	*SD* 3.38
Mental effort while learning (max. score = 18)	*M* 9.43	M_{rel} 0.52	*SD* 3.41	*M* 12.54	M_{rel} 0.70	*SD* 3.27
Mental effort while solvingthe items of the posttest (max. score = 18)	*M* 11.50	M_{rel} 0.64	*SD* 3.56	*M* 13.04	M_{rel} 0.72	*SD* 1.88

Note: There were 30 participants in the dynamic condition. Because of one missing value, the means and standard deviations of the dependent variables "all items" and "static items" only refer to 29 participants

addition to the posttest, self-reported mental effort was measured with respect to both learning about horse gaits and solving the items of the posttest. Mental effort was measured by an adapted and translated version of the "Cognitive load rating scale" (Paas, 1992; Paas, van Merriënboer, & Adam, 1994). Participants reported the mental effort that they had invested into learning and into solving the items of the posttest by rating its intensity from 1 ("very, very low mental effort") to 9 ("very, very high mental effort"). A second item asked participants to rate their perceived difficulty of the learning task and of the posttest, ranging from 1 ("very, very easy") to 9 ("very, very difficult"). These self-report measures were used in the current study because they would also be readily applicable in natural settings. However, a possible disadvantage of these subjective ratings might be the instability of the individual's framework of reference which can change during the course of learning due to adaptation processes (Schnotz & Kürschner, 2007). Participants were also tested for their prior perceptual experience with horse gaits and for their spatial ability because these two variables were assumed to be possible covariates (cf. Sanchez & Wiley, 2017, this volume).

Comparison of Animated and Static Graphics Regarding Performance Measures Table 4.1 shows the means and standard deviations of the relevant criteria of Study I. There was no significant difference between the dynamic and the static graphic conditions regarding their performances in the posttest ($t(55) = 0.62$, n.s.). Considering specifically only the dynamic and only the static items of the posttest, there were also no significant differences between the two conditions (dynamic items: $t(56) = 1.01$, n.s.; static items: $t(55) = 0.21$, n.s.). Performance in the dynamic condition was slightly better than in the static graphic condition with respect to all three criteria. Hypothesis H1 stating that learning about dynamics at a perceptual level should be more effective from dynamic visualizations than from static graphics, could therefore not be confirmed. This indicates that participants who had learnt from static graphics were also able to build up fine-grained perceptual representations of the different horse gaits. Their perceptual representations did not seem to strongly differ qualitatively from those representations that were built up by participants having learnt about horse gaits from dynamic visualizations so that both groups yielded much the same scores in the posttest. However, it remains unclear if, and to what extent, cognitive processes such as drawing inferences were involved in perceptual learning about horse gaits from static graphics.

There was no significant interaction between the presentation format of the learning condition (dynamic vs. static graphics) and the presentation format of the items in the posttest (dynamic vs. static items; $F(1, 55) = 0.97$, n.s.). Learning from static graphics did not lead to better performance with static items and neither did learning from dynamic visualizations lead to better performance with dynamic items. Rather, both groups showed significantly better performance with dynamic items ($F(1, 55) = 10.99$, $p = 0.002$, $\eta_p^2 = 0.17$). This may be a further indication that participants who had studied horse gaits from static graphics also built up dynamic perceptual representations of the corresponding movement patterns that were well suited for answering dynamic items.

In order to test for possible interactions (i) between learning condition and prior perceptual experiences with horse gaits and (ii) between learning condition and spatial ability, we conducted linear multiple regression analyses. Learning condition was included as a categorical contrast-coded variable into the analyses whereas prior perceptual experiences and spatial ability were centered metric variables (cf. Richter, 2007). There were neither any significant interactions between learning condition and prior perceptual experiences nor between learning condition and spatial ability focusing participants' performances in the posttest, in the dynamic items and in the static items as dependent variables (prior perceptual experience – posttest: $B = -0.06$, $SE = 0.43$, $t = -0.14$, n.s.; dynamic items: $B = -0.17$, $SE = 0.24$, $t = -0.72$, n.s.; static items: $B = 0.07$, $SE = 0.22$, $t = 0.33$, n.s.; spatial ability – posttest: $B = -0.17$, $SE = 0.13$, $t = -1.34$, n.s.; dynamic items: $B = -0.05$, $SE = 0.07$, $t = -0.78$, n.s.; static items: $B = -0.11$, $SE = 0.07$, $t = -1.67$, n.s.).

Comparison of Dynamic and Static Graphics Regarding Mental Effort Measures Whereas no significant differences were found between the two learning conditions regarding their performances in the posttest, there was a highly significant difference between the dynamic and the static graphic condition regarding the mental effort that had been invested into learning about the movement patterns of horse gaits ($t(56) = -3.53$, p(one-tailed) < 0.001, $d = 0.93$). Participants who studied from dynamic visualizations reported investing significantly less mental effort into learning than those who studied from static graphics. This result is consistent with the prediction of hypothesis H2 that perceptual learning about dynamics from dynamic visualizations should require less effort than learning from a set of static graphics.

Based on z-standardizations of posttest performance scores and scores of mental effort invested into learning, an index of learning efficiency was derived by interrelating both values. Average performance associated with average effort was interpreted as indicating average learning efficiency. Increasing the performance score by a certain z-value should be associated with the same increase in the mental effort score to further indicate average learning efficiency. According to these considerations, learning efficiency was significantly higher in the dynamic than in the static graphic condition ($t(55) = 2.61$, $p(\text{one-tailed}) = 0.005$, $d = 0.69$). Building up a perceptual dynamic representation seems to be mentally less taxing when done from dynamic visualizations than when done from static graphics.

In addition, participants' performances in the posttest were considered in relation to the mental effort they invested while solving the items of the posttest. Based on z-standardizations, both scores were related to each other in the same way as described for learning efficiency. This led to an index for the efficiency of recall of the perceptual dynamic information in the test situation. The index represents the ease with which information can be retrieved from the perceptual dynamic schema and be applied for solving appropriate tasks. Learning from dynamic visualizations again led to a significantly higher efficiency in recalling perceptual dynamic information than learning from static graphics ($t(55) = 1.68$, $p(\text{one-tailed}) = 0.0495$, $d = 0.45$). Retrieving and applying information from a perceptual dynamic representation seems to be mentally less taxing when done from dynamic visualizations than when done from static graphics.

Conclusion The results of the current study revealed that participants who studied movement patterns from dynamic visualizations or from static graphics could recognize these movement patterns equally well. This appears to indicate that dynamic perceptual schemata can be built up satisfactorily not only from dynamic visualizations but also from a set of static graphics. These schemata also seem to contain more extensive information about a specific movement pattern because participants who studied static graphics were better at solving dynamic posttest items. Learning from dynamic visualizations and learning from static graphics both seem to lead to qualitatively comparable dynamic perceptual schemata. However, the construction of these schemata seems to be more mentally taxing when studying static as opposed to dynamic graphics. According to participant reports, learning from static graphics also leads to more effortful retrieval and application of the dynamic information. Therefore, studying dynamic patterns from dynamic visualizations offers an advantage with respect to learning efficiency and the application of dynamic information.

4.2.3 Cognitive Learning About Dynamics

As mentioned above, individuals who are expected to learn about how and why something works need to engage in cognitive processing that leads to a mental model of the underlying dynamics. In the field of neurobiology, for example, students might be expected to understand from pictorial information how signal transmission at neuronal synapses works. Learners will first represent the pictorial information by a visual image in working memory (Kosslyn, 1994; Mayer, 2014; Schnotz, 2014), which will then serve as input for further semantic processing leading to the construction of a mental model (Johnson-Laird, 1983; Johnson-Laird & Byrne, 1991). Generally speaking, mental models are internal cognitive structures that represent an external subject matter based on analogy relations, that is, by mapping external structures on internal mental structures (Gentner, 1989; Schnotz, 1993). Whereas perceptual representations are sensory specific, mental models are more abstract as they can integrate information from different sensory modalities. Mental models are not necessarily dynamic: If a subject matter includes only stable spatial relations, there is no reason to include any dynamic characteristics into a mental model. However, if the subject matter is dynamic, that is, undergoes processes of change, the corresponding mental model needs to have dynamic characteristics as well (cf. Lowe & Boucheix, 2008). In this case, the structure mapping includes both spatial and temporal structures. Such models can be "run". That is, they can be triggered to simulate external processes by corresponding sequences of mental model states. We call such models of dynamic subject matter 'dynamic mental model'.

Following Schnotz and Lowe (2008), we do not expect dynamic mental models to include a full continuous sequence of internal 'snapshots' of high temporal granularity corresponding to the sequence of states occurring in the external process. Instead, we assume dynamic mental models to consist of a finite number of highly informative internal key states which represent corresponding external key states.

The concept of key states needs further consideration. Most processes can be segmented into meaningful units. Visiting a restaurant, for example, can be segmented into 'entering', 'being seated', 'ordering', 'drinking and eating', 'paying', and 'leaving'. Similarly, the hop of a kangaroo can be segmented into 'preparing', 'jumping', 'flying', and 'landing'. Such segmentation defines units or temporal chunks usually called 'events' or 'episodes' that can be subsumed under temporal categories. As with all other natural categories, temporal categories have fuzzy borders, but are nevertheless cognitively useful due to high conformity among individuals (Rosch, 1978). Events have a beginning and an end. Referring to an example from Schnotz and Lowe (2008), we consider the cutting of tomatoes as part of preparing a meal. At the beginning, all tomatoes are uncut. At the end, all tomatoes are cut. Between these two states, there is a middle state which can be considered as the best representation of all other states and, thus, as a prototype state of the event. We refer to the beginning state, the prototypical middle state, and the end state as the event's key states. According to Schnotz and Lowe (2008), key states are the most informative states within a process which therefore enable this process to be represented in a highly parsimonious fashion.

Following the general assumption of cognitive economy, we assume that dynamic mental models consist of internal cognitive representations of such highly informative key states because this is the most parsimonious kind of representing the corresponding events. Schnotz and Lowe (2008) argue that 'key states capture a maximum of relevant information about their corresponding events and are mentally represented as cognitive schemas in long-term memory' (p. 342). Accordingly, dynamic mental models consist of an ordered set of internal representations of a finite number of key states. ‘Running’ such a model means a stepwise mental simulation of the dynamic subject matter in which these internal key state representations become sequentially activated and brought into working memory step by step for further analysis. Although such a sequence has a relatively low temporal granularity, it may nevertheless be sufficient to answer questions about why and how something works.

Dynamic subject matter always has both spatial and temporal characteristics. Accordingly, its internal representation by a dynamic mental model can put more or less emphasis on one or the other kind of information. The specific emphasis might be strongly influenced by the kind of visual display used to present the subject matter. Spatial characteristics can best be shown by simultaneously presented static graphics because these graphics are stable over time. Learners can inspect them for as long as they need and can direct their attention to all spatial details. While studying static graphics, temporal relations between the single states need to be inferred by learners. Temporal information about a process can best be shown by a dynamic visualization due to its sequential and fluent character. When studying a dynamic visualization, however, it is more difficult to capture the spatial characteristics of the depicted content because learners would have to do something akin to ‘mentally freezing’ the dynamic visualization so that the single states could be interrogated in working memory, while they have to be prevented simultaneously from being ‘overwritten’ by new visual stimuli. Between these extreme forms of presentation, there are also intermediate forms such as static graphics presented one-by-one at a slower or at a faster pace. Perception is necessarily selective and cognitive processing capacity is limited (Baddeley, 1986; Chandler & Sweller, 1991; Cowan, 1997) which implies that one kind of cognitive processing has frequently to be performed at the expense of another kind of cognitive processing. It follows that there may be a trade-off between processing of spatial patterns and processing of temporal patterns (Lowe & Pramono, 2006). In other words, information about spatial patterns could be processed at the expense of information about temporal patterns, and vice versa.

If dynamic mental models are assumed to consist of an ordered set of mental representations of key states, it seems plausible to follow the assumption of Schnotz and Lowe (2008) ‘that a series of static graphics could sometimes be a better basis for constructing a dynamic mental model than an animation’ (p. 342). One argument for this assumption might be that a carefully selected ordered set of external static graphics showing key states within a process could directly provide ‘ready-made’ depictive building blocks for the construction of the dynamic mental model. Dynamic visualizations, on the contrary, due to their different processing affordances (Lowe, Schnotz, & Rasch, 2011) require learners to identify the relevant key states by themselves ‘on the fly’. Due to the transient nature of the visual display, this is highly challenging and requires sufficient prior knowledge about the depicted subject matter. Although dynamic visualizations provide more information than static graphics, the mere availability of this additional information does not necessarily mean that it is appropriately processed by the learners. Based on these considerations, one can formulate the following hypotheses:

(H3) If individuals construct dynamic mental models in order to explain how something works, learning from an ordered set of static graphics illustrating key states of the process should be more effective than learning from a dynamic visualization.

(H4) In this context, constructing a dynamic mental model from a series of static graphics illustrating key states should also require less mental effort than learning from dynamic visualizations.

Furthermore, if dynamic mental models are expected to put the main emphasis on temporal rather than on spatial patterns, one can formulate the following additional hypotheses:

(H5) If individuals are expected to construct dynamic mental models in order to explain how something works, displays that emphasize the temporal characteristics of the dynamic sufficiently should allow more effective learning than displays emphasizing mainly the spatial characteristics.

(H6) In this context, displays emphasizing the temporal characteristics should also require less mental effort than displays emphasizing the spatial characteristics.

These hypotheses were tested in an experiment about the acquisition of knowledge about the neurobiological sequence of signal transmission by synapses requiring the construction of a dynamic mental model (cf. Wagner, 2013).

4.2.4 Study II: Cognitive Representations of Functional Sequences

Method 80 university students of social and educational sciences participated in this study (age: $M = 22.1$ years; $SD = 2.5$ years; sex: 81.3% female). Students of psychology and biology were excluded from participation due to the likelihood of their having too much prior knowledge of the topic of the learning material. The learning task was to develop a knowledge of the synaptic signal transmission process by studying a dynamic visualization or a set of static graphics. Participants were expected to encode and understand the synaptic processes in order to answer knowledge and inferential questions after the learning phase. Participants were randomly distributed to four different experimental conditions. One group learned about signal transmission by synapses with the help of a dynamic schematic drawing. In this dynamic condition, the whole process of signal transmission was shown seven times in succession. The other three groups learned the same content via ten schematically drawn static graphics that showed key states of the process. The static graphics were arranged differently amongst the remaining three conditions: In the second condition, the static graphics were presented sequentially at a relatively fast pace. Each of the ten single graphics was shown for 3.4 s so that the whole process of signal transmission was presented seven times. In the third condition, these same static graphics were also presented sequentially but at a relatively slow pace (12 s per graphic) so that the whole process of signal transmission was presented twice. In the fourth condition, the static graphics were arranged in five lines each showing two consecutive graphics presented simultaneously on the screen. Total presentation time was 4 min. in all four conditions. Figure 4.3 shows a snapshot taken from the learning material in the dynamic condition. Figure 4.4 shows the arrangement of the simultaneously presented graphics in one of the static conditions.

The structural depiction used in the learning material was rather simple in that it showed only those parts of synapsis that are directly relevant for the transmission process. Nevertheless, the conditional and causal relations within the synaptic processes running on this structure were rather complex. Constructing a corresponding dynamic mental model could therefore be considered as a challenging task in this respect.

After the learning phase, participants took a pencil and paper test about the signal transmission by synapses. This test consisted of 12 knowledge and 18 inferential tasks with an open-ended response format (after an item analysis, 4 items were excluded) that were intended to measure the quality of the dynamic mental model. If a mental model of a dynamic subject matter is well-constructed in that it represents the essential elements and structures of the process, knowledge items can be answered directly by reading off the necessary information from the model. Knowledge items asked for information that was explicitly depicted in the learning material. In addition, the items focused the temporal or sequential information presented by both dynamic visualizations and static graphics. An example item was: "What is the main function of calcium ions during the synaptic transmission process?". A well-constructed dynamic mental model can also be run and manipulated in order to generate the novel information required to answer inferential items. In the current study inferential items asked for information that was not explicitly shown in the learning material but had to be inferred by participants. In the case of synaptic signal transmission, inferential items could ask for possible effects of diseases and poisons on the synaptic transmission. An example item was: "The Alzheimer's disease causes neurons to produce a too little acetylcholine. Why are Alzheimer's patients often prescribed cholinesterase inhibitors by a physician?" Knowledge items and inferential items were presented within the posttest in a mixed order. In addition to the posttest, the mental effort invested both in learning about synaptic signal transmission and in solving the items of the posttest was measured as two further criteria of the study. As in study I, mental effort was measured by an adapted and translated version of the "Cognitive Load rating scale" (Paas, 1992; Paas, van Merriënboer, & Adam, 1994). Furthermore, participants were asked to rate their perceived difficulty of the learning task and of the posttest, ranging from 1 ("very, very easy") to 9 ("very, very difficult"). Participants were also tested for their prior knowledge and for their spatial ability because these two variables were assumed to be possible covariates.

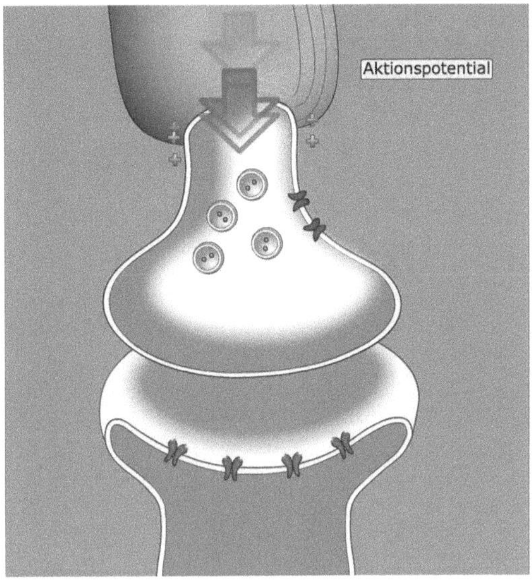

Fig. 4.3 Example of a snapshot from the presentation of signal transmission by synapses in the dynamic condition

Comparison of Dynamic and Static Graphics Regarding Performance and Mental Effort Measures Table 4.2 shows the means and standard deviations of the posttest scores and the mental effort scores of Study II. Hypotheses H3 and H4 were tested by calculating a-priori contrasts that compared the dynamic condition with the other static graphics conditions. Because there was no interaction between the presentational format in the learning condition (dynamic visualizations vs. static graphics) and the type of item in the posttest (knowledge vs. inferential item) ($F(1, 78) = 0.61$, n.s.) the total score of the posttest was considered as the main criterion for learning success of the study. There were no significant differences between the dynamic and the static graphic conditions regarding performances in the posttest ($t(78) = 0.35$, n.s.) or the mental effort that participants had invested into learning ($t(78) = -0.23$, n.s.) and into answering the questions of the posttest ($t(78) = 0.56$, n.s.). Therefore, hypotheses H3 and H4 that assumed learning from static graphics would be more effective and would require less effort than learning from dynamic visualizations if the learning aim was to construct a dynamic mental model were not supported.

In order to test for possible interactions both between learning condition and prior knowledge and between learning condition and spatial ability, we again performed linear multiple regression analyses. Because of heterogeneous sample sizes, learning condition was included as a categorical weighted effect coded variable into the analyses whereas prior knowledge and spatial ability were centered metric variables (cf. Richter, 2007). There were no significant interactions between learning condition and prior knowledge ($B = -0.44$, $SE = 0.61$, $t = -0.72$, n.s.) or between learning condition and spatial ability ($B = -0.14$, $SE = 0.35$, $t = -0.40$, n.s.) focusing participants' performance in the posttest as dependent variable.

Fig. 4.4 Arrangement of the simultaneously presented static graphics

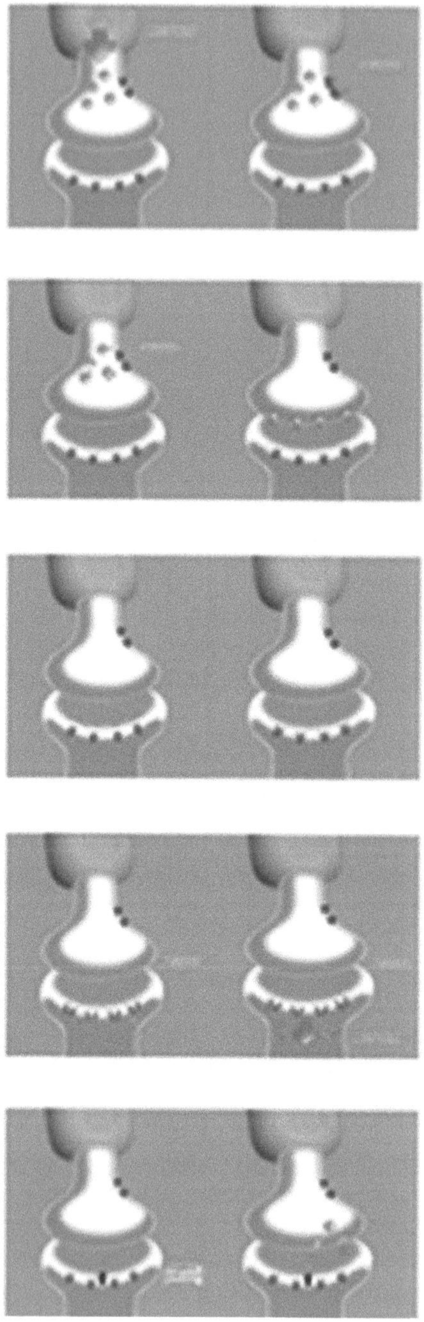

Table 4.2 Study II: means and standard deviations of the dependent variables

Criterion	Dynamic visualization ($n = 22$)			Static graphics: fast sequential ($n = 19$)			Static graphics: slow sequential ($n = 19$)			Static graphics: simultaneous ($n = 20$)		
Posttest (max. score = 52)	*M* 32.27	M_{rel} 0.62	*SD* 12.05	*M* 32.63	M_{rel} 0.63	*SD* 13.16	*M* 33.63	M_{rel} 0.65	*SD* 11.79	*M* 27.50	M_{rel} 0.53	*SD* 12.41
Mental effort while learning (max. score = 18)	*M* 10.77	M_{rel} 0.60	*SD* 2.39	*M* 10.74	M_{rel} 0.60	*SD* 3.54	*M* 11.26	M_{rel} 0.63	*SD* 2.33	*M* 10.80	M_{rel} 0.60	*SD* 2.88
Mental effort while solving the items of the posttest (max. score = 18)	*M* 13.91	M_{rel} 0.77	*SD* 1.93	*M* 13.55	M_{rel} 0.75	*SD* 2.74	*M* 13.61	M_{rel} 0.76	*SD* 2.06	*M* 13.55	M_{rel} 0.75	*SD* 3.00

Comparison of Simultaneous and Sequential Graphics Regarding Performance and Mental Effort Measures Hypotheses H5and H6 were tested by calculating a-priori contrasts that compared the condition in which the synaptic transmission processes were shown by simultaneously presented static graphics (simultaneous condition) with the other conditions, in which the synaptic transmission processes were shown either by a dynamic visualization or by sequentially presented static graphics (sequential condition). In the simultaneous condition, spatial information about synapses is given more emphasis, whereas in the sequential condition temporal information about synaptic processes is more emphasized. Because there was again no interaction between the presentation format in the learning condition (simultaneous vs. sequential graphics) and the type of item in the posttest (knowledge vs. inferential items; $F(1, 78) = 1.35$, n.s.) only the total posttest score was considered. The sequential condition yielded a significantly higher posttest score than the simultaneous condition ($t(78) = 1.69$, p(one-tailed) $= 0.048$, $d = 0.43$). Putting the emphasis on temporal information in the sequential condition was clearly beneficial for the construction of a dynamic model that enabled knowledge and inferential items of the posttest to be answered correctly. Accordingly, hypothesis H5 was supported. Regarding the mental effort invested by participants into learning about synapses and into answering the posttest questions, no significant differences were found between the simultaneous and the sequential condition (mental effort for learning: $t(78) = 0.16$, n.s.; mental effort for answering the posttest questions: ($t(78) = 0.24$, n.s.). Therefore, Hypothesis H6 was not supported. A possible explanation is that in the simultaneous condition more effort had to be invested into drawing temporal inferences whereas in the sequential condition more effort had to be invested into identifying spatial information. Both cognitive processes may have been equally effortful for participants. Therefore, there were no significant differences between the conditions regarding learning efficiency ($t(78) = 0.92$, n.s.) and application efficiency ($t(78) = 0.84$, n.s.). Furthermore, there were no significant interactions between learning condition and prior knowledge ($B = -0.09$, $SE = 0.23$, $t = -0.38$, n.s.) or between learning condition and spatial ability ($B = -0.19$, $SE = .13$, $t = -1.43$, n.s.) focusing participants' performance in the posttest as dependent variable.

Conclusion The results presented above suggest that an emphasis on temporal aspects was highly important to the construction of a dynamic mental model. However, emphasis could be given to temporal aspects not only from dynamic visualizations, but also from sequential presentation of static graphics. Lack of sequential information resulted in lower performance. Thus, the presentation of key states through external static graphics is not beneficial *per se* for the construction of a dynamic mental model. On the one hand, it is possible that under specific conditions the key state information can also be extracted from dynamic visualizations to some extent. On the other hand, the mental representations of key states must be linked to each other in a temporally adequate manner so that a complete dynamic mental model of the process can be constructed. This temporal information can probably be delivered most effectively by sequential presentation formats such as a dynamic visualization or sequentially presented static graphics. In the current study, participants had to know about the temporal and causal relations between the single states of the synaptic transmission process in order to answer the items in the posttest correctly. It was therefore very important that temporal information in particular was integrated during the process of mental model building. This meant that better results could be achieved by learning from either sequentially presented static graphics or from a dynamic visualization, both of which facilitated such integration.

4.3 Aligning Spatial and Temporal Information in Learning About Dynamic

4.3.1 Balancing Temporal and Spatial Processing

Based on the arguments of Schnotz and Lowe (2008), we hypothesized that a series of static graphics presenting the key states within a dynamic process should provide a better basis for constructing a dynamic mental model than a dynamic visualization. Study II presented above did not support this hypothesis. According to the results, it seems that making a sharp distinction between static graphics and dynamic visualizations was not helpful for a meaningful interpretation of the empirical findings. In all static experimental conditions, the static graphics presented key states of the underlying process. However, the static conditions were either not or only slightly superior to the dynamic visualization in terms of learning outcomes. This finding suggests that the traditional dichotomy between learning from dynamic visualizations versus learning from static graphics is dubious and should perhaps be reconsidered.

Because cognitive processing capacity is limited (Baddeley, 1986; Chandler & Sweller, 1991; Cowan, 1997) it is widely accepted that in many cases one kind of cognitive processing may be performed only at the expense of another kind of cognitive processing. We have already noted that both dynamic visualization and sequences of static graphics can display temporal and spatial information. In the case of dynamic visualization, the relative emphasis of temporal versus spatial information depends on the speed of presentation (Fischer, Lowe, & Schwan, 2008). High presentation speed gives the learner less time for analyzing the various spatial structures following each other than a slow presentation speed and, thus, puts more emphasis on temporal information. In the case of a sequence of static graphics, the relative emphasis given to temporal versus spatial information depends on the speed of presentation and on the temporal granularity which is defined by the number of states presented per time unit (Lowe, Schnotz, & Rasch, 2011). The distinction between dynamic visualization and static graphics seems self-evident at first sight but is thrown into doubt when scrutinized more closely. Technically speaking, there is no clear-cut answer to the question as to where a sequence of static graphics ends and where a dynamic visualization begins. There seems to be a somewhat fluent passage between the two categories – a dynamic visualization can be played frame by frame which turns it into a sequence of static graphics. However, there are perceptual thresholds in terms of temporal granularity and presentation speed that make humans perceive continuous motion instead of a series of still frames (Schnotz & Lowe, 2008). Accordingly, the distinction between dynamic visualization and static graphics results from an interaction between temporal granularity, presentation speed, and human temporal visual discrimination capabilities.

Although the distinction can be made from a psychological (perceptual) point of view, the findings presented above indicate that instead of focusing on the distinction between static graphics and dynamic visualizations, it might be more fruitful to analyze how much emphasis has been put on the temporal information within the display (but at the expense of the spatial information) and how much emphasis has been put on the spatial information within the display (but at the expense of the temporal information). In other words, one can distinguish visual displays of dynamics in terms of their emphasis on temporal-sequential versus spatial information rather than continuing the traditional dichotomy of static versus dynamic graphics.

Dynamic visualizations seem to be only one way amongst various possibilities for conveying the temporal information needed to construct a dynamic mental model. A sequence of static graphics presented at an adequate rate could also provide sufficient temporal information for this purpose. A lower presentation rate is more conducive to analyzing the spatial structure of the various states that follow each other, but puts correspondingly less emphasis on temporal (i.e. sequential) aspects of the content. On the one hand, a set of simultaneously presented static graphics, even if they are arranged in a clear order, seems to provide a relatively weak prompt to read temporal or sequential information from the display. This may be because it requires learners to activate prior knowledge about presentation orders to read the static display sequentially and superimpose a temporal structure on the static display. On the other hand, simultaneously presented static graphics can be repeatedly interrogated by the learner at his/her own pace. Thus, a set of static graphics provides the opportunity for detailed analyses of differences between the spatial structures of the depicted states (which does not of course guarantee the learner will actually carry out such analysis). In contrast, dynamic visualizations do not allow the same kind of detailed analysis of specific states due to their transient nature.

With respect to analysis of spatial structures, the processing demands imposed by dynamic visualizations would be considerably higher because the learner has either to hold a previous state in working memory or to retrieve it from long-term memory in order to make the required comparison with the subsequent state (Lowe, 1999). Comparison of different states would also be more difficult in case of a successive presentation of static graphics than with their simultaneous presentation (cf. Lowe, Schnotz, & Rasch, 2011; Ploetzner & Lowe, 2014, 2017, this volume). It seems to us that there is no best way of constructing a dynamic mental model either from static graphics or from dynamic visualizations. Instead, the crucial issues are how much emphasis should be put on the temporal and how much emphasis should be put on the spatial aspects of the learning content and what type of tasks are to be performed after the learning experience.

4.3.2 Managing Presentation Speed

As already mentioned above, human perception is characterized by a limited range of sensitivity for change. Humans cannot see the growing of a plant – this type of change becomes noticeable only over days, weeks or months. They cannot see either how a bullet leaves a gun or how a chameleon 'shoots' an insect with its tongue, because these events take place within milliseconds. In order to become visible, changes that are normally too slow to be detected have to be speeded up and changes that are too fast have to be slowed down. Thus, presentation speeds need to be tailored to perceptual sensitivity. According to our abovementioned considerations, it should be noted that this applies both for dynamic visualizations and for sequences of static graphics.

As mentioned above, sequences of events can be subsumed under superordinate higher-order events. Such embedding relations create hierarchies of events, which allow superordinate macro-events and subordinate micro-events to be distinguished. The embedding of micro-events within macro-events results in a hierarchy of dynamics ranging across different levels, from temporal microstructures through to temporal macrostructures. A dynamic visualization and a sequence of static graphics usually include multiple levels of change. However, these various levels are not necessarily of equal importance. For some purposes, the macro-changes are more important than the micro-changes. For other purposes, the micro-changes are more important.

Complex processes are usually hierarchically structured: They consist of more comprehensive macro-events (whose participating elements undergo macro-changes) that can be subdivided into less comprehensive micro-events (whose elements undergo micro-changes). As micro-events are embedded into macro-events, it follows that micro-movements are superimposed on macro-events. Accordingly, the micro-events are faster than the macro-movements. In view of humans' limited range of sensitivity for change, it follows that in the case of a higher presentation speed, macro-events tend to be more salient (i.e. have higher relative perceptibility) than micro-events because macro-events' presentation speed will be closer to the optimal perceptual sensitivity than the micro-events' presentation speed which in this case might be too high resulting in lower perceptual salience. Conversely, if presentation speed is low, micro-events become more salient than macro-events because in that case micro-events are more easily perceivable than macro-events whose presentation speed is relatively low which results in low perceptual salience. In line with these assumptions, Fischer, Lowe, and Schwan (2008) as well as Meyer, Rasch, and Schnotz (2010) found that higher speeds emphasize global (i.e. macro) events, whereas lower speeds accentuate local (i.e. micro) events. It seems that the salience of macro-events can be increased at the expense of the salience of micro-events, and vice versa.

4.4 Instructional Consequences

The results of the learning experiments presented above provide some preliminary indications only of how dynamic visualizations and static graphics may be related to each other. However, they do raise the possibility that black-and-white discussions about dynamic visualizations versus static graphics are not likely to be very fruitful. We suggest that a more useful alternative may be to consider dynamic visualizations and static graphics as instances on a continuum of temporal granularity. Depending on the display rate of dynamic visualizations as well as sequences of static graphics, there can be more emphasis on the visuo-spatial aspect or on the temporal aspect of the displayed dynamic subject matter at the expense of the other. Because both aspects can be important, it is unlikely that research will be able to uncover a universally applicable rule of thumb about using dynamic visualizations or static graphics for the display of dynamic content. Rather, one should think of combining dynamic visualizations (high temporal granularity) with sequences of static graphics (lower temporal granularity) to take full advantage of distinctive affordances offered by these different forms of information display. This could involve the development of hybrid learning environments that flexibly combine dynamic visualizations and static displays in order to ensure optimal alignment of dynamic learning content, individual learning prerequisites, and learning aims.

Dynamic visualizations and sequential displays of static graphics can be presented at different rates according to the learning goals and the required perceptual and cognitive processing. Very fast processes could be made more accessible by presenting the process more slowly (temporal zooming-in) by a dynamic visualization, but also by a sequential presentation of static graphics. Very slow processes could be speeded up (temporal zooming-out) in order to make the changes more readily apparent. Depending on the speed of presentation, the level of macro events or the level of micro events will receive more attention (Fischer, Lowe, & Schwan, 2008; Meyer, Rasch, & Schnotz, 2010).

When dynamic visualizations are used to present dynamic information, the key states within the process could be signaled to the observer by having the dynamic visualization stopped at appropriate frames by the learning environment. Such stops turn the dynamic visualization into a static picture. This should be beneficial when deeper interrogation of those especially informative states within the whole sequence is required. Pauses in a dynamic visualization should therefore be placed strategically to facilitate such intensive processing where it is required. Depending on the learner's zone of proximal development (Vygotski, 1963), the selection of key states presented visually by static pictures and the number of presented key states should at the beginning of learning be determined by the system (i.e. the instructional designer). Only after acquisition of sufficient domain-specific background knowledge, should learner-controlled selection of states for deeper analysis be allowed.

When specific states within a dynamic visualization are selected – either by the instructional system or by the learner – in the form of snapshots, this can be considered as kind of "freezing" of parts of the dynamic visualization into static graphics. These frozen extracts from the dynamic visualization can be transferred to a gallery of static graphics for a more detailed analysis of key states. With such a gallery, learners could perform detailed analyses of the spatial relations and visual features displayed by these graphics (provided they have enough time). However, this may well come at the cost of obtaining less information about the subject matter's underlying dynamics. Running (i.e. "thawing") a dynamic visualization, on the contrary, displays the temporal changes explicitly. "Thawing" a key graphic means continuing the dynamic display to the next key state graphic. With such a learning environment, learners could move from an emphasis on spatial relations to an emphasis on temporal relations, and vice versa, and modulate the relative weight of the two kinds of information accordingly. This should assist them in building a well-elaborated and integrated mental model of the dynamic content at hand. Similarly, faster presentations of a sequence of static graphics combined with higher temporal granularity would put more emphasis on the temporal features, but at the expense of the visuo-spatial features, as compared to slower presentations combined with lower temporal granularity. Accordingly, learning could in addition alternate between lower and higher temporal granularity and between different speeds of presenting sequences of static graphics of dynamic content, whereby speed and granularity should be system-controlled for novices and learner-controlled only for more advanced students. Such hybrid systems that flexibly combine displays of dynamic visualizations and of static graphics in order to ensure optimal alignment of dynamic learning content, individual learning prerequisites, and learning aims might be a promising further development.

Acknowledgements We would like to thank Radu Georghiu for the technical production of the learning material that was used within the studies. We are also grateful for the help of our student assistants Sabine Boysen, Lena Buescher, and Katharina Allgaier in collecting the data for our studies. Furthermore, we would like to thank Dr. Christoph Mengelkamp for critical discussions and helpful suggestions concerning the studies.

References

Baddeley, A. D. (1986). *Working memory*. Oxford: Clarendon Press.

Chandler, P., & Sweller, J. (1991). Cognitive load theory and the format of instruction. *Cognition and Instruction, 8*, 293–332.

Cowan, N. (1997). *Attention and memory: An integrated framework*. New York: Oxford University Press.

Fischer, S., Lowe, R. K., & Schwan, S. (2008). Effects of presentation speed of a dynamic visualization on the understanding of a mechanical system. *Applied Cognitive Psychology, 22*, 1126–1141.

Gentner, D. (1989). The mechanisms of analogical learning. In S. Vosniadou & A. Ortony (Eds.), *Similarity and analogical reasoning* (pp. 197–241). New York: Cambridge University Press.

Höffler, T. N., & Leutner, D. (2007). Instructional animation versus static pictures: A meta-analysis. *Learning and Instruction, 17*, 722–738.

Johnson-Laird, P. N. (1983). *Mental models. Towards a cognitive science of language, inference, and consciousness*. New York: Cambridge University Press.

Johnson-Laird, P. N., & Byrne, R. M. J. (1991). *Deduction*. Hillsdale, NJ: Lawrence Erlbaum.

Kosslyn, S. M. (1994). *Image and brain*. Cambridge, MA: MIT Press.

Lowe, R. K. (1999). Extracting information from an animation during complex visual learning. *European Journal of Psychology of Education, 14*, 225–244.

Lowe, R. K., & Boucheix, J.-M. (2008). Learning from animated diagrams: How are mental models built? In G. Stapleton, J. Howse, & J. Lee (Eds.), *Diagrammatic representation and inference* (pp. 266–281). Berlin: Springer.

Lowe, R., & Boucheix, J.-M. (2017). A composition approach to design of educational animations. In R. Lowe & R. Ploetzner (Eds.), *Learning from dynamic visualization — Innovations in research and application*. Berlin: Springer (this volume).

Lowe, R. K., & Pramono, H. (2006). Using graphics to support comprehension of dynamic information in texts. *Information Design Journal, 14*, 22–34.

Lowe, R. K., & Schnotz, W. (Eds.). (2008). *Learning with animation: Research implications for design*. New York: Cambridge University Press.

Lowe, R. K., & Schnotz, W. (2014). Animation principles in multimedia learning. In R. E. Mayer (Ed.), *Cambridge handbook of multimedia learning* (2nd ed., pp. 513–546). New York: Cambridge University Press.

Lowe, R. K., Schnotz, W., & Rasch, T. (2011). Aligning affordances of graphics with learning task requirements. *Applied Cognitive Psychology, 25*, 452–459.

Marey, E. J. (1874). *Animal mechanism: A treatise on terrestrial and aerial locomotion*. New York: Appleton and Co.

Mayer, R. E. (2014). Cognitive theory of multimedia learning. In R. E. Mayer (Ed.), *Cambridge handbook of multimedia learning* (2nd ed., pp. 43–71). New York: Cambridge University Press.

Meyer, K., Rasch, T., & Schnotz, W. (2010). Effects of animation's speed of presentation on perceptual processing and learning. *Learning and Instruction, 20*, 136–145.

Neisser, U. (1976). *Cognition and reality*. San Francisco, CA: Freeman.

Paas, F. G. W. C. (1992). Training strategies for attaining transfer of problem-solving skill in statistics: A cognitive-load approach. *Journal of Educational Psychology, 84*, 429–434.

Paas, F. G. W. C., van Merriënboer, J. J. G., & Adam, J. J. (1994). Measurement of cognitive load in instructional research. *Perceptual and Motor Skills, 79*, 419–430.

Ploetzner, R., & Lowe, R. (2014). Simultaneously presented animations facilitate the learning of higher-order relationships. *Computers in Human Behavior, 34*, 12–22.

Ploetzner, R., & Lowe, R. (2017). Looking across instead of back and forth – How the simultaneous presentation of multiple animation episodes facilitates learning. In R. Lowe & R. Ploetzner (Eds.), *Learning from dynamic visualization — Innovations in research and application*. Berlin: Springer (this volume).

Richter, T. (2007). Wie analysiert man Interaktionen von metrischen und kategorialen Prädiktoren? Nicht mit Median-Splits! [How to analyze interactions of metric and categorical predictors: Not with median splits!] *Zeitschrift für Medienpsychologie, 19*, 116–125.

Rosch, E. (1978). Principles of categorization. In E. Rosch & B. B. Lloyd (Eds.), *Cognition and categorization* (pp. 27–48). Hillsdale, N.J: Erlbaum.

Sanchez, C. A., & Wiley, J. (2017). Dynamic visuospatial ability and learning from dynamic visualizations. In R. Lowe & R. Ploetzner (Eds.), *Learning from dynamic visualization — Innovations in research and application*. Berlin: Springer (this volume).

Schnotz, W. (1993). On the relation between dual coding and mental models in graphics comprehension. *Learning and Instruction, 3*, 247–249.

Schnotz, W. (2014). Integrated model of text and picture comprehension. In R. E. Mayer (Ed.), *Cambridge handbook of multimedia learning* (2nd ed., pp. 72–103). New York: Cambridge University Press.

Schnotz, W., & Kürschner, C. (2007). A reconsideration of cognitive load theory. *Educational Psychological Review, 19*, 469–508.

Schnotz, W., & Lowe, R. K. (2008). A unified view of learning from animated and static graphics. In R. K. Lowe & W. Schnotz (Eds.), *Learning with animation: Research implications for design* (pp. 304–356). New York: Cambridge University Press.

Spangenberg, R. W. (1973). The motion variable in procedural learning. *AV Communication Review, 21*, 419–436.

Tversky, B., Morrison, J. B., & Bétrancourt, M. (2002). Animation: Can it facilitate? *International Journal of Human-Computer Studies, 57*, 247–262.

Tversky, B., Heiser, J., Mackenzie, R., Lozano, S., & Morrison, J. (2008). Enriching animations. In R. K. Lowe & W. Schnotz (Eds.), *Learning with animation: Research implications for design* (pp. 263–285). New York: Cambridge University Press.

Ullman, S. (1984). Visual routines. *Cognition, 18*, 97–159.

Vygotski, L. S. (1963). Learning and mental development at school age. In B. Simon & J. Simon (Eds.), *Educational psychology in the U.S.S.R* (pp. 21–34). London: Routledge & Kegan Paul.

Wagner, I. (2013). *Lernen mit Animationen: Effekte dynamischer und statischer Visualisierungen auf die Bildung perzeptueller und kognitiver Repräsentationen beim Erwerb von Wissen über dynamische Sachverhalte.* [Learning from animation: Effects of dynamic and static visualizations on the construction of perceptual and cognitive representations for the acquirement of knowledge about dynamic subject matters.] Ph.D. thesis. University of Koblenz-Landau, Faculty of Psychology. Retrieved January 15, 2015, from http://kola.opus.hbz-nrw.de/volltexte/2013/858/

Chapter 5
The Role of Craft-Based Knowledge in the Design of Dynamic Visualizations

Jodie Jenkinson

5.1 Introduction

Illustration of medical and scientific subject matter is a practice that dates back centuries, advancing in lockstep with scientific discovery (see also McGill, 2017, this volume). Much of the early work in this area focussed on detailed documentation of comparative anatomy, by depicting cadaveric specimens in various stages of dissection. Closely linked to this was the visual translation and mapping out of basic physiology and theories of disease processes. Even in its earliest form, medical illustration was a practice that faced the challenge of capturing the dynamic nature of living things. Biomedical visualization as a profession is relatively new, evolving over the last century through a formal system of training and practice-derived heuristics. The first of these training programs was established by Max Brödel, in 1911 at Johns Hopkins University School of Medicine in Baltimore, Maryland (Crosby & Cody, 1991). The focus of these training programs is on the development of competencies that encompass understanding of basic and clinical sciences and skill in visual communication. Attention is also devoted to the mastery of various illustration techniques that are applied to the depiction of biomedical subject matter, much of which addresses complex structural and functional or dynamic relationships. Many of the strategies used to represent different subject matter are well documented in instructional textbooks (e.g., Hodges, 2003; Jastrzebski, 1985; Wood, 1994). These illustration techniques, many of which find their origins in fine art, have evolved, often in response to the practical affordances and limitations of the printing process at the time. While print provides an adequate platform for representing static subject matter, it presents a particular challenge for depicting complex, dynamic interrelationships.

J. Jenkinson (✉)
University of Toronto Mississauga, Mississauga, ON, Canada
e-mail: j.jenkinson@utoronto.ca

R. Lowe, R. Ploetzner (eds.), *Learning from Dynamic Visualization*,
DOI 10.1007/978-3-319-56204-9_5

Since its foundation over a century ago, the practice of biomedical visualization has grown considerably, in particular with the introduction of dynamic visualization techniques. At its core, it remains a problem-solving activity involving the deconstruction of the structures, processes, or mechanisms to be depicted and the identification of visual communication strategies that convey the subject matter with clarity and accuracy (see also McGill, 2017, this volume). Throughout this chapter we will examine the decision-making involved in this problem-solving process primarily as it relates to the depiction of dynamic subject matter. We begin by outlining a general approach to visualizing dynamic events used in static representations. Next, we will focus on animation as a visualization strategy, and the inherent complexity that this introduces to the decision-making process. With the addition of the temporal dimension and the design choices this introduces, decisions faced by animators may at times be informed by techniques borrowed from film, intuition, or pure speculation. Finally, we will discuss innovations in the representation of biomedical subject matter, many of which are driven by advances in technology, and the implications of technological advancements for future practice in dynamic visualization design.

5.2 Illustrating Dynamic Processes in Static Depictions

The opportunity to escape the limitations associated with representing a dynamic three-dimensional world on a static, two-dimensional printed page has resulted in the development of a number of graphic devices, many of which are still in use. Depicting process or change can be achieved using a variety of rendering techniques. Visual explanations may be represented across a range of styles, from the highly visually complex to more simplified depictions, depending on the end communication goal or intended audience of the work. For example, in depicting a surgical procedure it is most common for the illustrator to depict this dynamic process as a series of sequential illustrations rendered essentially as line drawings (much like a comic). Surgical illustration is commonly delineated into a series of discrete steps. Each illustration focuses on a key event in the procedure, depicted from a perspective that maximizes the clarity of the procedure. It is most often oriented from the point-of-view of the surgeon or from an anatomical viewpoint. The surgical field, an obstructed view of retractors, towels, and hands is impossible to decipher visually. Illustration is essential for providing clarification and drawing attention to the relevant features of this complex scene. Whereas tonal illustrations (drawn using a range of greyscale values) are helpful in identifying the fine details of what one might see in an operating theatre, line illustrations, by abstracting and clarifying each step, provide the necessary cues to support the student's conceptual understanding of the procedure. In a successful illustration, information about texture, depth, and spatial interrelationship is clearly communicated. The illustrator uses a number of techniques, including the use of cast shadows, occlusion, transparency, variation in line weight, and linear patterns to convey form and selectively

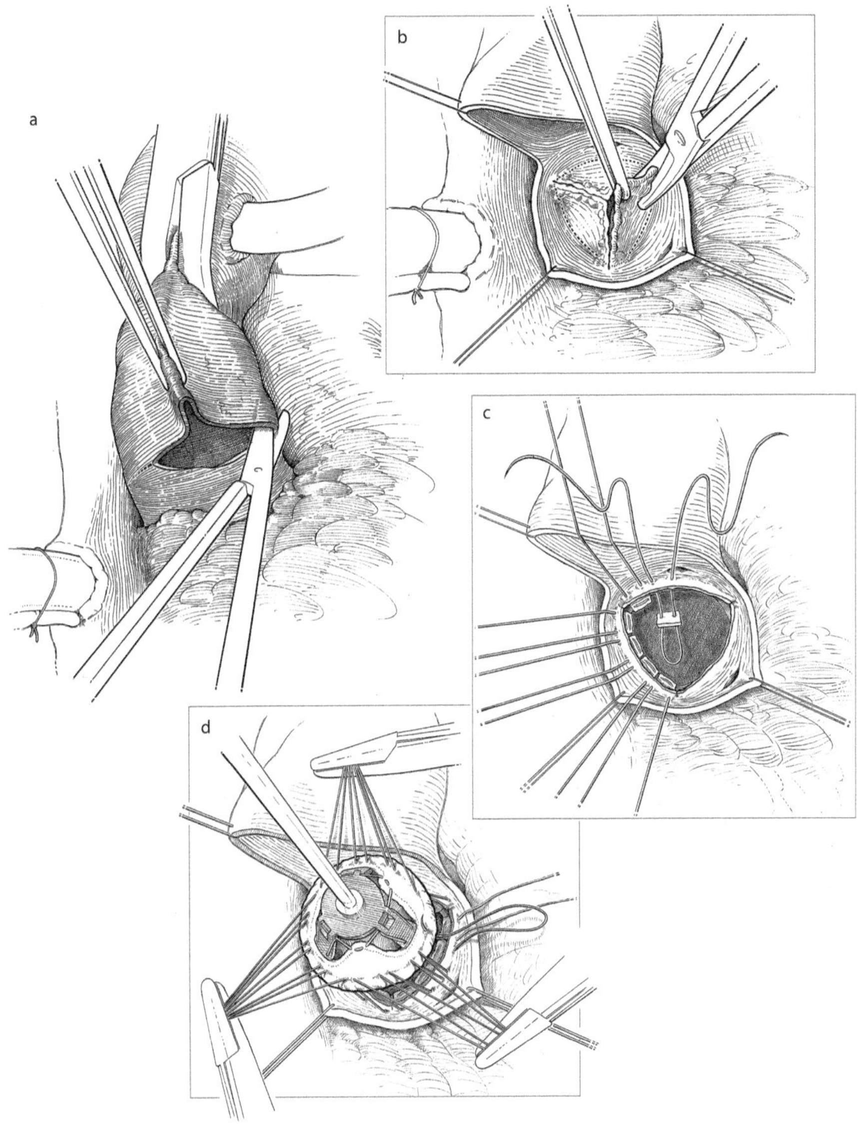

Fig. 5.1 This sequential series depicts aortic valve replacement surgery beginning with (**a**) dissection of the aorta, (**b**) excision of the aortic valve leaflets, (**c**) interrupted horizontal mattress suture technique, and (**d**) valve replacement (Copyright 2012 by Michael Corrin. Reprinted with permission)

emphasize or de-emphasize structures. These techniques create the necessary illusion of depth to depict the complex spatial and functional inter-relationships of anatomical structures without the use of color. Figure 5.1 shows a series of surgical illustrations demonstrating an approach to aortic valve replacement surgery.

In addition to clarifying the visuospatial aspects of the process, and providing adequate context in relation to anatomical landmarks, Fig. 5.1 shows how the illustrator has also broken the process down into a series of discrete steps that link together to represent the process as a whole. The illustrator has identified the key moments of this process in four illustrations. He creates focus in each step by using instrumentation (surgical scissors, clamps, forceps, etc.) and suture lines to draw attention to the area of greatest thematic relevance and to convey the actions occurring in each surgical step. As well, the use of value contrast (increased variation in the depiction of light and dark values) heightens our awareness of these areas. In many ways these sequential depictions form a sort of storyboard equivalent to the pre-visualization process involved in creating animations (a process by which the animated narrative is described in a sequence of annotated static images). Many of these well-established design heuristics used by illustrators have been confirmed and legitimized by recent research in visual perception. This research has provided formal empirical evidence about features that, in the early state of perception, "pop out" in a way that distinguishes them from their surroundings. Illustrators have long called on the collective experience of accumulated craft knowledge to manipulate the visual "weight" of elements by heightening or reducing visual properties such as contrast in size, luminance, color, or form. An understanding of pre-attentive processing, the visual system's rapid, early detection of these visual properties, can translate into principles for organizing and displaying information in such a way that focuses the viewer's attention (Ware, 2004).

5.3 Depicting Dynamics via Static Illustrations

The features that often distinguish depictions of dynamic from non-dynamic phenomena include representations of the passage of time, change over time, variation in spatial scale, and the illustration of multiple viewpoints. The representation of time within a static medium presents a particular challenge to the illustrator. Time may be represented across a series of panels as exemplified in the surgical series (Fig. 5.1) or depicted within one single continuous image. When using panels or frames the rate at which time is shown as passing may be manipulated by increasing or decreasing the number of panels used, the negative space (the space surrounding an object or an image) between the panels, and the shape or size of each panel (McCloud, 1993). These strategies are used in comics and occasionally in storyboard development. In contrast, when depicting the passage of time within a *single* continuous frame the illustrator relies instead on a number of visual cues to convey temporal change and to guide the viewer's eye around the composition. Figure 5.2 depicts a dynamic process (specifically the sequence of events associated with immune response regulation) in which the illustrator demonstrates the use of additional resources such as color, variation in line weight, and graphic devices (arrows and numbered labels) to depict change and focus the viewer's attention on the most relevant information. The illustration simultaneously depicts events occurring at

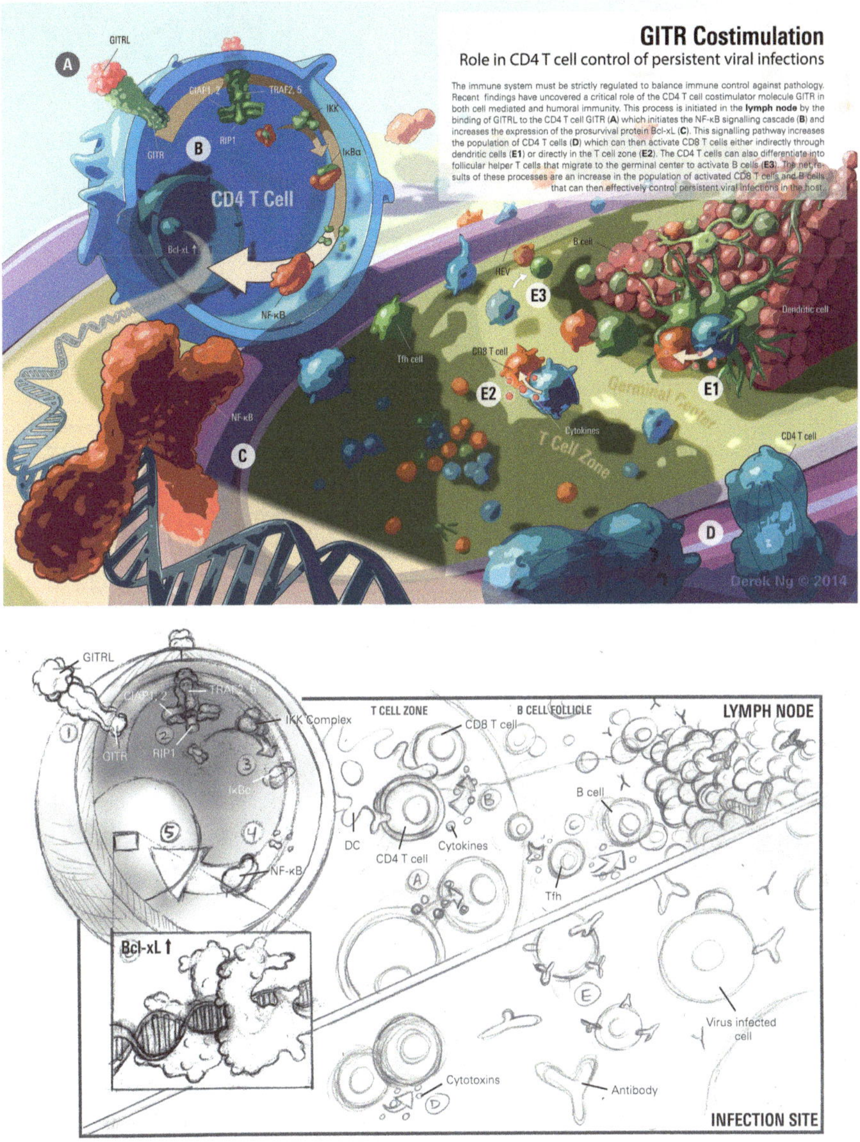

Fig. 5.2 This illustration and accompanying preliminary sketch use a variety of visual cues in depicting the role of GITR molecules, which contribute to the activation of T cells, in regulating immune response (Copyright 2014 by Derek Ng. Reprinted with permission)

both the molecular and cellular level that contribute to the development of an effective immune response.

The illustrator has organized the story presented in the illustration to be read from left to right. The narrative is established with a depiction of molecular processes

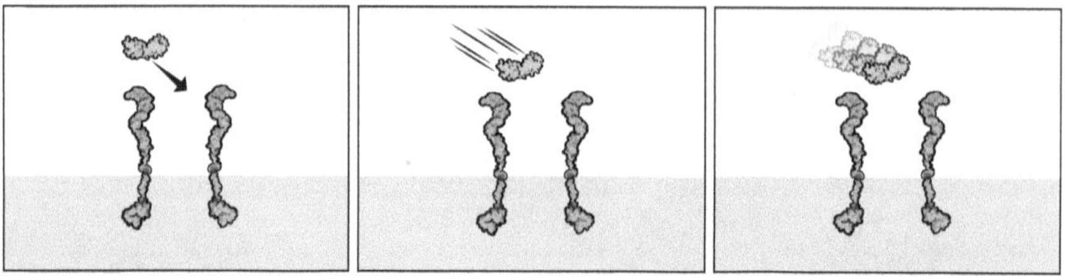

Fig. 5.3 Arrows, action lines, and ghosting are three techniques commonly used in illustration to depict motion

occurring within the cell. The illustrator manipulates color properties and contrast to increase the salience of these events and create the illusion of 3-dimensional space. Color may serve a number of functions in in the depiction of dynamic processes. It can aid in the segmentation of complex scenes, the ordering of events, and the categorization of structures. It may also be used to great effect to highlight subtle changes when multiple images are displayed at once (Tufte, 1997). Both design layout and color may be used to guide the eye through a natural progression of information (or in this case, the dynamic unfolding of a story). In Fig. 5.2 the illustrator has used high levels of color saturation and contrast (in color and relative size) to draw attention to the main features of the narrative (the role of co-stimulator molecule GITR in regulating the immune system). Reduced contrast in color value creates the illusion of atmospheric perspective (the effect that the atmosphere has on an object when viewed from a distance), establishing depth relationships. For example, if we compare the T cells in the foreground with the T cells in the background of the image, the background cells are depicted with reduced color saturation, and lighter shadows, contributing to the illusion of depth. This is further supported by introducing variation in line weight to describe the contours of each element in the scene. Note that foreground objects are rendered with a heavier contour, while background objects are delineated using finer lines. This standard graphic approach to manipulating line weight has the effect of creating a depth hierarchy within the picture plane (Wilson-Pauwels, 1997; Woolridge, 2013).

Throughout this illustration dynamic change is depicted with the use of arrows and supporting text labels. Arrows are used to suggest causality (the activation of a molecule or cell by another) and to illustrate with directionality, a cascade of events (change over time). They are intended to highlight isolated events within the narrative. Text labels are used to guide the viewer around the illustration orienting them to the process as a whole. These are very common visualization strategies in medical and scientific illustration. As exemplified in Fig. 5.3, arrows, action lines or multiple layered views (often incorporating transparency to stratify information or to convey spatial inter-relationships) are techniques used to suggest motion or change.

Similarly, (as previously described), sequential illustration may serve the same instructional goal (e.g., Wagner & Schnotz, 2017, this volume). Arrows and action lines are also used to convey the motion of an entity through space. Another visual cue that is used to convey motion is ghosting or onion skinning (overlaying several adjacent frames simultaneously), which mimics the slow shutter speed of a camera as it illustrates the path of an object. While the same types of design strategies may be used to express a variety of different movements or relationships, they are generally used in conjunction with textual descriptions, without which their specific meaning could be easily misconstrued. As the dynamic interactions that are to be depicted become increasingly complex, the need for additional explanatory text increases, as does the overall complexity of the illustration. This presents a challenge to the illustrator who must balance the visual readability of the representation with the provision of adequate information about the phenomenon depicted. In this respect, animation is a medium that is far better equipped to communicate complex interactions. It has a far richer vocabulary for describing the effects of physical forces upon objects.

In addition to the depiction of temporal change, illustrators are often tasked with depicting changes in spatial scale. The simultaneous representation of multiple scales is a challenge to communicate in a single static depiction. When switching between two or more orders of magnitude, a visual call-out technique (an inset figure that signals the close-up of an area of detail) is frequently used to illustrate change or magnification at various levels. In Fig. 5.2, the illustrator has used an exaggerated perspective to convey spatial scale. Interestingly, if we are to compare the final illustration with the preparatory sketch shown in Fig. 5.2, there is a notable shift in the illustrator's approach to depicting the difference in spatial scales. Where originally he had conceived of a visual call-out to create the magnifying effect of viewing activity occurring within the nucleus of the CD4 T cell, he redesigned the image using perspective instead to convey this change in scale. He describes his decision as an attempt to "ensure a continuous flow visually and informationally, across the image", acknowledging that the exaggerated perspective may present a problem in that it "… is not scientifically accurate with regard to the scale of the different cellular and molecular elements" (D. Ng, personal communication, May 21, 2015). This raises a very important issue in the visualization of medical and scientific information, and that is the role of artistic license in the design of visual explanations. An illustrator or animator will frequently make decisions about how best to visualize a process that extend beyond the direct mapping of information into a visual format and ultimately influence the outcome of the final representation. This decision process takes into consideration factors such as the context in which the work will be viewed (format, audience, and learning objectives), budgetary constraints, as well as more elusory considerations relating to the visual impact or style of the representation (aesthetic merit and potential to engage). We will explore these aspects of decision-making throughout the visualization process in greater detail in our discussion of animated representations.

5.4 Depicting Dynamics via Animations

Sequential static illustration has for years been a staple learning resource in both medical and life sciences education. As exemplified by the surgical series illustrated in Fig. 5.1, it can be an effective strategy for depicting a stepwise process. It is also often used in explaining physiological, chemical, or biological processes. While sequential illustration has been a helpful tool for communicating change, it relies heavily on the student's ability to fill in the gaps between the individual illustrations. This process of mental interpolation demands much on the part of the end viewer and depending on the domain often requires a substantial degree of subject matter expertise in order for the depiction to be effectively understood. Animation would appear to provide a more intuitive and fluid mechanism for establishing clear connections between a sequence of static images, effectively filling in the gaps for the viewer. However, while developers and consumers of animation perceive the potential communicative value of this medium to be great, this is not necessarily borne out by the research assessing its impact upon learning. Many studies have demonstrated that animation is no better than (e.g., Rieber, 1989; Rieber & Hannafin, 1988; Sanger & Greenbowe, 2000), or in many cases less effective than, static imagery (e.g., Lewalter, 2003; Lowe, 2003; Tversky, Morrison, & Betrancourt, 2002). Regardless of the negative or neutral effects reported in the literature, general enthusiasm for animated media persists. Further, recent advances in technology, including the availability of low-cost consumer-level software applications, have fueled a dramatic increase in the development and subsequent adoption of animated media in the classroom.

Scientific animation, as a sub-domain of medical and scientific illustration, derives many of its guiding principles from the heritage of formal training programs in these areas and also more recently from advancements in the mainstream film and animation industries. On one hand, the design of dynamic visualization borrows many of the established strategies (for example, attentional cues such as color coding, arrows, call-outs, and text labels etc.; see De Koning & Jarodzka, 2017, this volume) that have guided the development of static illustrations. On the other hand, innovation in animation (in particular, the evolution of stylistic conventions) have contributed to broadening the practice of illustration. While some illustration strategies translate well to an animated environment, many do not. In a review article assessing the transferability of illustrated cues to an animated environment, De Koning, Tabbers, Rikers, and Paas (2009) have classified cues by function according to Mayer's (2001) theory of multimedia learning. Cues are classified as (1) attention guiding (drawing attention to essential information); (2) organizational (emphasizing a visual hierarchy); and (3) relational (emphasizing the relationships between structures). The authors conclude that while attentional cueing may be successful in directing focus it may not be successful in inferring causality or function. Organizational and relational cues appear not to translate as well from a static medium to a dynamic one and when used in isolation are less successful in improving understanding. More recent research by Boucheix, Lowe, Putri, and Groff

(2013) suggests that relational cueing in animation is more effective when combined with attentional cueing in so-called *Relational Event Unit* cueing, a form of dynamic cue that draws greater attention to the relationships between entities and their associated behaviours (see also De Koning & Jarodzka, 2017, this volume).

While animation and illustration may share several perceptual characteristics, there are so many more factors introduced in the development of an animated visualization that it is difficult to measure how features interact with one another in contributing to the whole. At the most basic level, where animation distinguishes itself from static representation is in its capacity to use motion to depict change explicitly and directly; particularly in its ability to convey dynamic change in both temporal and spatial scale. In this respect, comparisons between the two mediums may not prove productive. Increasingly, researchers who study the impact of animation are looking at it less in terms of its absolute effectiveness or how it compares with static imagery, and more in terms of the influence of animation within very specific learning contexts. The issue here is not whether animation is superior to other modalities of communication, but rather that we need to alter our characterization of the animated medium (Schnotz & Lowe, 2008), consider when and why animated instruction may be effective (Ainsworth, 2008; Betrancourt & Chassot, 2008) and how the design of animations may impact their effectiveness (Schnotz & Lowe, 2008; Tversky, Heiser, MacKenzie, Lozano, & Morrison, 2008). Within the mainstream film industry, animators tasked with visualizing medical and scientific subject matter will find a wealth of ideas that can feed into and enrich the many design decisions undertaken during production. Visuospatial techniques borrowed from those used in static illustrations have been complemented by increasing adoption of spatiotemporal techniques developed by "Hollywood" filmmakers and animators). For the most part techniques borrowed from this well-established production pipeline have advanced the practice of dynamic visualization immeasurably. However, while the goal of mainstream animation (such as Pixar's Toy Story or Finding Nemo) is to entertain and engage viewers in the narrative, the goals of scientific visualization are most often to inform and educate. This means that at times these techniques are not transferable from one purpose to the other. Whether adopting techniques from static illustration or from mainstream animation, developers should exercise caution in applying these techniques to the development of scientific animation.

5.4.1 The Design Process

The design of scientific animations involves a complex decision-making process in which considerations such as purpose and communication objectives, complexity of subject matter, narrative structure, and aesthetic appeal must all be balanced (see also McGill, 2017, this volume). As described by Sharpe, Lumsden, and Woolridge (2008), scientific animation commonly serves one of three purposes: (1) Explanation (communication, education, or public outreach); (2) Simulation (drawing upon empirical data relating to structure or function within a scientific domain); or (3) Speculation (more exploratory or hypothetical in nature). Dynamic visualizations created for educational purposes are most often of the explanatory variety. Hence, in this chapter we will focus on design processes used for developing explanatory animations. From a design perspective, an explanatory animation is commonly conceived of as an interpretive narrative through which explanation of the referent subject matter is be progressively presented to the target audience in a coherent and comprehensible manner. It integrates traditional storytelling strategies (storyline, plot structure, pacing, narrative arc, etc.) in order to engage and hold the viewer's attention.

Given the considerations described above as well as the cost and investment of effort involved in the development of animation, developers typically adopt a tightly organized workflow borrowed from production heuristics used in filmmaking. This process is often initiated by the client stakeholder who issues a "request for proposal" (RFP) describing the scope and objectives of the proposed animation. Once job specifications are firmly established, production can begin. The production process can be broken down into three distinct phases: (1) Pre-production (script, treatment, storyboard, narration, animatic); (2) Production (layout, modeling, keyframes, animation); and (3) Post-production (compositing, editing, and encoding). This three-phase workflow is an iterative process of generation and refinement in which the narrative is re-visited and adjusted in order to increase the 'readability' of the visualization. It is guided largely by traditional storytelling conventions, communication goals, and learning objectives, which are balanced against the needs of the target audience for whom the visualization is intended. Figure 5.4 illustrates the development process involved in designing 3D computer animation with an explanatory note on each phase. Throughout this process the animator is engaged in a decision-making process that centres around managing the viewer's attention while telling an engaging and readable story.

5.4.2 Pre-production

During the pre-production phase of development, the animator collects reference material in order to understand the subject matter, while at the same time defining the visual appearance or "look" and the narrative flow (coherence and pacing) of the animation. The pacing of the narrative is initially established by a script that is written in advance of storyboard development. The script serves the purpose of moving the story forward by identifying plot points (events that may signal a change in pace or take the action in a new direction) and establishing a tone or mood for the narrative. Together the script and storyboard are an integral part of the design process and serve as a blueprint for the animation.

Staging the Narrative The development of the storyboard is organized around a hierarchy of shots, which are assembled into scenes that may be further organized

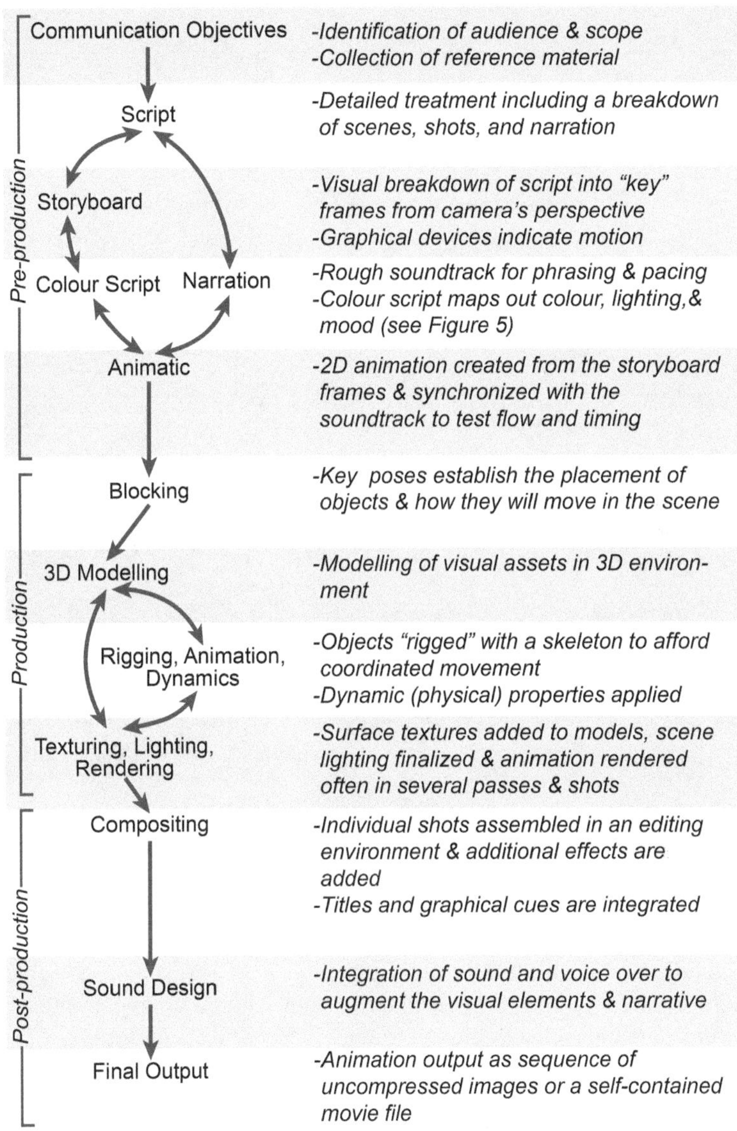

Fig. 5.4 The typical workflow for 3D computer animation design (From Jantzen, Jenkinson, and McGill (2015). Copyright 2015 by Stuart Jantzen. Adapted with permission)

into the sequences that comprise an explanatory story. Throughout this process the animator is exploring scenarios for framing or staging the narrative. This involves decisions about composition (both within the display area and across multiple frames) that provide adequate context, while focusing the viewer's attention on the most relevant details, and producing a coherent structure to the story. This process

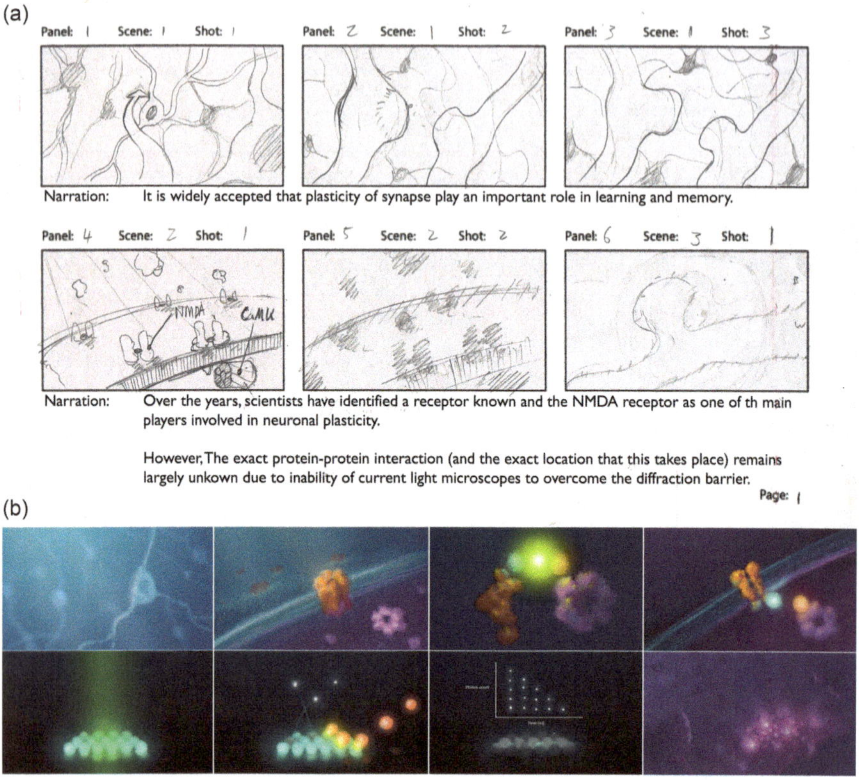

Fig. 5.5 (**a**) Six panels from a storyboard illustrate a breakdown of the proposed animation as a sequence of scenes containing numerous shots and include narration. (**b**) The color script provides an overview of visual treatment (color, lighting, and mood) of the animation (Copyright 2015 by Jerry Won. Reprinted with permission)

is very similar to the approach taken in illustrating the surgical series depicted in Fig. 5.1, insofar as key changes in action and point of view are represented in individual frames, requiring interpolation. Figure 5.5 provides an example of a storyboard excerpt and subsequent color script. The visual style of an animation is usually determined by its communication goals and end target audience. For example, an animation targeting a grade school audience would naturally have different communication goals than one designed to meet the needs of the scientific research community.

Visual Style There is no prescribed formula for determining how a scientific animation should look (i.e., its visual style), although there are a number of popular conventions for depiction, spanning the photorealistic to the schematic. The schematic rather than photorealistic depiction of scientific concepts may sometimes be a reflection of budgetary constraints because a photorealistic treatment typically takes far longer to implement. However, more often it reflects a deliberate decision to simplify the representation for the sake of clarity (much in the same way that the depiction of surgical procedures using line drawing techniques lends clarity to the process). It may be argued that a photorealistic approach to depicting scientific phenomena is a more faithful and true-to-life form of representation. However, it may also be counter-argued that a highly realistic depiction of complex phenomena presents subject matter with such deceptive clarity that students convince themselves they understand the subject matter and take away only a superficial understanding of the concepts depicted (Linn, Chang, Chiu, Zhang, & McElhaney, 2010). It is true that there are advantages and disadvantages associated with either, but it is dependent upon the learning context. There are many instances of students misinterpreting or interpreting literally schematic representations of scientific phenomenon, just as there are instances of students being overwhelmed by the level of detail included in some highly realistic animated representations. The issue is not whether schematic or photorealistic visualizations are better for learning, but rather one of identifying the specific circumstances under which learning with one or the other is more advantageous. The visual saliency of features within an animated display should support the thematically relevant elements, which in turn, should fulfill the learning objectives (cf. Lowe, Schnotz, & Rasch, 2011).

In addition to staging the narrative and assigning a visual style there are many other decisions undertaken during the pre-production phase of development that impact greatly upon the outcome of the visualization. It is at this early stage that the animator will decide where to position the camera (in a 3D modeling environment a virtual camera is used to frame the scene; see also Schwan & Papenmeier, 2017, this volume) and what point of view to adopt. This decision-making process is guided by determining how to most clearly represent information and how to visually transition between the various environments that the animator is tasked with depicting. Often scientific visualization involves providing access to the inner detail of a structure, in which case the animator will also rely upon illustrative techniques such as transparency, cross-sectional, cutaway (whereby interior features are made visible by removing a portion of the exterior surface), or exploded views in order to provide access to this environment. As well, since a great deal of scientific visualization involves depicting phenomena existing within a spatial scale that cannot be appreciated with the naked eye, it is necessary to provide adequate landmarks to orient the viewer within an unfamiliar landscape. This is a particular challenge in representing a crowded cellular environment and animators will typically adopt either an immersive or a cross-sectional perspective in order to depict both the extracellular and intracellular spaces. As illustrated in Fig. 5.6, the immersive point-of-view places the viewer within the environment, providing an engaging and highly focused scene. This can be a powerful tool for drawing attention to very specific activities within the cell but as McGill observes (2008) this method requires that we reduce the concentration of molecules that is actually present within the environment in order to provide a vista through which the camera can travel. Alternately the cross-sectional perspective provides the viewer with a more comprehensive view inside the environment. It provides a mechanism for depicting the full complexity of the cell but at the expense of depicting depth-associated spatial relationships.

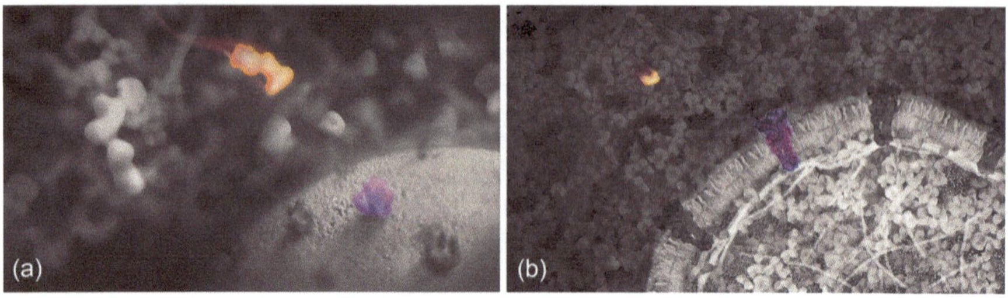

Fig. 5.6 Depiction of a molecule approaching a receptor on the surface of a cell from an immersive perspective (**a**) and a cross-sectional perspective (**b**) (Copyright 2014 by Naveen Devasagayam. Reprinted with permission)

This is typical of decisions faced by scientific animators, who must balance accuracy against effective communication with the target audience. All of the decisions regarding the style, pacing, and staging of the narrative made during the pre-production process provide a necessary scaffold for supporting subsequent phases of development.

5.4.3 Production

Whereas the pre-production phase may draw upon many techniques used in static illustration, the production process, which involves modelling, lighting, and animation, draws upon the established practices of the mainstream film and animation industries. Block (2008) describes the key to visual structure within a film as balancing the "contrast and affinity" of the various components. The principle of Contrast and Affinity posits that "The greater the *contrast* in a visual component, the more the visual intensity or visual dynamic increases. The greater the *affinity* in a visual component, the more the visual intensity or dynamic decreases" (Block, 2008). If, for example, we compare the two frames contained in Fig. 5.6, the first (Fig. 5.6a) is higher contrast in terms of the focus on a single molecule (using shallow depth-of-field) as well as size contrast between the molecule and the cell membrane in the lower righthand corner; whereas Fig. 5.6b contains many small objects of the same size increasing their affinity and reducting its overall dramatic intensity. Visual components such as space, line, shape, value contrast, color, movement, and rhythm may all be described in terms of their contrast and affinity. With the exception of narrative pace these are all pre-attentive features that may be used in the design of a cinematic display to increase the salience of specific features and ultimately help to focus the viewers' attention upon important attributes of the narrative. Relationships between various visual elements are approximated during pre-production but ultimately refined as decisions about lighting, camera and object movements, and various other visual effects are made during production.

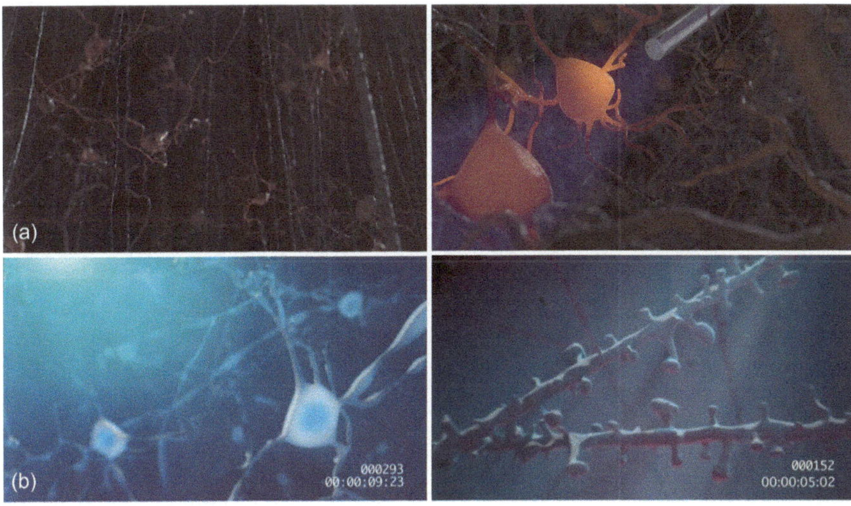

Fig. 5.7 Depicts the impact of lighting design on two shots depicting a neuronal network as (**a**) a dense forest and (**b**) an aqueous environment (**a** Copyright 2013 by Stuart Jantzen. Reprinted with permission. **b** Copyright 2015 by Jerry Won. Reprinted with permission)

Approaches to lighting and shading may impact greatly upon the mood or style of a shot and may be used to achieve a wide variety of visual effects. Lighting may be designed to mimic reality or evoke visual metaphor. For example in Fig. 5.7 two animators have taken different approaches to designing a shot that depicts a neuronal network. One animator, Stuart Jantzen, has created the illusion of a dense forest by lighting the scene from behind (rim lighting) and from the upper right. The intent here is to convey the crowded inaccessibility of such an environment. By contrast, another animator, Jerry Won, has designed neuronal network that appears to exist beneath the surface of a body of water. He uses diffuse volumetric lighting, producing illusory beams of light, and incorporates caustic effects (mimicking the refractory properties of water) to illuminate the entire scene with the intent of helping the audience visualize the seemingly foreign environment by connecting it to visual elements that are more commonly observed in the real world.

As demonstrated in both of these approaches, lighting design may also contribute to evoking mood or feeling, and to the visual structure (contrast and affinity) of visual components. As well, lighting serves a similar purpose to color, in that it may be used to encode or categorize elements, or focus attention. Finally, lighting and shading may serve the more practical purpose of clearly demonstrating structure and highlighting associated spatial and structural inter-relationships.

Creating the illusion of motion within a scene may be achieved either by animating objects within the scene or by animating the 'camera' (cf. Schwan & Papenmeier, 2017, this volume). Within the animation software authoring environment virtual cameras share the same attributes as their real world counterparts. Like real world cameras, they contain aperture and focal length settings and may be used to guide attention by providing scale, depth, and motion cues. While the visual-callout is sometimes used in animation to depict scale, camera movement (e.g., a dolly shot in which the camera moves smoothly along a horizontal axis) and zoom (enlarging or shrinking objects in frame) will have the impact of moving through space and providing a continuum of scale. When creating the illusion of object motion, an animator may also reference established principles of animation in order to assign dynamic attributes to a character. Developed by Disney animators in the 1930s and more formally described in the 1980s by Thomas and Johnston (1981), these 12 basic principles have been adopted by many computer animation studios. The primary purpose in establishing these principles was to provide a reference for how characters should move in adherence with the laws of physics, essentially the depiction of familiar 'everyday' subject matter. For example, if we were to animate the trajectory of a bouncing ball we would depict the ball compressing as it makes contact with the ground and stretching as it becomes airborne, recognizing that flexible objects respond to forces such as gravity (while retaining the same mass or volume). However, when depicting phenomena occurring within a *scientific* domain, we are often tasked with representing the movements of entities to which these principles do not apply. One molecule colliding with another would not experience macroscopic scale forces and thus would not be depicted with the distortion of a 'bouncing ball'. Properties such as temperature, gravity, and friction operate very differently at the cellular mesoscale (existing between the micrometer and nanometer scales), often producing quite disparate effects (Goodsell, 2010). In conveying spatiotemporal behaviors, interactions, and locomotion at this level the animator will rely upon varied references that may be quantitative in origin (published experimental data), qualitative (descriptive), or even speculative. Indeed, visualizing detailed molecular interactions is a great challenge for many animators; a proverbial black box that is drawing increasing attention within the scientific visualization community.

5.4.4 Post-production

The depiction of temporal change is also addressed during the post-production phase of development and may be realized using any number of techniques. During post-production the animator will compose the output from the 3D authoring environment and edit the film into a coherent narrative.

Editorial Transitions Editorial transitions in animation (e.g., cuts or dissolves) are introduced to suggest the passage of time, changes in environment, perspective, mood, or simply to provide a distinct break in the narrative. In this respect the animated transition shares much in common with the panel-to-panel transitions that are its static counterpart in a sequence of static illustrations. Within a sequence of static illustrations transitions are communicated by perceptually linking panels through composition and closure through which the viewer must mentally simulate time, motion, and change in order to construct the narrative. McCloud (1993) has categorized panel-to-panel transitions according to five distinct classifications: (1) moment-to-moment (demanding minimal closure); (2) action-to-action; (3) subject-to-subject; (4) scene-to-scene; (5) aspect-to-aspect (exploring aspects of an environment or mood); and (6) the non-sequitur (offering no connection between two panels). Animated transitions serve many of the same functions in linking individual shots. Strategic use of editing techniques is essential for managing the viewer's attention, controlling pace, and heightening emotional intensity. In essence the editorial "cut" (the juxtaposition of two shots in time without transition) is the equivalent to the panel-to-panel transition and often is all that is needed to suggest a change in motion, timeline pace, or narrative (Woolridge, 2012). When two scenes are connected the cut should occur "on the action". For example, if you were shooting a baseball player from two different angles the first shot would end mid-swing and the action would resume in the second shot to provide visual continuity.

Editorial cuts may be further described as "match cuts" (where an effort is made to match the graphic qualities between two clips) and "jump cuts" (in which there is intentional discontinuity between two clips, often used to suggest the passage of time). Editorial "dissolves" are used to soften the transition between cuts. One shot is progressively blended into another. This may be used to suggest the passage of time or to capture a particularly vast environment. A third form of editorial transition is the "mise-en-scène", in which camera movement reframes the action as it unfolds, using long uninterrupted shots. Woolridge (2012) remarks that novice animators sometimes feel that it is necessary to depict change using this technique because they believe it resembles their experience of the world. However, this can be confusing for viewers who may be overwhelmed by the perceptual effort required to orient themselves within this changing landscape while attempting to hang on to the narrative thread. Editorial decision-making is perhaps one of the most impactful aspects of the animation process with respect to establishing relationships, furnishing the viewer with adequate context, and providing a smooth readable narrative with which to build a robust mental model of dynamic events.

Manipulating Time In addition to any number of editorial transitions, an animator may also use other visualization strategies in depicting change over, or within, time. Temporal transition may be achieved with a process known as time-stretching or time-remapping during post-production. This is often used in animation to focus attention. It may be used to bring the action to a halt, as in a freeze frame shot or to alter the perceived temporal scale of a process. Events that ordinarily occur too quickly to be perceived by the naked eye can be seen more clearly when the speed of the event is slowed down using an effect informally known as bullet time cinematography, so-called after the Matrix films (Block, 2008). The opposite to this would be using a time lapse or time compression approach to speed up events occurring at a rate that is too slow to be perceptible. The pacing of animation has a significant impact on its readability and has been the subject of research examining the relationship between the learners' perceptual and cognitive resources and presentation speed (e.g., De Koning, Tabbers, Rikers, & Paas, 2011; Fischer, Lowe, & Schwan, 2008; Fischer & Schwan, 2010). Animation design can integrate temporal manipulation both as a means of guiding attention and of managing the viewer's perceptual resources (for example freezing or stretching time to focus on detail without the distraction of motion to divide the audience's attention).

Regardless of the target audience and communication goals, many dynamic visualizations are created using the same basic three-stage workflow described here. Throughout these phases of production the animator will draw upon many resources in a process of experimentation and refinement in order to craft a readable story. Many decisions are undertaken along the way with respect to the framing of the narrative, what details to include or exclude for the sake of clarity, where, when, and how to focus attention, and how to visually represent concepts where the evidence may be lacking or more hypothetical in nature. Artistic license plays a significant role in the design of a compelling and explanatorily effective communication tool.

5.5 Artistic License in the Design of Dynamic Visualizations

Static scientific visualization spans a continuum where at one end exists raw data (unbiased/empirical mappings) from which one hopes to extract meaning or insight. At the opposite end of the spectrum lies visualization for educational outreach (designed for learning in informal contexts where this type of animation is more editorial in nature). In the design of *dynamic* visualization, artistic license operates on the same continuum. When the goal of the visualization is peer-to-peer communication (presenting findings to scientific colleagues) there clearly needs to be a very tight fit between the depiction and the data it represents. Even so, there is a small level of artistic license in depicting theoretical constructs or dynamic events that cannot be appreciated with the naked eye. Color and shading do not exist at sub-microscopic level, but without their inclusion in the depiction of these structures they would be far too complex to understand. Colour and shading extend the scientific language by labelling, categorizing, and delineating structures.

If the communication goal is to educate, this most often demands a level of clarification and abstraction that requires some artistic judgement. Sometimes leaving out information may facilitate understanding. Through selective disclosure, the animator can simplify a topic by stripping away all non-essential or distracting information to focus attention. Goodsell and Johnson (2007) compare this to the reductive approaches used in science. The decision of "what to include" is a critical component of visualization design in so far as the outcome has the potential to mislead (through oversimplification) or overwhelm the viewer (by providing an extraneous level of detail). As an example, in a study examining the role of visual detail in teaching undergraduate life sciences students molecular interactions (in this case a receptor-ligand binding event), Jenkinson and McGill (2012) found that while a more simplified visualization (cf. Fig. 5.8) was effective in communicating the basic concept (surface-level understanding) of "A binds to B" it failed to convey any of the more abstract concepts (deep-level understanding) related to molecular binding events. Learners who viewed the more visually complex or detailed representations

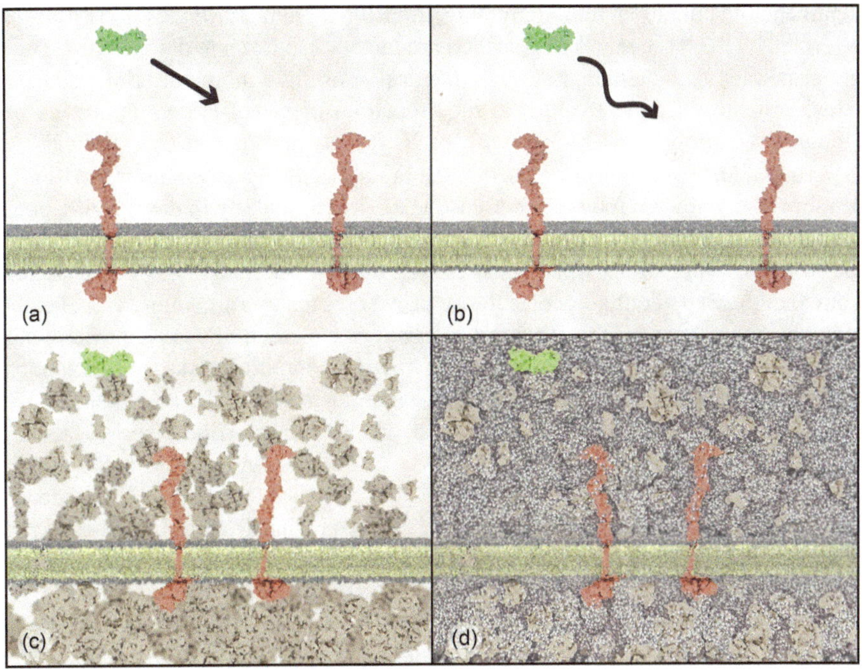

Fig. 5.8 Representative frames from four animations representing the same receptor–ligand binding event: (**a**) depicts the event using smooth, directed motion of the ligand toward the receptors; (**b**) introduces random, non-directed (Brownian) motion of the ligand and the receptors; (**c**) builds upon treatment "**b**" and introduces molecular crowding; (**d**) builds upon treatment "**c**" and introduces molecular water

held correspondingly deeper understanding of the nature of molecular binding. In this particular example, the dynamic event depicted addressed subject matter that was both complex and emergent in nature. What might at first appear to be extraneous detail (certainly more than is traditionally depicted in educational animations on the topic), did not pose a barrier to understanding. On the contrary, it constituted a scaffold to deeper intuition and understanding of the interactions that contribute to binding events. The suggestion here is that we need to establish, on a per scenario basis, what constitutes extraneous information in the design of a coherent whole. In so doing, careful attention should be paid to designing representations that will hold meaning for the audience, who may assume incorrectly that a highly schematized or simplified depiction holds some physical reality and stands alone as an explanation (Johnson & Hertig, 2014).

Finding an appropriate and consistent form of representation is particularly important if the aim of the visualization is outreach. This is a special case, in which the audience is not 'captive', and so engaging this audience is generally considered to warrant a much higher degree of artistic license (Johnson & Hertig, 2014). Since the primary objective of outreach is to engage an uninformed and potentially uninterested audience, capturing their attention takes priority. Decisions about the visual treatment of animation (color, lighting, transitions, camera angles etc.) may be informed by the basic tenets of visual communication, but where there are few guiding principles or exemplars to inform the design of these more editorial visualizations, animators frequently turn to the animation or visual effects industries for guidance.

Artistic license is also often used to fill in knowledge gaps when information is missing or unknown (Goodsell & Johnson, 2007). Particularly in the growing practice of molecular visualization, an animator may need to make decisions that guide the design of animations when fully definitive reference material is not available. For example, in depicting a molecular scale event, such as a signalling cascade, the animator must consider the structure and behavior of each of the elements within that environment. Visual elements may be based on empirical data, interpretations, or speculation. If the target audience is at the research end of the continuum, missing, uncertain, or ambiguous data may be depicted as dashed lines, circles, or other simple geometric shapes (Johnson & Hertig, 2014). However, as we move toward the outreach end of the continuum, using artistic license to speculate in the depiction of missing structures can minimize distractions that would otherwise detract from a readable narrative. Again, a balance must be struck between communicating clearly to the audience and visualizing a hypothetical model with deceptive clarity, possibly biasing the viewer's understanding. Iwasa (2010) suggests that when communicating these processes, different rendering styles (from the schematic to the highly detailed), may be effective visualization strategies for communicating the degree to which the process is understood and validated by the scientific research community.

5.6 Innovations in Dynamic Visualization

Visualization has been a transformative influence on our understanding of the intricate dynamic interactions occurring at all levels of life. In turn, depicting the complexity of newly discovered vistas demands from the designer, the exploration of new strategies for describing these phenomena. Innovation in visual representation is driven by necessity and afforded in part by advances in imaging technology. Present-day visualization design makes extensive use of large datasets derived from biomedical research. High resolution volumetric Computer Axial Tomography (CT) and Magnetic Resonance Imaging (MRI) data are commonly used as a starting point for modelling biomedical structures. As well, imaging techniques in structural and cell biology provide access to an untold wealth of data that affords the representation of molecular scale phenomena. Hence, the challenge faced by biomedical animators is two-fold. In addition to demanding the exploration of new visual strategies for depicting and disseminating novel research findings, from a practical perspective it also requires the development of software toolkits capable of integrating data with existing animation software. As Sharpe et al. (2008, p. 13) have astutely remarked, "… even the most amazing idea for new ways of seeing the world is powerless without the tools and technical means of bringing the new vision into practical use."

Fortunately, on the software front a number of advances have been made. Recent progress, including the availability of low-cost consumer-level graphics, three-dimensional, and post-production software applications, have led to a dramatic increase in the development and availability of dynamic visualizations. Open-source image processing software such as OsiriX have made it possible to process output from imaging equipment (CT, MRI, etc.). From these data three-dimensional models are constructed and may be imported into a 3D modeling environment, refined and otherwise manipulated. This is a particularly helpful resource for animators modeling anatomical structures. Affordable animation software such as Maxon Cinema 4D, Autodesk Maya® (a mainstay of the Hollywood animation industry), and Blender (a free open-source software) have provided the biomedical visualization community with powerful modeling and animation tools (see also Schwan & Papenmeier, 2017, this volume). As well, a number of specialized toolkits have been developed that allow artists working with molecular structures to import structural data directly into one of these animation environments. Molecular Maya (mMaya) is one such toolkit that extends Maya's capabilities by allowing users to import and animate molecular structures. The embedded Python Molecular Viewer (ePMV) is another plug-in for Maya, Cinema 4D and Blender, providing molecular viewer functionality inside the software environment. BioBlender is a similar software package built on the Blender 3D engine. These software innovations make it possible to visualize cellular environments using a number of different rendering styles, with greater accuracy, and at a level of detail not conceivable until now.

5.6.1 *Visualization Challenges*

Some of the greatest challenges in dynamic visualization may be found in the representation of biological concepts. Capturing the sheer complexity of these environments, while depicting phenomena occurring over several orders of spatial and temporal magnitude, is a herculean task. A typical cell is about 1000 times smaller than the smallest observable object in our everyday world, and within each cell we encounter the world of molecules, entities so tiny that they are smaller than the wavelength of light (Goodsell, 2010). Even more challenging than the depiction of spatial scale is the representation of temporal scale. From atomic level thermal vibrations to the behavior of different cell types, scientific animators are tasked with representing temporal phenomena operating over several levels of magnitude. For the biomedical animator striking a balance between scientific accuracy at this scale and clarity is difficult to manage. In particular, when the process to be depicted is either highly complex or emergent in nature, visualization designers are challenged to devise new techniques and approaches to depiction. Many of the design decisions undertaken at this level are without precedents (either in research or practice) and are guided largely by intuition.

5.7 Bridging the Gap Between Research and Practice

Research about dynamic visualization could be described as falling into one of two categories: (1) research *into* animation; and (2) research *through* animation. In the first case, this research is centered on assessing the effectiveness of animations in a variety of contexts and in comparison with various modalities of representation. The latter is concerned with the credibility of animated resources as models of insight and discovery. With respect to research that examines the effectiveness of animation, there would appear to be a considerable gap between educational research and craft-based practice. In part this is to be expected as both groups operate in siloed environments. This gap may also be linked to poor dissemination on the part of the research community. Animators working in industry are not generally aware of findings in the educational research domain. Whether this is because of the inaccessibility of primary literature when working outside of an academic setting, or more simply a perception that educational literature does not fit into the animation process (within the well-established practice of animation, research in the design and educational impact of animation is still in its infancy). On a more practical note, tight production timelines present a barrier to investing time exploring the growing body of literature in this area. On the other side of this breach, there is very little documentation to describe the decision-making process that underlies the design of animations. It is a practice has evolved through a process of training and mentorship, guided in part by apprenticeship-derived conventions drawn from medical and scientific illustration.

Identifying this gap underscores a need for greater inter-disciplinary collaboration between the two domains. Within the animation community there is a growing interest in educational research as a vehicle for supporting visualization decisions and justifying the utility of animation as an effective investment. Animation, 3D animation in particular, is painstaking to develop and very expensive to produce, and there is a great need for research that addresses the impact of decisions made within each phase of the production pipeline. More fine-grained assessment of narrative design, framing, visual treatment, and editorial decisions undertaken would contribute much to our understanding of effective dynamic visualization in a way that would reflect back on real world practice.

5.7.1 Improving the Credibility of Dynamic Visualization

Because visualizations may be very compelling and convincing in the depiction of events, they present a challenge to the end viewer to determine to what degree the development of any particular visualization has been guided by artistic license. Currently, there is no mechanism in place for describing how the design and behavior of features within an animated display may have been informed. Animators rely upon a wide variety of references that include personal communications with domain experts as well as information from database queries, images, simulations, videos, articles, and textbooks. At each stage of pre-production, production, and post-production, the development of visualizations makes use of these resources in a number of ways. For example, in animating a process occurring at the molecular level, the animator must take into consideration the physical attributes of the molecular "actors", their behavior (in particular, changes they may undergo when interacting with other molecules), and both the appearance and behavior of the surrounding environment. It is unlikely that the animator will find reference material to support all of the features included within a dynamic display. Many decisions may be based upon estimates, descriptive sources, or pure speculation. This presents a challenge for the end viewer who has no means of discerning how the design of the dynamic visualization is informed. While there currently is no mechanism in place for identifying the sources that inform the development of scientific animation, Jantzen et al. (2015) propose that an integrated system of citation (providing a more detailed account of references) would increase the credibility of scientific animations. As well, the establishment of a citation system would encourage viewers to be more critical consumers of animated media.

A second credibility issue concerns whether or not the widely assumed intrinsic power of animations to make dynamic subject matter comprehensible is realized in practice. Evidence reported in the literature assessing the educational impact of dynamic visualization would suggest otherwise. A number of factors may be involved in the failure of animation to "deliver". On one hand, given the ubiquity of consumer level software for creating dynamic representations, one needn't have received formal training in order to produce scientific animation. The rigorous workflow described in this chapter would not necessarily be familiar to self-taught animators. On the other hand, as previously remarked, there is a wide disconnect between the design and evaluation communities. The evaluation community would benefit from the design of high quality purposefully designed stimuli and visualization practitioners would benefit from insights that inform and reflect the work that they are producing. Concerted efforts and collaboration on both fronts would contribute greatly to establishing animation as a credible educational medium. While we have yet to fully understand the educational impact of dynamic visualization, it is nevertheless generally recognized as a potentially powerful and compelling means of communicating visually that which is difficult or impossible to communicate lexically. The visual examples included in this discussion exemplify a very small fraction of the many different visual variations that can be employed in the design of dynamic representations. The profession of scientific animation is relatively new (within the extended history of animation), but the practice has advanced rapidly within the last decade. As instruments of research, education, and outreach, there remains much to explore about the effective design of dynamic representations.

Acknowledgements The author is grateful for the generous contribution of illustrated works by graduate students and faculty in Biomedical Communications at University of Toronto. Research highlighted here is supported in part by grants NSF #DUE- 1220512 from the National Science

Foundation (USA) and SSHRC #SIG-13/14 from the Social Sciences and Humanities Research Council (CAN).

References

Ainsworth, S. (2008). How do animations influence learning? In D. H. Robinson & G. Schraw (Eds.), *Recent innovations in educational technology that facilitate student learning* (pp. 37–67). Charlotte, NC: Information Age Publishing.

Betrancourt, M., & Chassot, A. (2008). Making sense of animation. In R. K. Lowe & W. Schnotz (Eds.), *Learning with animations: Research implications for design* (pp. 141–164). New York: Cambridge University Press.

Block, B. (2008). *The visual story: Creating the visual structure of film, tv and digital media* (2nd ed.). Burlington, MA: Focal Press/Elsevier.

Boucheix, J.-M., Lowe, R. K., Putri, D. K., & Groff, J. (2013). Cueing animations: Dynamic signaling aids information extraction and comprehension. *Learning and Instruction, 25*, 71–84.

Crosby, R. W., & Cody, J. (1991). *Max Brödel, the man who put art into medicine*. New York: Springer.

De Koning, B. B., & Jarodzka, H. (2017). Attention guidance strategies for supporting learning from dynamic visualizations. In R. Lowe & R. Ploetzner (Eds.), *Learning from dynamic visualization—Innovations in research and application*. Berlin: Springer (this volume).

De Koning, B. B., Tabbers, H. K., Rikers, R. M. J. P., & Paas, F. (2009). Towards a framework for attention cueing in instructional animations: Guidelines for research and design. *Educational Psychology Review, 21*, 113–140.

De Koning, B. B., Tabbers, H. K., Rikers, R. M. J. P., & Paas, F. (2011). Attention cueing in an instructional animation: The role of presentation speed. *Computers in Human Behavior, 27*, 41–45.

Fischer, S., Lowe, R. K., & Schwan, S. (2008). Effects of presentation speed of a dynamic visualization on the understanding of a mechanical system. *Applied Cognitive Psychology, 22*, 1126–1141.

Fischer, S., & Schwan, S. (2010). Comprehending animations: Effects of spatial cueing versus temporal scaling. *Learning and Instruction., 20*, 465–475.

Goodsell, D. S. (2010). *The machinery of life* (2nd ed.). New York: Copernicus Books Springer.

Goodsell, D. S., & Johnson, G. T. (2007). Filling in the gaps: Artistic license in education and outreach. *PLoS Biology, 5*(12), e308.

Hodges, E. (Ed.). (2003). *The guild handbook of scientific illustration* (2nd ed.). Hoboken, NJ: Wiley.

Iwasa, J. H. (2010). Animating the model figure. *Trends in Cell Biology, 20*, 699–704.

Jantzen, S. G., Jenkinson, J., & McGill, G. (2015). Transparency in film: Increasing credibility of scientific animation using citation. *Nature Methods, 12*, 293–297.

Jastrzebski, Z. (1985). Scientific illustration: A guide for the beginning artist. Englewood Cliffs, NJ: Prentice-Hall, Inc.

Jenkinson, J., & McGill, G. (2012). Visualizing protein interactions and dynamics: Evolving a visual language for molecular animation. *CBE–Life Sciences Education, 11*, 103–110.

Johnson, G. T., & Hertig, S. (2014). A guide to the visual analysis and communication of biomolecular structural data. *Nature Reviews Molecular Cell Biology, 15*, 690–698.

Lewalter, D. (2003). Cognitive strategies for learning from static and dynamic visuals. *Learning and Instruction, 13*, 177–189.

Linn, M. C., Chang, H. Y., Chiu, J. L., Zhang, H. Z., & McElhaney, K. (2010). Can desirable difficulties overcome deceptive clarity in scientific visualizations? In A. Benjamin (Ed.),

Successful remembering and successful forgetting: A festschrift in honor of Robert A. Bjork (pp. 239–262). New York: Routledge.
Lowe, R. K. (2003). Animation and learning: Selective processing of information in dynamic graphics. *Learning and Instruction, 13*, 157–176.
Lowe, R., Schnotz, W., & Rasch, T. (2011). Aligning affordances of graphics with learning task requirements. *Applied Cognitive Psychology, 25*, 452–459.
Mayer, R. E. (2001). *Multimedia learning*. New York: Cambridge University Press.
McCloud, S. (1993). *Understanding comics*. New York: Kitchen Sink Press/Harper Collins.
McGill, G. (2008). Molecular movies… Coming to a lecture near you. *Cell, 133*, 1127–1132.
McGill, G. (2017). Designing instructional science visualizations in the trenches: Where research meets production reality. In R. Lowe & R. Ploetzner (Eds.), *Learning from dynamic visualization—Innovations in research and application*. Berlin: Springer (this volume).
Rieber, L. (1989). The effects of computer animated elaboration strategies and practice on factual and application learning in an elementary science lesson. *Journal of Educational Computing Research, 5*, 431–444.
Rieber, L., & Hannafin, M. J. (1988). Effects of textual and animated orienting activities and practice on learning from computer-based instruction. *Computers in Schools, 5*(1/2), 77–89.
Sanger, M., & Greenbowe, T. J. (2000). Addressing student misconceptions concerning electron flow in aqueous solutions with instruction including computer animations and conceptual change strategies. *International Journal of Science Education, 22*, 521–537.
Schnotz, W., & Lowe, R. K. (2008). A unified view of learning from animated and static graphics. In R. K. Lowe & W. Schnotz (Eds.), *Learning with animations: Research implications for design* (pp. 49–68). New York: Cambridge University Press.
Schwan, S., & Papenmeier, F. (2017). Learning from animations: From 2d to 3d? In R. Lowe & R. Ploetzner (Eds.), *Learning from dynamic visualization—Innovations in research and application*. Berlin: Springer (this volume).
Sharpe, J., Lumsden, C., & Woolridge, N. (2008). *In silico: 3D animation and simulation of cell biology with Maya and MEL*. New York: Morgan Kaufmann.
Thomas, F., & Johnston, O. (1981). *The illusion of life: Disney animation*. New York: Hyperion.
Tufte, E. (1997). Visual explanations. Cheshire, CT: Graphics Press Inc.
Tversky, B., Heiser, J., MacKenzie, R., Lozano, S., & Morrison, J. B. (2008). Enriching animations. In R. K. Lowe & W. Schnotz (Eds.), *Learning with animations: Research implications for design* (pp. 263–285). New York: Cambridge University Press.
Tversky, B., Morrison, J. B., & Betrancourt, M. (2002). Animation: Can it facilitate? *International Journal of Human-Computer Studies, 57*, 247–262.
Wagner, I., & Schnotz, W. (2017). Learning from static and dynamic visualizations: What kind of questions should we ask? In R. Lowe & R. Ploetzner (Eds.), *Learning from dynamic visualization—Innovations in research and application*. Berlin: Springer (this volume).
Ware, C. (2004). *Information visualization: Perception for design*. San Francisco: Morgan Kaufmann.
Wilson-Pauwels, L. (1997). Bringing it into focus: Visual cues and their role in directing attention. *Journal of Biocommunication, 24*(3), 12–16.
Wood, P. (1994). *Scientific illustration* (2nd ed.). New York: Van Nostrand Reinhold.
Woolridge, N. (2012). To cut, or not to cut. *Journal of Biocommunication, 38*(1), 27–30.
Woolridge, N. (2013). Drawing a line in the mind: Some reflections on the fundamental nature of linear depiction. *Journal of Biocommunication, 39*(1), 26–31.

Chapter 6
Designing Instructional Science Visualizations in the Trenches: Where Research Meets Production Reality

Gaël G. McGill

6.1 Introduction

Multimedia is ubiquitous in today's instructional materials, especially in science education (see also Jenkinson, 2017, this volume). These materials are either embedded within curricula delivered through learning management systems, provided within textbook supplements, or accessible 'a la carte' on the web from a multitude of content creators. In support of the development of these dynamic visualizations is a healthy industry that caters to the needs of publishers, museums, scientists and teachers who use these materials and approaches to communicate their scientific messages. Dynamic visualizations are also commonly used in the somewhat different context of marketing and communication campaigns by biotechnology, pharmaceutical and medical device companies that also face the challenge of explaining complex science (disease biology, drug mechanisms of action or technology platforms to name a few) to a variety of non-expert audiences. Engaging and scientifically-accurate visualizations are in high demand and there is great interest in better understanding how we learn from such materials in the context of the classroom, laboratory, hospital, boardroom, museum exhibit, or browser.

There is great variety in the types of people tasked with creating these materials—professional and amateur alike. In the life sciences—my field of expertise and that from which I will draw most of the examples and experiences in this chapter—anyone from students, research fellows, research and teaching faculty to artists, illustrators and animators with little or no science background are involved. There is also a growing number of those dually- or triply-trained as scientist-artist-programmers who can help to bridge the communication gap between the fields of science and those of art, design and multimedia programming. Even within this last group of professionals (i.e., those who bring some degree of scientific expertise to

G.G. McGill (✉)
Harvard Medical School & Digizyme Inc., Brookline, MA, USA
e-mail: mcgill@crystal.harvard.edu

R. Lowe, R. Ploetzner (eds.), *Learning from Dynamic Visualization*,
DOI 10.1007/978-3-319-56204-9_6

the multimedia design process), there are those trained as medical illustrators, those who started as PhD-level research scientists and faculty but who shifted their focus to the creation of visualizations, and those who started from a pure art and/or animation background and, over a period of years, built up their science knowledge to the point of deserving 'honorary PhDs' in particular topics. This professional diversity is wonderful in that it has resulted in a great variety of approaches, treatments, styles and other sensibilities in multimedia. However, it also represents a challenge if we are interested in understanding the process of 'good design' and extracting commonalities in its underlying driving principles. Indeed, understanding the context of where, how and who engages in instructional multimedia design is critical if we are to improve the circumstances under which these materials are born and the outcomes for learners are assessed.

Historically, there have been those who create dynamic visualizations and those who study them. Designers, illustrators, animators and programmers make up the first group and typically base their design decisions on practical experience, individual aesthetics and personal intuition. There is a creative freedom that is implied in the work and approach of a designer. The second group devises research studies to understand how individuals process the products of designers and measure their effectiveness in educational and other settings. This research community is rarely the home of designers—at least to the extent that principal investigators who lead these groups and drive the fundamental questions in the field are typically not trained as designers. The underlying assumption for researchers is that good design should seek to address clearly defined learning objectives and they are typically concerned with uncovering and/refining empirically verifiable and generalizable design principles that can be deployed according to the learning objectives. Ideally, designers' decisions should always be aligned with the specified learning goals with this focus superseding all other issues. In reality, however, designers and their clients are not always primarily focused on learning objectives. As noted above, design instinct and intuition are the typical driving forces, rather than strategically chosen principles derived from academically-based research studies.

In practice, designers and researchers work in very different contexts that are characterized by very different affordances and constraints. For example, the production intensity that results from aggressive deadlines and dwindling budgets rarely if ever allows designers to pick apart individual design variables in ways similar to controlled studies. Not only is a methodical, research-based approach unrealistic from a timing perspective, but there are often too many variables to address in combination. This is no doubt one of the reasons design 'intuition' plays a key role in real-world contexts. Ironically, some designers would argue that the principles 'uncovered' from academically-based research studies are not only ones they have instinctively known and applied for years, but are the very ones originally initiated by practitioners and subsequently tested in a research context. Nevertheless, designers may be able to benefit from consciously and explicitly integrating relevant research-based knowledge in their work. The impact on design is likely to reach its full potential only if key research conclusions are summarized and made more readily available to the design community. These conclusions cannot remain in the depths of the academic literature if they are to be used effectively and be deployed in the heat of the moment on a project with a deadline. Although certain handbooks and other publications attempt to address the craft of scientific animation from a professional and practical perspective (De La Flor, 2004; Frankel & De Pace, 2012; Hodges, 2003; Weinschenk, 2011), few if any integrate and summarize leading studies and results coming out of the academic community. Another potential benefit to designers is that results from published peer-refereed research studies could also be used as 'ammunition' in discussions with clients who have strong but misguided views about design. Indeed, designers may be more effective in offering alternative design approaches to clients if they can appeal to an external impartial authority like the research community. The likelihood of increased success in design choices in these situations remains to be tested, however, as clients may often simply 'go with their gut' despite proof of the contrary—whether the opinion is presented by an experienced designer or extracted from a controlled research study!

A key concern of this chapter is the issue of how we might go about integrating the craft- and research-based knowledge from both sides (see also Jenkinson, 2017, this volume). Without first-hand design experience deployed in the context of client-driven projects or sufficient familiarity with practitioners' craft knowledge, how can researchers become more aware of the processes and pressures that drive design decisions? Conversely, without meaningful exposure to existing studies and methodologies in instructional design research, how can designers draw upon this wealth of information to become more effective in their everyday work? The goal for some researchers such as Mayer & Moreno (2002) is to "develop a cognitive theory of multimedia learning that will guide designers in effectively using animation in multimedia presentations". The claim is that "the future of instructional animation is bright to the extent that its use is guided by cognitive theory and research". Although no doubt a worthy goal, it must be tempered by consideration of the complexities of the client-designer relationship and how this can impact design decisions. I will explore ways in which the collaborative process between client and designer can be affected by seemingly trivial project management inefficiencies and how design can be disrupted by the dominance of priorities (such as aesthetics and seductive technology among others) that are inconsistent with communication objectives.

This chapter provides a glimpse of the forces at play in the animation design and production process, and asks how we might improve the network of interdependencies between researcher, designer and client. What are the pressures that influence the design process when interacting with a client? How can we integrate research-based knowledge and the principles derived from it to enhance the quality and effectiveness of projects? In what novel ways could designers and researchers interact in order to bridge the cultural gap between these communities and gain a better mutual understanding of what is needed in both camps to increase the quality of learners' experience? Fundamentally, how do we leverage the craft knowledge of design practitioners and the insights derived from research to guide design considering the complex choreography by which projects are born, managed and completed (Fig. 6.1). In this chapter, I will not rehearse aspects such as neurobiology of the human visual system, cognitive psychology (including theories of perception and

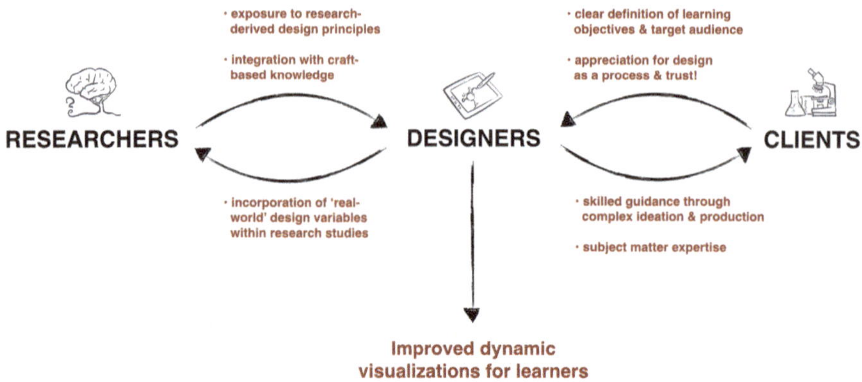

Fig. 6.1 Idealized researcher-designer-client relationship

mental image formation), the science of learning and how these findings impact approaches to multimedia and visualization design. These are dealt with in some detail by other contributors to this book. Instead, my focus will be upon issues that tend not to be addressed in the research literature such as a consideration of the client-designer interaction. I will review not only common disruptions that occur from both sides but also design variables that transcend the relationship. I will then offer strategies not only for navigating that interaction (in a way that minimizes conflict and negative impact on design) but also provide specific examples of how one can take advantage of some of the most recent multimedia and dynamic visualization technologies to successfully integrate multiple media modalities (text, static imagery, animation, user-paced interactives) for the benefit of the learner.

6.2 A Wealth of Studies to Guide the Design of Dynamic Visualizations

Undoubtedly one of the most fundamental shifts in our understanding of the working of the visual system is that our eyes and brain do not passively transmit and decode images collected from the world around us. Seeing is now understood to be an active and highly selective process in which attentional resources are limited (Ware, 2008, 2013). In carrying out this process, we are all to some extent 'visual virtuosos' (Hoffman, 1998) who selectively sample the visual fields around us and reconstruct them along two basic information processing axes: (1) bottom-up processing which is primarily dependent on the nature and quality of the stimuli around us (in the case of dynamic visualizations, the information display) and (2) top-down processing that looks at how a viewer's prior knowledge (or lack thereof) shapes how he or she attends to the display, extracts information from it and integrates this new information with existing concepts and mental models (Kriz & Hegarty, 2007).

A theory for understanding the nature of internal representations involved in visual mental imagery also suggests to us that perception and representation are intrinsically linked (Kosslyn, 1994). In practice, integrating perceptual exercises within the learning process has been explored with powerful results in terms of engagement and transferable creative problem solving in both young audiences (Yenawine, 2013) and higher education (Gremmler, 2014). As it relates specifically to dynamic visualizations in instructional media and how we effectively learn from them, key questions and themes have been explored over the past few decades. Some emphasize the notion that effectively managing the amount and frequency of information we present to learners is paramount (Sweller, 1988). How much information is 'too much'? What is the nature and frequency of visuals that lead to 'cognitive overload' and diminishing pedagogical returns (Hasler, Kersten, & Sweller, 2007)? Rather than being pre-occupied with just the overall amount of information presented, a more nuanced approach is needed. As succinctly summarized by Kirby (2008, p. 175), "if where we look is where we should look, and if we know what to do with what we see, animation is supporting and perhaps enhancing learning".

Parallel efforts and advances in the analysis and crafting of data visualizations have been ongoing, especially given the recent advent of 'big data' and its associated challenges (Munzner, 2014). Increasingly the data are being collected using powerful new methodologies (such as crowd-sourcing) that have the potential to significantly grow the number of study participants along with the statistical confidence with which trends are detected (Strobelt, Oelke, Kwon, Schrek, & Pfister, 2015). Further, at a more fine-grained level, techniques such as eye-tracking are helping researchers to characterize and appreciate the potentially powerful consequences of individual differences on visual processing. A rich literature summarizes the principles and elements of design as well as provides memorable examples of how failed design can have catastrophic outcomes (Tufte, 1997). A dynamic field of research has emerged, one focused on how to design effective instructional materials in the context of animated and/or interactive graphics (Ainsworth, 2008; De Koning, Tabbers, Rikers, & Paas, 2009; Lowe, 2008; Mayer, 2014; Plass, Homer, & Hayward, 2009; Ploetzner & Lowe, 2012). In the context of education, there has been much interest and progress in deploying teaching strategies that are responsive to the great variety of learning styles, as well as the detection and remediation of preconceptions, and opportunities for adaptive learning (Hattie, 2009; Hattie & Yates, 2014). Interestingly, efforts aimed at understanding how best to design animations, have shown that students who author, share and evaluate their own animations display improved learning outcomes (Hübscher-Younger & Narayanan, 2008).

As we will see, however, there are many factors that influence design in the 'real world' and that result in a deviation from otherwise widely accepted good design practices, let alone the findings from empirical research. In fact, in the context of dynamic visualizations in educational media, the very assumption that projects are methodically constructed and strategically executed around learning or communication objectives is unrealistic. Competing influences present in the design context may not always be reconcilable and designers will undoubtedly have to balance the application of empirically-derived principles with client-driven preferences. It is also not uncommon for certain instructors—even ones who use visualizations and multimedia as instructional tools—to think of the value of animation purely as a student engagement tool, an aesthetic embellishment that will earn them 'attention credits' from the students and keep them engaged and awake. Although in the context of this chapter, we will focus on design that flows from a clear understanding of target audience and learning objectives, I will offer the need to inspect the client-designer interaction (understanding that 'client' is used in a very broad sense that means 'he/she who commissions a visualization') and we will touch on the variety of individuals and conditions that may disrupt the design of dynamic visualizations for science instruction.

6.3 A Look at the 'Typical' Client Interaction

Dynamic visualizations are commissioned in a variety of settings, via interaction with a diversity of clients, and for a wide range of reasons. Our own work has spanned the needs (and wants) of scientists, physicians, teachers, publishers, museum exhibit designers, documentary film producers, as well as biotechnology, pharmaceutical and medical device company marketing, communications and sales executives. Each client and project is unique, and varies according to target audience, funding level, number of reviewers involved in feedback cycles and, of course, the nature of the content and specific learning or communication objectives. As designers we must adapt our communication style and design approach to these varied settings and there is obviously no such thing as a 'typical' client interaction. However, for the purposes of trying to understand broader themes in the design and production of dynamic visualizations, we can extract common characteristics across many such projects and discuss a few representative case studies that help illuminate the rest. In particular, we can focus our attention on the development of dynamic visualizations and multimedia for science instruction.

The typical stages of the visualization and multimedia development pipeline are borrowed from those with the greatest experience in implementing this complex process: the entertainment and gaming industries. Overall, the activities can be broken down into three broad phases: preproduction, production and postproduction (Fig. 6.2). During preproduction, an initial client-designer meeting will focus on understanding the target audience, the main learning or communication objectives, a discussion of likely medium (animation, interactive, game etc.) and setting for the use of the instructional materials (online, in the classroom, in a museum exhibit etc.). Will narration be used or will the media play alongside a live presentation? With these basic parameters in hand, a second meeting might involve the review of an initial text outline that captures the basic content progression (in the case of a linear story treatment), interfaces (in the case of an interactive, menu-based system), gaming levels (in the case of a 'gamified' treatment) or even simulation parameters (in the case of an educational simulation to be controlled by learners). At this stage, a draft script can also be created since the timing of narration and

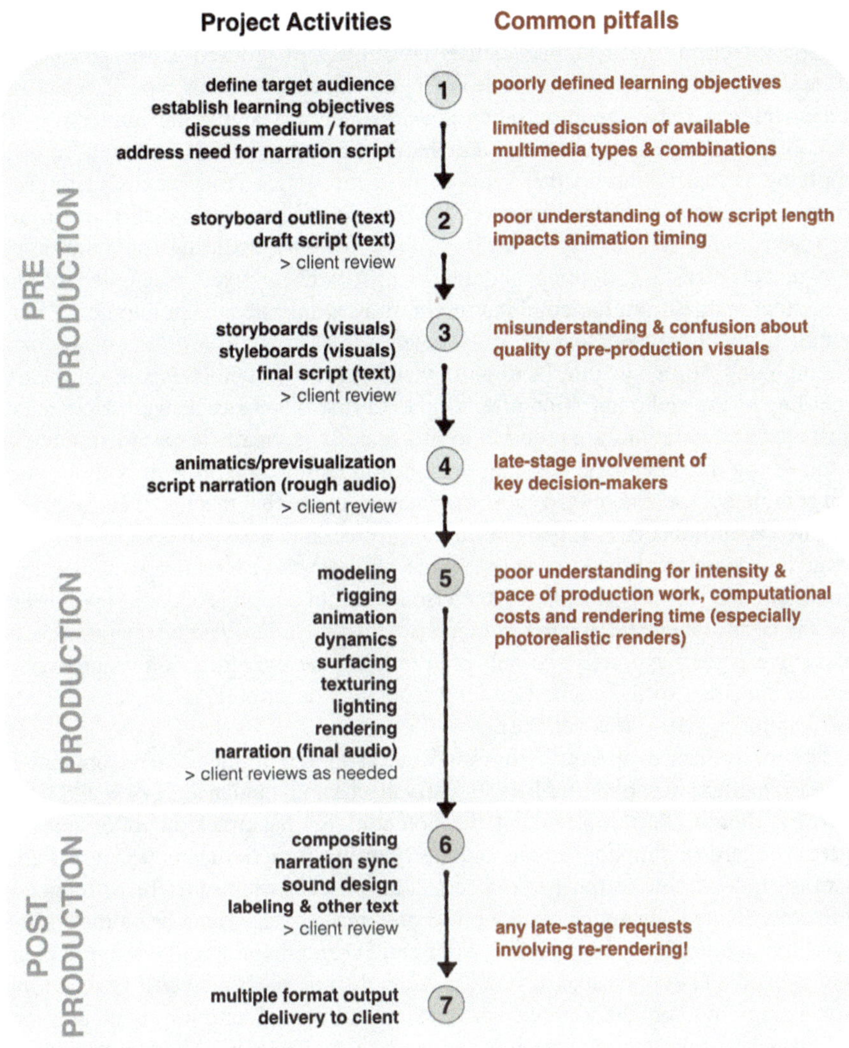

Fig. 6.2 The animation production pipeline and its pitfalls

imagery is a critical aspect of production that is best addressed as early as possible in the process. Generating even a rough audio recording of the script can also be a good idea at this stage. With agreement on these materials, the designer (or designer team) will now progress to the storyboard stage where a series of hand-drawn visuals will capture key frames or shots within the story progression. Depending on the project, styleboards (typically a collection of materials that evoke the visual feeling of the project, including color palette) can also be created at this stage and complement the linear storyboard treatment. Ideally, the narration script is also finalized at this stage so as not to leave uncertainty in the specific timing of animation and synching with audio. With a good sense of story progression, narration and even stylistic direction, the designer team is now ready to proceed to the production phase. The specific steps in this phase are highly dependent on the type of visualization or interactive presentation being developed. For example, an immersive 3D animation will often proceed through modeling (creating 3D objects), rigging (applying virtual armatures that support these models), animation (using the rig/armatures to move models), dynamics (particle or rigid body simulations), surfacing (determining how object surfaces respond to light), texturing (painting detail onto object surfaces), lighting and rendering (pixel-by-pixel calculation of the image that results from the combination of all previous steps). An interactive may include all of these steps (to produce visual assets) in addition to the design and programming of menus, interfaces and other interactive components. Professional recording of the audio narration often occurs during this phase as well. How many 'postings' and iterations are required to successfully navigate the production phase is something that client and designer must negotiate for each project. With production completed, there is often a 'post-production' phase that ties together the different threads and materials generated during production: shots are composited (i.e., image layers are combined to create the final visuals) and edited together, narration is overlaid and properly timed with the visuals, sound design is added if applicable, and any additional text and label elements are also included. The final animation or interactive is then exported and delivered to meet the dissemination requirements initially specified by the client. For a more detailed description of this process, see also Jenkinson (2017, this volume).

This phase-based approach to visualization and multimedia development is intended to maximize opportunities for early discussion, guidance, review and feedback by clients while minimizing the potential for backtracking as production moves forward (a time-consuming and costly proposition in the context of 3D animation). Indeed, unlike most live action film projects where the final product is ultimately shaped during the editing phase and most footage remains on the cutting room floor, animation (and especially high-end 3D animation) requires careful planning of each and every frame. If one considers the computational cost of rendering even a single photorealistic frame (or one derived from a complex simulation that took days or weeks to compute), it is no wonder that designers and animators strive to keep clients within a structured approach to defining story, shots and ultimately style. Downstream changes to the original production plan can have a disproportionately large impact on deadlines and cost, a reality that few clients truly appreciate. The right side of Fig. 6.2 lists some pitfalls commonly encountered in this process.

A broad breakdown of the client-designer work relationship yields two main categories of activity: (1) process management and (2) media design and production. The first category is concerned with the logistics of how one productively engages with the client and his/her team in order to elicit high quality information and timely feedback. The second focuses on the kinds of design and ensuing technical decisions that are made in response to the chosen target audience and learning objectives. One might imagine that these categories can be somewhat divorced, but in reality they are closely interconnected and poor implementation in one can have devastating consequences to the other. In extreme cases, the process is experienced as one that pits clients against designers, the former considered as opponents to be carefully managed and sometimes forced into submission when it comes to making the 'right' design choices. Finding effective design solutions (even empirically-derived ones featured in controlled and peer-reviewed research studies) under conditions where 'tensions run high' can be difficult and otherwise objective decisions can veer into agendas—either pushed from a client with a pet peeve or preference for a particular aesthetic, or from a designer who loses sight of learning objectives and elevates his or her work to the status of 'art for its own sake.' Despite best intentions and assuming that the interaction with a client is not approached as an outrightly adversarial one, there nevertheless remain numerous factors that influence the client-designer relationship. In the following sections, we will discuss the more common types of disruptions in the process of visualization design and organize these according to whether they arise from clients, designers or transcend the client-designer divide.

6.4 Client-Side Disruptions

A critical first step in developing a productive working relationship with a client is to introduce them to the creative opportunities available with 'design thinking.' As already noted, perception is an active process and the human visual system is highly selective in what it captures. Skilled designers consciously or intuitively understand this process and seek to manipulate it to their advantage. They do so via a number of strategies including managing the learner's attention, guiding visual queries and taking care not to overload visual working memory in the process (especially with visuals in motion). "Effective design should start with a visual task analysis, determine the set of visual queries to be supported by a design, and then use color, form, and space to efficiently serve those queries" (Ware, 2008, p. 21). Clients need to be aware of this. A client who appreciates that design decisions are made in the context of these perceptual goals is primed to engage in a meaningful conversation about design. The bar is not set so high as to expect clients to understand these design principles in detail, but to realize that they exist and that the design process is about leveraging these principles in service of the learning objectives. Without a basic understanding of this process, designers and clients cannot begin to discuss the merits of specific design solutions based on whether they work for or against the pedagogical goals of a given scene.

Clients also sometimes have difficulty understanding that in addition to carefully managing the amount of visual information on screen at any given time (and the frequency with which it changes), the modality of multimedia presentation can also have a significant impact on the success of the design. Consider the following production examples created by the author's team, chosen both for the density of information they have to convey as well as the fact that they depict events occurring at

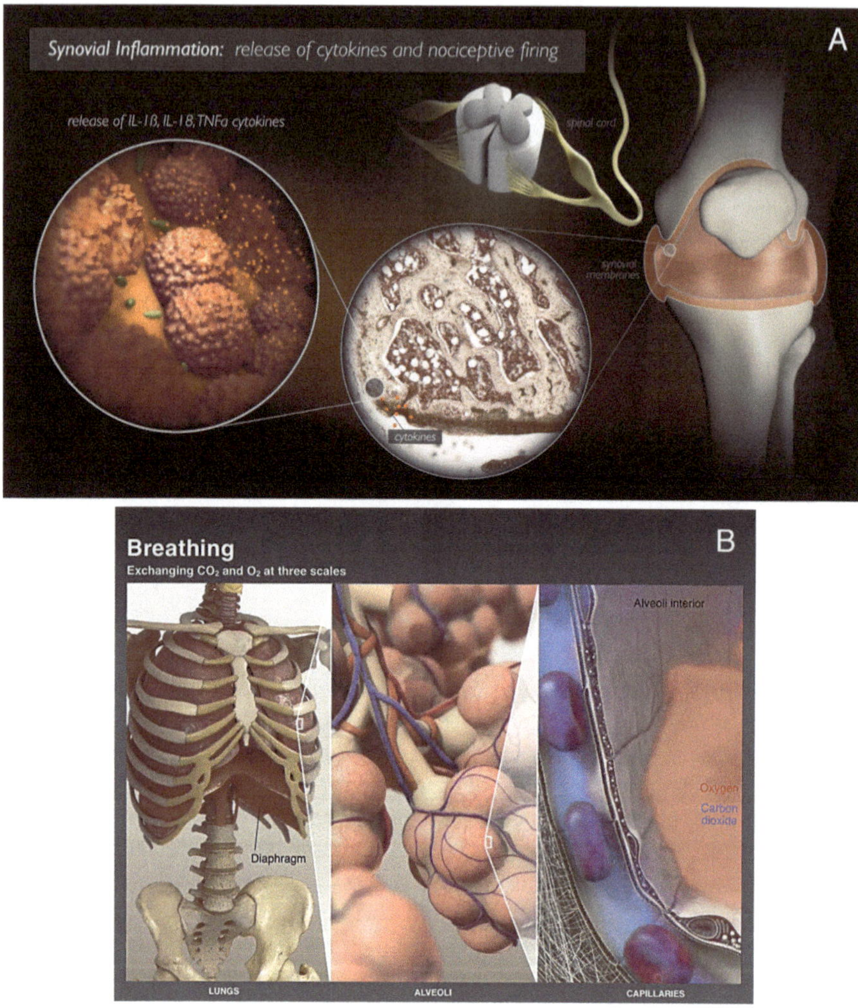

Fig. 6.3 A frame from a 3D animation on knee pain (Panel **A**) and a looping 3D animation from E.O. Wilson's *Life on Earth* iBook (Panel **B**). Copyright 2011 by Digizyme Inc. Reprinted with permission

multiple scales simultaneously (Fig. 6.3). In the case of the 3D animation on knee pain (cf. Panel A), the client was trying to explain a series of organ, tissue and cellular scale processes that are all interrelated. It was decided that these different levels of scale should be present at once on the screen which, despite care to connect these levels visually with lines and 'zoom-in' insets, likely results in split-attention problems for the viewer. This is an example where the choice of a linear animation—where information is necessarily fleeting—is problematic. Not only is there potentially too much information on screen at once, but because it is a linear animation, that information is not persistent enough to allow time for the viewer to meaningfully process the different parts. This issue of simultaneous versus successive presentation of complex information is a critical one that has been the subject of recent research studies (Lowe, Schnotz, & Rasch, 2011; Ploetzner & Lowe, 2014, 2017, this volume). In Panel B, the task was to depict another complex process—breathing—that similarly occurs at multiple levels simultaneously (i.e., organ, tissue and cellular). However, the design strategy for this movie was to make it a loop so that viewers would have time to inspect and re-inspect the display as needed over a period of time. Incidentally the breathing process lent itself well to that treatment and despite the density of visual information on the display and the perpetual motion onscreen, our judgment was that the learning objectives were better served with that treatment than with linear animation.

Note that in both cases the underlying decision and assumption was that viewers would benefit from seeing how the different scales are interrelated not only by making these visible on screen together, but also by connecting the imagery with series of lines that imply one inset zooming out of another. Despite similar densities of content, however, the difference in type of animation (i.e., linear, narrative-style movie versus looping movie) potentially shifts the balance away from likely confusion to understanding.

Another common issue can arise when clients are also content experts—a situation that is not unusual in the context of multimedia projects in the sciences. The 'curse of knowledge' that Pinker (2014) discusses in relation to ineffective writing or teaching is entirely relevant to the design choices that experts make when creating instructional multimedia for novices. Indeed, experts are temporally removed from their own learning experience and this can place them at a disadvantage when attempting to create pedagogical materials that are effective for novice learners. Especially in the context of the sciences, experts are comfortable parsing and using all kinds of visual representations to inspect and communicate the data they generate. As such, however, they may fail to consider whether learners are equipped to understand these specialized conventions and whether they should be used in instructional media. An example of this is found in molecular visualization where scientists have developed a host of specialized representations to depict specific molecular characteristics. However, outside the realm of structural cell and molecular biology, these representations are far from intuitive and require proper training to interpret. This can be said of many 'specialist representations' across the sciences. Even if the ultimate goal is for designers to help transition learners from simple to more advanced 'specialist' representations, they must guard expert clients against the temptation to use these without proper context and explanation.

In a broader sense, this speaks to the issue of visual literacy, or 'graphicacy,' in science and whether learners are properly equipped to interpret complex scientific images without the requisite scaffolding (Lowe, 2000). In other words, expert clients may not only forget that their audience is not familiar with a specific representational style otherwise commonplace in their field, but they are also likely to underestimate the learner's general ability to interpret scientific imagery since this is not a skill that is adequately taught at any level of science education (Metros, 2008). Furthermore, a learner's ability to interpret scientific imagery can also be influenced by their experience of creating such imagery. Indeed, the process of designing visualizations is a powerful way not only to become intimately familiar with the subject matter, but also to integrate multiple data types, become aware of 'missing data,' experiment and become familiar with multiple representations (Marx, 2013; McGill, 2014; McGill, Nowakowski, & Blacklow, 2017). Amidst these concerns and opportunities, the designer must first and foremost remind the client that the use of specialized scientific representations with learners who lack prior exposure and practice in parsing them will result in confusion and poor learning outcomes.

Finally, having discussed some of the conceptual and design-related pitfalls contributed by clients, we should not forget that more mundane reasons are also pervasive. In fact, probably the single most common obstacle to starting a productive design project is the client's lack of experience. One of the first questions we ask when starting a project is: 'Have you ever been involved in an animation project?' If the client answers 'no' then the designer (or animator as may be the case) has the responsibility to educate the client about the process they are about to embark upon (cf. Fig. 6.2). Although a polite explanation and warning about the associated pitfalls can also be given, in practice such warnings bear little impact on whether such pitfalls eventually occur. In light of this complex process, it is probably most important that clients understand the nature and timing of feedback that is expected of them. In my experience, and despite their neglect in the literature, project management logistics may easily be one of the leading causes for disruptions to the design process. Especially in the context of multi-tiered, hierarchical organizations, key stakeholders and decision makers will often be shielded by project managers from what is considered to be somewhat inconsequential, early phase work (in other words critical pre-production phase work). Intermediate work product—sometimes as far into the process as production itself—will trigger the desire to show progress and the project manager will only then share the latest version with their superior, sometimes with little to no context for how the design process arrived at this visual solution. The ensuing feedback can be misguided or disruptive, at best, to downright catastrophic, often resulting in the proverbial 'back to the drawing board.' It is therefore critical to communicate to project managers that key stakeholders and decision makers be included from the outset of a project and kept active in their design feedback throughout.

We already touched upon the notion that designers must educate clients about the nature of the design process and be willing to explore design avenues that result in effective engagement and learning. However, another common issue encountered with certain clients is the 'we pay you to implement, not to design' syndrome—where the vendor is seen purely as a production studio, one that receives a 'specification' and is simply tasked with implementing it. This is in contrast to the view of 'vendor as partner' who not only understands subject matter, but also takes into consideration the learning objectives and offers design solutions to the client while advising them on the pros and cons of each design path. A client's misunderstanding that they are paying for a process rather than a specific deliverable can be one of the reasons for 'sticker shock' (i.e., surprise with regard to high project costs). This may sound like a rather inconsequential financial issue, but in fact it has a significant impact on the client-vendor relationship, the production process (technology choices, number of iterations etc.) and, as a result, the design process. This also connects to the fundamental issue of trust—something that shapes the client-designer relationship and is critical if you consider that the effectiveness of a vast majority of instructional science visualizations are never assessed. Because the visualization's impact on learning is almost never measured in most real-world production settings, clients are left without any concrete metric of success and must *trust* that designers have the right priorities and are basing their decisions on sensible design parameters. All of these factors taken together—design, business and financial-related ones—shape the client-designer experience and, in doing so, have a very real impact on the context in which key design decisions are made.

6.5 Designer-Side Disruptions

On the other side of the client-designer relationship, a number of variables can impinge on the success of a designer, especially an inexperienced one. For example, designers will often tend to overestimate clients' ability to properly read and use storyboard outlines, storyboards, styleboards, animatics and/or previsualizations. Although they are all tools that are commonly used by designers and animators to progressively discover and refine the arc of story and the most effective ways to present it visually, they can be rather foreign to clients. Clients often have a difficult time making inferences that designers make with regard to how the finished product will look, and unfortunately this limits the usefulness of these planning and production techniques. Each is developed to address a specific phase of this creative process and, ideally, their use productively restricts the scope of possible designs as one makes progress through the pipeline. However, clients must first be taught how to interpret a storyboard and expectations must be set as to their benefits and also limitations. Despite warning clients that animatics (showing a paced succession of storyboard frames) or previsualizations (showing important camera moves and shot framings but using unshaded and simplistically animated models) lack the color and lighting information—let alone the polish they expect of a final animation—they will commonly proceed to ask if 'this is the way it will look?' This is why setting and continuously managing client expectations is an art and one that can spell the difference between a successful project and one that not only misses the mark from the clients' point of view, but also results in financial losses on the designers' side (due to unforeseen additional cycles of review and edits and a vanishing profit margin as a result).

We already discussed how client subject matter expertise can sometimes interfere with the design process in that experts may chose scientific representations that novice learners are not trained to understand. One way to alleviate this problem, is to have designers who are highly trained in the sciences and therefore able to weigh the scientific communication objectives alongside the representational choices. This is a growing trend in the industry as we witness examples of scientists who shift their efforts full time to scientific visualization or young designers who select a career path that incorporates graduate-level scientific training (as is the case with medical illustration programs for example). At a time when the increasing complexity of science is challenging even to scientists in neighboring fields, designers who bridge the gap of scientific knowledge with their clients are likely to enjoy a more collaborative process. Regardless of background, designers, artists, and animators who serve the sciences should undoubtedly see it as an intrinsic part of their role to listen carefully and educate themselves not only about the subject matter they depict, but also the relevant aspects of perceptual and cognitive psychology. These are not only skills that are important when interacting with clients, but they would also abate researchers' criticism that designers sometimes drive viewer attention to parts of the display that do not contain information critical to learning. "The technicians and programmers who are most able to design and produce these animations are the ones least trained to predict or understand their effects." (Kirby, 2008, p. 167). Although this may be the case in theory, a lack of appreciation on the researcher's side for both the technical and practical design implementation barriers are just as detrimental to the process. Ideally designers who are able to weigh both the learning goals and understand the design strategies that best serve these goals are most likely to craft effective dynamic visualization for instructional media. We will discuss in more depth later how further supplementing and embedding assessment technologies within the animation or interactive can inform the designer as to the prior knowledge and preconceptions of learners. This information can then serve as the basis for an adaptive learning system that triggers different follow-up media to the learner depending on how well they are currently doing.

In this context, and given the increasing variety and complexity of multimedia software and programming languages now available to designers, it is perhaps not surprising that a major challenge for designers, animators and programmers is to remain up to date and technically proficient. It is worth considering that limitations in the designer's ability to implement certain visual solutions for purely technical reasons can significantly affect the visuals, especially in content areas that require high levels of technical skill. These limitations are very real and are exacerbated in small studio environments where there are fewer human resources and therefore more uneven coverage of the tool landscape. This limited breadth of skill is reminiscent of the proverbial Law of the instrument (Kaplan, 1964, p. 28): "if all you have is a hammer, everything looks like a nail." In other words, a designer's search for the best design solution (which encompasses both visual and technical solutions) may be swayed towards implementations that they are already most familiar with (rather than ones that are most effective). For example, a learning module like that in Fig. 6.4 benefits learners because it combines immersive 3D animation (Panel A) with user-paced interactive anatomy exploration (Panel B) and simulation (Panel C). The primary learning objective of the module is to understand how beetles from the arid Namib Desert (also known as fog-harvesting beetles) condense water vapor onto their bodies and channel it to their mouths. To most effectively depict this incredible adaptation, the module combines the benefits of a narrated animation (the row of frames shown in Panel A are from a short introductory photorealistic animation with narration that shows the beetle climbing a small hill and lifting its rear into

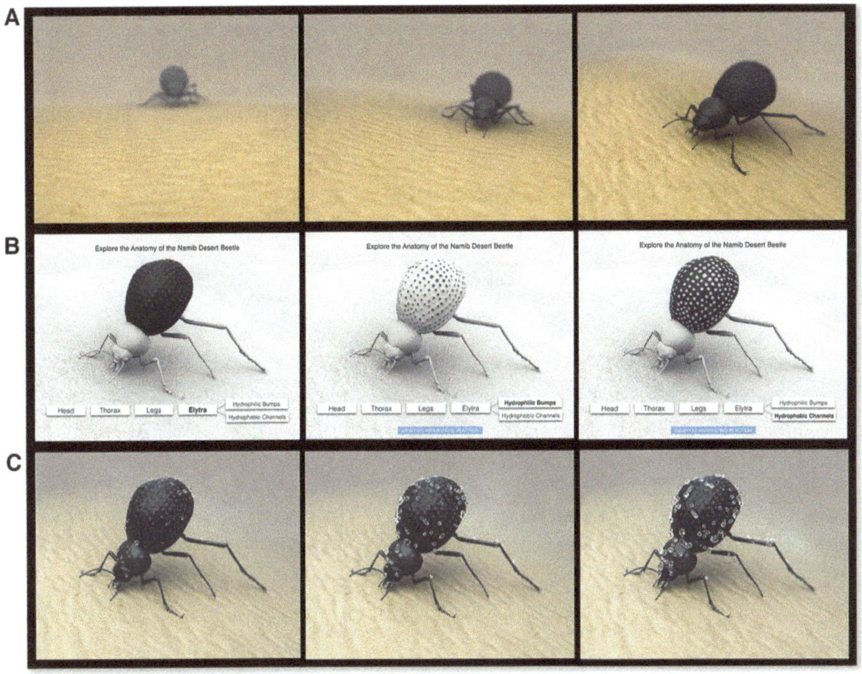

Fig. 6.4 The fog-harvesting beetle. Interactive iBook widget from *Nature's Toolkit* that combines multiple media types for optimal pedagogical treatment of the content. Copyright 2016 by Digizyme Inc. Reprinted with permission

the air), self-paced inspection by viewers (the buttons in Panel B highlight key anatomical structures of the beetle), visual cueing (by switching the entire display to greyscale while focusing the viewer's attention on the selected structures), as well as advanced simulation (the frames in Panel C are taken from a simulation of fog being deposited on the super-hydrophilic bumps of the beetle's back followed by a channeling of the growing water droplets towards the mouth along super-hydrophobic valleys). Therefore, the widget balances and combines the benefits of these different treatments to optimize the viewer's understanding of environmental context, key anatomical structures and chemical process of fog harvesting. In the hands of a designer or animator who is either unfamiliar with the possibilities of interactive programming or lacks the knowledge to implement them, this learning module would have likely been created as a simple linear animation and thereby missed a potential opportunity for improved pedagogy.

The previous example speaks both to the incredible opportunities offered by the latest multimedia technologies, but also to the burden placed on designers and animators to remain informed in a rapidly moving field and sharp in their technical implementation skills. Conversely, one must also beware of these very same technologies should they become too 'seductive' to the designer. Given the richness of what technology enables, it is no surprise that design decisions can sometimes

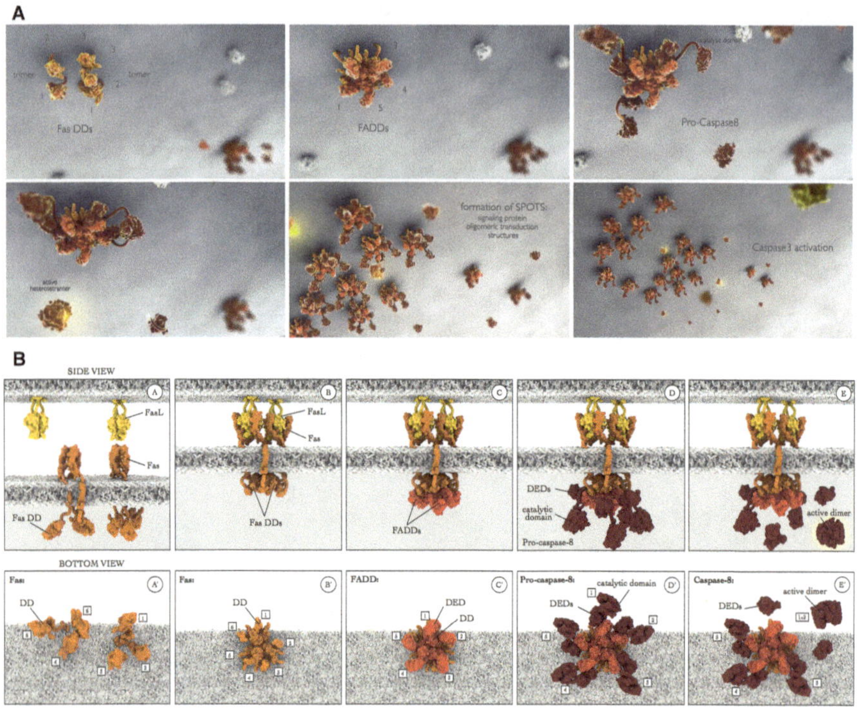

Fig. 6.5 Assembly of the Death-Inducing Signaling Complex (DISC). Panel **A** shows selected frames from a 3D animation of the dynamic process and Panel **B** shows a temporal progression of static images. Copyright 2010 by Digizyme Inc. Reprinted with permission

veer towards 'what is possible' (technologically and/or aesthetically) rather than remain focused on 'what is needed.' We are right back where we started with the 'Law of the instrument' alluded to earlier, except this time it is not because the designer lacks a broad menu of solutions to draw from (and solves design challenges with a limited few) but because the designer is enamored with a particular technical approach and tries to apply it to every design challenge available. Although this kind of seductive power of multimedia technology can be an issue for both designers and clients alike, it is mentioned here as an issue primarily with designers and animators because they are often the ones with a better knowledge of 'what is possible' and therefore perhaps most likely tempted by it.

A simple example of this concept is shown in Fig. 6.5 where we show two different treatments of the same content—a complex, multi-step dynamic assembly of a large molecular complex. The particular complexity of this content lies in both the spatial disposition of the subunits in the molecular complex (the layers of protein progressively docking to this complex exhibit multiple types of symmetry) as well as the dynamic assembly process. The client's assumption was that using immersive 3D animation would be the natural choice to facilitate learning of this structure and its assembly. This decision can certainly be supported in light of the fact that this is a highly dynamic process and motion could be recapitulated with this approach.

However, the choice of immersive 3D in this case was clearly also driven by the 'cool' factor and the excitement of being able to fly in and observe the process in all its dynamic complexity (see also Schwan & Papenmeier, 2017, this volume). The camera would be free to roam and observe the process from any optimal angle as well as zoom in and out as needed either to highlight details or provide wide shots (a few selected frames from this animation are shown in Panel A). Although the 3D animation does give the viewer a great sense of the highly dynamic nature of this assembly process, by refocusing on the learning objective, we realized that a series of static graphics with carefully chosen, consistent camera vantage points (in this case side and bottom views in Panel B) would most effectively focus the viewer's attention on the different symmetries of the structure and the step-by-step chronology of its assembly. This is reminiscent of empirical research showing that static depictions can sometimes be more effective than dynamic ones when detailed visual analysis of the subject matter is required (Lowe et al., 2011).

6.6 Design Disruptors that Transcend the Designer/Client Divide

'Naive realism' is something that both designers and clients have a tendency to mistakenly embrace and is rather pervasive in modern media. In part fueled by the technological race of entertainment media to create highly realistic and immersive imagery (in the context of special effects for feature film and/or games), naïve realism generally assumes that the more 'true to life' (from a visual and behavioral point of view) a visual depiction, the more likely the engagement and learning value is to be high for learners (Lowe, 2006, Smallman & Cook, 2011). This faulty assumption is based on the idea that a realistic depiction is one that is easy, accurate and complete from a perceptual standpoint. A number of studies have shown that while realism and complexity of depiction are often preferred by viewers, users do not necessarily know what's best for them from a learning outcomes perspective (Andrew & Wickens, 2011). But what if we could enjoy the benefits of immersive, photorealistic media (in terms of viewer engagement) along with the pedagogical power of cued visual explanations provided in the context of the former? This is what was developed in the example in Fig. 6.6. Intended for a middle school audience, we felt it was important for students to feel immersed and engaged in the environment hence the choice of photorealism (Panel A). However, to highlight the parrotfish's unique mechanism of grinding down coral skeletons, we temporarily stopped the animation (Panel B) and by using transparency, reveal the key skeletal features of the fish: a second of set of jaws in their throat (Panel C)!

A similar approach was used in Fig. 6.7, a 3D animation that explains the stinging mechanism of coral polyps that use specialized cells to paralyze and capture passing prey. In this animation we chose to begin with photorealistic, immersive imagery (Panel A) but then transition to cutaways (Panel B) and other representations that include overlaid data (Panel C) to better support the learning objectives of the piece.

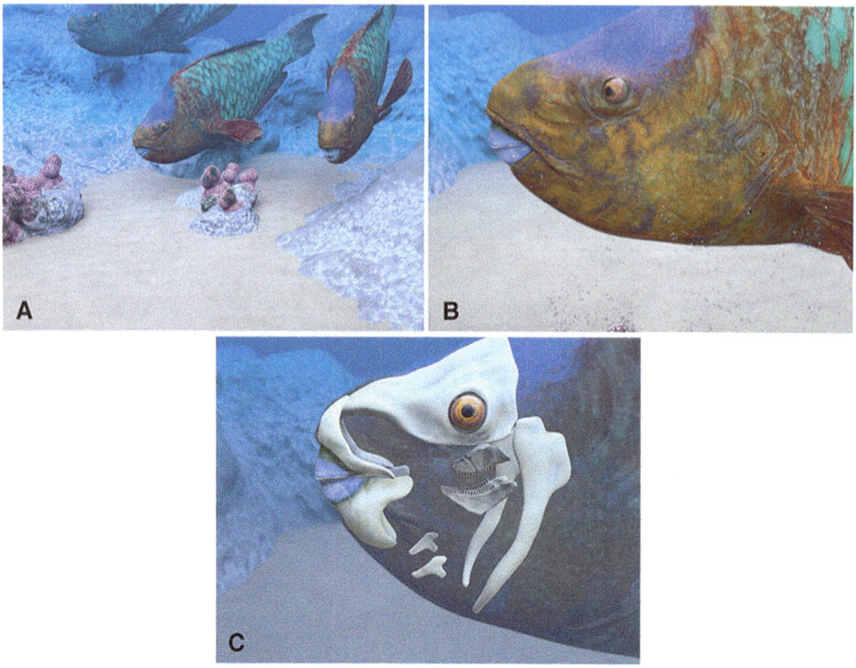

Fig. 6.6 Photorealistic 3D animation of a parrotfish. iBook widget from *Coral Reefs* that combines multiple media types for optimal pedagogical treatment of the content. Copyright 2016 by Digizyme Inc. Reprinted with permission

Fig. 6.7 3D animation of coral polyps and their nematocysts. iBook widget from *Coral Reefs* which uses multiple representations within the same animation. Copyright 2016 by Digizyme Inc. Reprinted with permission

The theme here is that animations need not adhere to a particular style the entire way through—they can easily leverage different representational styles depending on what the specific learning objectives call for. As long as the burden of decoding unfamiliar representations is kept to a minimum, then this approach provides endless flexibility to combine different design solutions within a single learning module.

Some of our own collaborative studies in this area aim to understand exactly what features of a representation can be kept or taken away in order to meet learning objectives. The ants in Fig. 6.8 are part of an ongoing experimental effort to understand

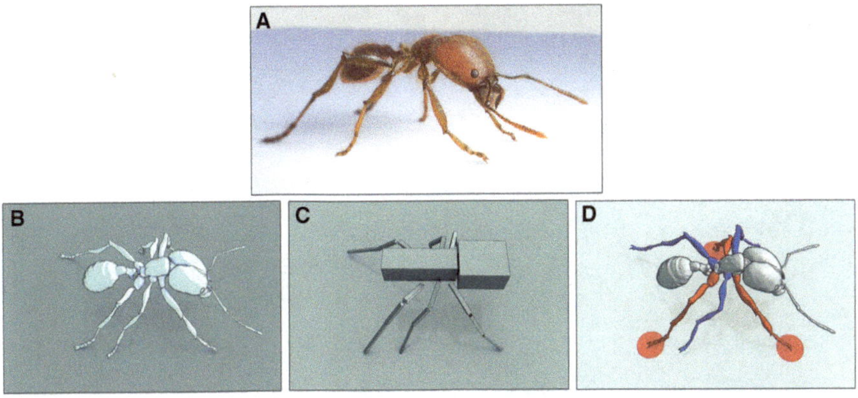

Fig. 6.8 3D models of ants developed for an ongoing study about representations most effective to teach ant locomotion. Panel **A**: photorealistic 3D ant; Panel **B**: toon-shaded ant model; Panel **C**: 'box ant'; Panel **D**: toon-shaded ant with color cueing on legs to highlight 'alternating tripods' insect walkcycle mechanism. Copyright 2016 by Digizyme Inc. Reprinted with permission

what features (structural, textured and animated) of an ant are critical in the design of an effective animation or interactive that teaches students the 'alternating tripods' concept in insect locomotion (Lowe, Boucheix, & Fillisch, 2017, this volume; Lowe, Jenkinson, & McGill, 2014).

Incidentally, a related pitfall with naïve realism is the client's or designer's prioritization of aesthetics over whichever representations best support learning objectives. This is another common issue in the development of instructional media whereby clients assume that beauty is a key component during the learning process. This is not to imply that aesthetically memorable and intrinsically engaging media is not a goal to strive for. However, as with every other design variable we have discussed here, the primary representational characteristics of the animation or interactive content should derive from careful consideration about what best serves learning objectives, as opposed to being decisions made upstream—and often at the expense—of the former. Indeed, Tversky, Morrison, and Bétrancourt (2002, p. 250) point out that "the advances in technology of producing attractive graphics often seem to drive and outstrip the development of tools and devices rather than research on their utility".

Another important factor that both designers and clients should consider when developing interactive visualizations or multimedia is that interactivity in and of itself, comes at a cost for the viewer (Boucheix, 2008; Lowe, 2008). Indeed, any new interface will necessarily require its own learning process in order for the user to become proficient with its use. Specifically, viewers must learn how the interface relates to the visualization they are presented with and this effort is accompanied by a cognitive cost. This concept is analogous to the idea of extraneous load in cognitive load theory (Schnotz & Rasch, 2008). In the same way that visual cues can be improperly used and lead viewers astray (i.e., when visual salience is misaligned with the pedagogical importance), interactivity can also be poorly deployed and

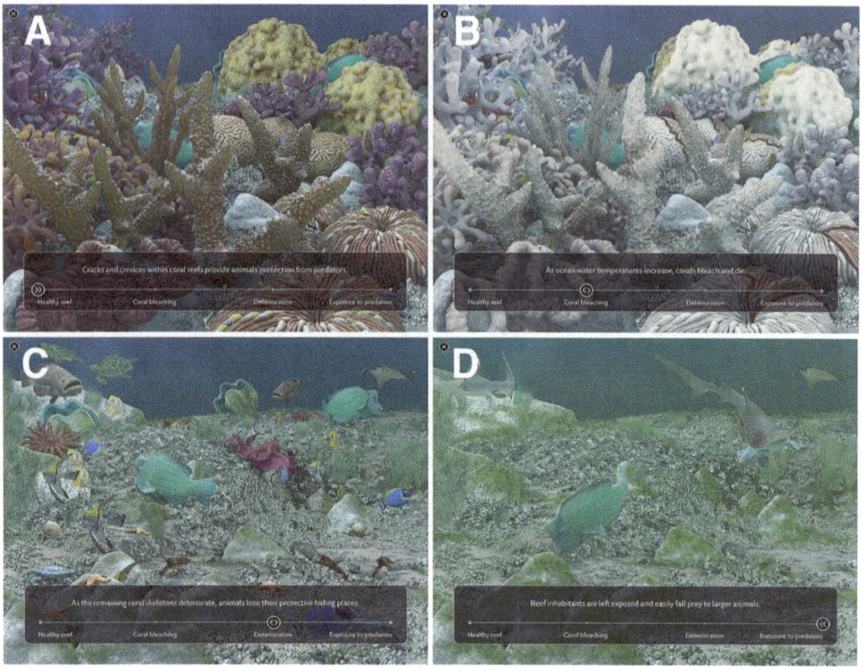

Fig. 6.9 iBook HTML5 widget from *Coral Reefs* that lets viewers horizontally drag across the bottom of the screen to control the reef's extent of bleaching. Coral are healthy in Panel **A**, bleached in Panel **B**, dead in Panel **C** and, as a result, smothered by algae in Panel **D**. Copyright 2016 by Digizyme Inc. Reprinted with permission

result in overly embellished graphic displays or, worse yet, interactive control features that have little or no relation to the content most critical to learning. On the other hand, if a simple and focused mode of interaction is 'mapped' to the right learning parameter, then interaction can become very effective. Figure 6.9 shows a simple interactive iBook widget programmed in HTML5 that lets learners control the evolution of coral bleaching and its impact on the surrounding ecosystem by swiping their finger along a slider at the bottom of the screen. Although the motion of the slider itself is smooth and continuous along the horizontal axis and this motion is controlling a smooth crossfade between four sets of images, we have defined and clearly labeled four states along this axis to help students realize that distinct ecosystem collapse stages do exist (i.e., Panel A: increase in ocean water temperature; Panel B: bleaching of the reef; Panel C: death of most of the coral and exposure of any animals that relied on the coral for their protection; Panel D: shift in the very food web relationships as the sea floor becomes smothered in algae and predators like sharks are able to prey on otherwise camouflaged animals). This discretization of the steps involved in coral bleaching is important for this audience (middle school science students) because part of the learning objective in this case is to explore cause-and-effect chains in nature.

To add to the complexity of variables to consider, we also encounter design challenges that are truly specific to the representational challenges of a particular field or area of content. The molecular realm, for instance, is a rich example of this in that the 'rules' of this scale of environment—from a representational standpoint—are quite different from those we experience at macroscopic scales. Color has essentially no meaning at the molecular scale since the size of most molecules is so small as to be below the wavelength of visible light. Yet we often use color to encode all kinds of features of molecules in our animations. Similarly, the density and speed of molecules in aqueous environments (such as cells) is such that placing a metaphorical 'nano camera' at that scale to film molecular events unfolding would result in seeing a compact wall of molecules moving at speeds far exceeding the limits of human perception. Perhaps most importantly, the agency and intent that we recognize in the movement of people and other living organisms at the macroscopic scale is gone from the molecular world, which is characterized by utter chaos of stochastic motion. These characteristics are fundamental to our understanding of how molecules work and most scientists and advanced students know this well. Yet graduate students creating molecular visualizations will sometimes inquire: 'What color should we make the DNA?' In other words, even those with deep knowledge of subject matter sometimes fail to apply that knowledge in the context of an unfamiliar task like the process of creating a visualization. So in response to the graduate student wanting to know how to color their DNA, the answer is not that we should avoid color altogether in molecular animation, but rather that we should be encouraged to think about how to use color most effectively to encode information within that perceptual channel. Studies aimed at gaining a better understanding of the design choices specific to these areas of science are underway and yielding some surprising insights into how we might integrate representations of complexity within instructional animations (Jenkinson & McGill, 2012, 2013).

6.7 From Research to Design Trenches

In light of the important factors that influence the design of dynamic visualizations, how can we increase the extent to which research results permeate design activities 'in the production trenches'? In parallel, how can designers' craft-based knowledge, technological savvy and experience managing the design process in collaboration with clients, influence the planning and analysis of research-based studies in order for them to better serve the complex web of design decisions characteristic of real-world projects? How should these culturally different communities intersect and what areas are most ripe for improvement?

A significant amount of client influence will always be at the heart of a project's design directions. In some instances, the parameters imposed by clients can seem almost unachievable. For example, we were recently tasked with explaining the 'central dogma' of biology (DNA > RNA > Protein) to a lay museum audience using no more than a few minutes of animation without the use of narration or onscreen text (other than simple labels)! In another project, the client's insistence on a specific molecular representation led to a highly inaccurate depiction of molecular scale, an unfortunate but realistic example of how 'look' and aesthetics occasionally dethrone accuracy. These client-initiated restrictions derive from the environment and context of use of the animation or interactive (i.e., online, classroom, museum exhibit, live presentation) and have a significant impact on the range of practical decisions available to the designer and client (not to mention the learning outcomes). Another source of influence can also come from regulatory or governmental bodies who recommend (or dictate in certain contexts) that multiple representations be available to learners with different learning styles and/or disabilities. As a result, starting the design process without first considering the tenets of Universal Design for Learning (UDL) can result in disruptive surprises further down the production pipeline (Burgstahler & Cory, 2010; Rappolt-Schlichtmann, Daley, & Rose, 2012). Finally, it should be noted that the research community rarely tackles the complexity of variables typical of live projects, preferring instead the analysis of isolated variables. More effort should be made to devise experiments that incorporate design variables and questions that more closely resemble those found in complex real-world projects.

Another important factor in the creation of instructional multimedia for the sciences is cost. When done well (i.e., based as much as possible on real scientific data and algorithms) and guided by teams of dually trained scientist-artists, the cost of design and production can become very expensive. What can be done to reduce these costs or at least leverage the assets and work developed in one project across other topic-related learning modules? The examples in Figs. 6.10 and 6.11 show how, when based on real data, the pre-production and even production work in one project can be successfully reused for other learning modules while serving the distinct learning objectives of each module and curricular level. In the first module (cf. Fig. 6.10), the goal was to expose introductory biology undergraduate students to the great structural diversity and dynamic nature of cellular membranes. A narrated, linear animation was chosen as the best format for this task (note that a basic introduction to membranes would have already been provided in an earlier module). Since we used real structural data and molecular simulation trajectories simulated in-house upon which to build these instructional materials, we were then faced with the challenge of linking the very simplistic diagrammatic representations of membranes typically found in textbooks with our more advanced visualization. First, we chose to remind students of the location of membranes in the context of a complex cellular environment (Panels A and B). We selected a color scheme that reinforced the palette usually applied to membranes in simpler diagrams (i.e., yellow; Panels A, B, and C). Only after this brief introduction did we proceed to increase the complexity of the visuals using both color (i.e., differentiating the lipid families found in membranes) and setting this crowded environment in motion to give students a sense for the molecules' highly dynamic and random motion (Panels D and E).

Unlike this first learning module aimed at introductory biology students, the second module shown in Fig. 6.11 addresses the same general topic—the structural diversity of cellular membranes—but this time for more advanced, upper level biology

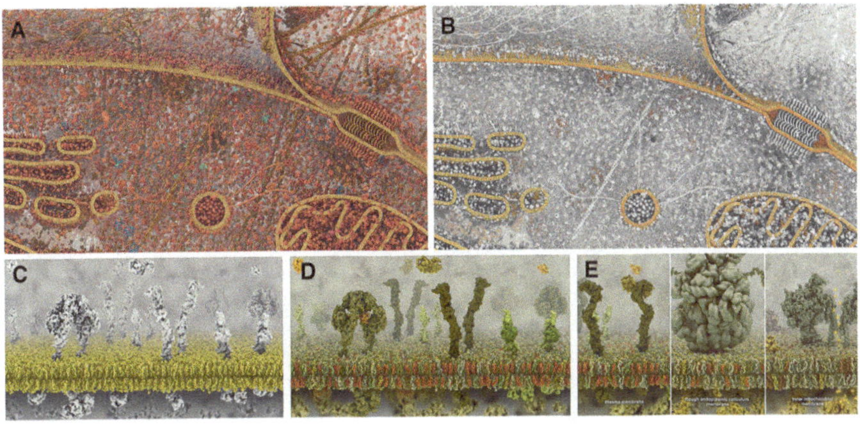

Fig. 6.10 3D animation depicting the structural and compositional diversity of lipids and proteins in cellular membranes for introductory biology undergraduate students. Copyright 2014 by Digizyme Inc. Reprinted with permission

students. The audience is now operating at a different curricular level and the learning objective has evolved as well. The goal is now to provide more detailed information about the actual quantitative differences between lipid families across membranes from different cellular compartments/organelles. Instead of presenting this content by means of a linear story using animation, we considered that the learning objective would best be served by giving students the ability to inspect the visuals at their own pace. The fleeting animated visuals that were meant to engage and leave no more than an intuition for molecular diversity and motion in the first year students now had to become more persistent visuals that could be inspected carefully and compared across membranes simultaneously (Lowe et al., 2011; Ploetzner & Lowe, 2012). Therefore, we created an interactive where the abundance of each lipid family across the three types of cellular membranes is triggered by mousing over the name of that lipid family in the upper right (upon which the entire image goes to greyscale expect for the selected lipid family which remains in full color). Not only is the format of each learning module best adapted to its learning objectives, but the underlying data and visual representations can be recycled across modules to minimize production effort (similar to what we observed in the DISC assembly animation and diagram shown in Fig. 6.5). This is possible because the data driving all the visuals is real to begin with and it is also advantageous because it provides visual consistency across curricular boundaries.

If our ultimate goal is for students who view and learn from our visualizations to be able to transfer this knowledge to other contexts, we must also consider strategies for embedding assessments within our multimedia materials. Too often, assessment is a separate consideration undertaken in isolation from the design phase of instructional multimedia creation. Not only is this a missed opportunity in that we are not maximizing the efficiency of our pedagogical interventions, but it also speaks to the disconnect between the level of engagement and interactivity found in multimedia

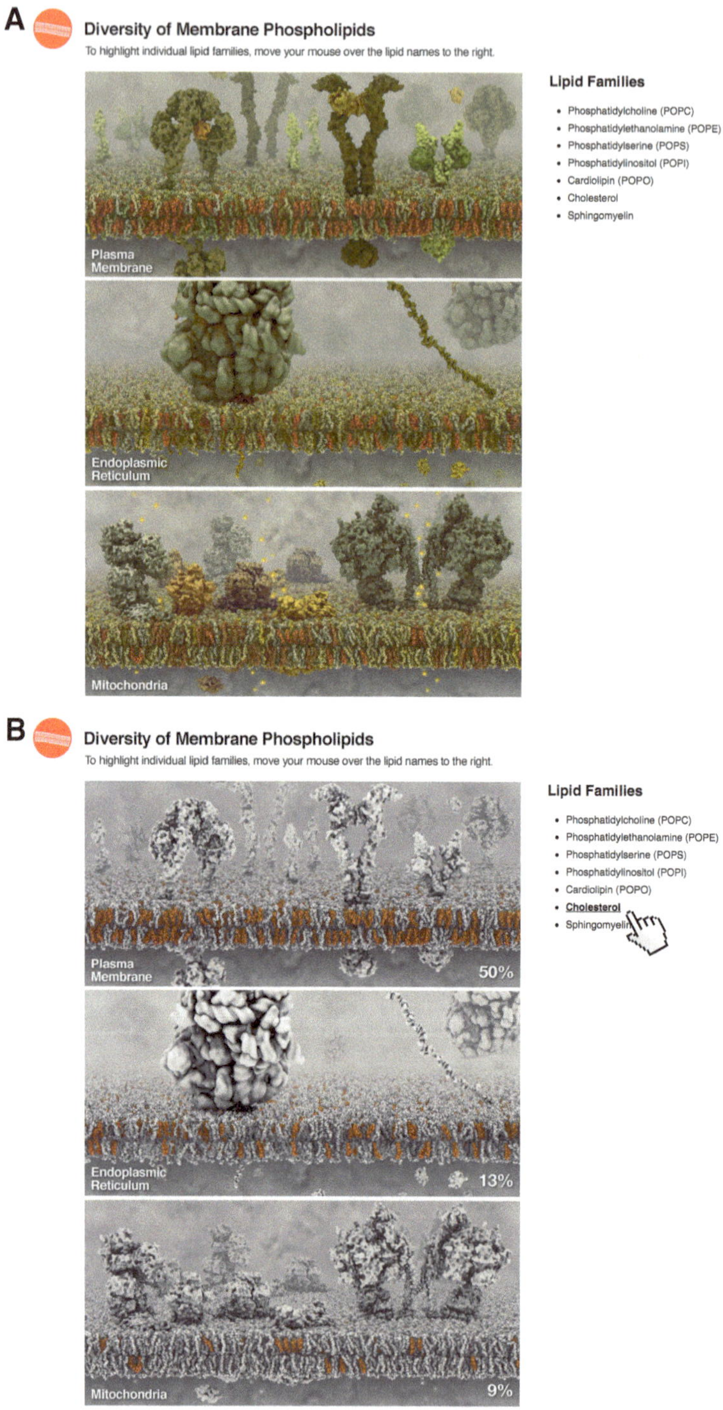

Fig. 6.11 Interactive module aimed at upper level biology undergraduates showing the structural diversity and quantitative difference in lipid components within three key cellular membrane systems. Panel **A** is the default state of the interactive and Panel **B** shows updated imagery in response to moving the mouse over individual lipid family names (in this case cholesterol) in the *upper right-hand column*. Copyright 2014 by Digizyme Inc. Reprinted with permission

and the simplistic multiple choice assessments that often follow them (Lowe et al., 2017, this volume). We know that tight integration and strategic timing of tests or 'booster' events can have a significant and lasting impact on information retention (Roediger, Agarwal, Kang, & Marsh, 2010; Roediger & Karpicke, 2006). Yet we see little discussion of this critical aspect of learning within the context of how animations and interactive media are designed, the broader theories of how we learn (Fry & Kolb, 1979) and implications for designing across different styles of learners (Kolb, 1976). Not only might we improve the overall effectiveness of our multimedia materials by blurring the line between instruction and assessment within our learning modules, but the principles we extract for the design of the former are undoubtedly relevant for the design of the latter.

Last but not least, it should be noted that designers sometimes question the validity of the sweeping conclusions put forth in the research literature because of the sub-par quality of visualizations used in these research studies. In order for the design community to be interested and influenced by the principles that emanate from research-based studies, there will need to be an improvement in the overall quality and variety of stimuli that are used. These will not only need to feel more relevant in terms of the quality of media expected in many real-world projects, but they also need to better represent the complexity of topics that are being shown. Many studies are based on rather simple animation examples and the conclusions drawn from such stimuli are therefore limited and do not always feel relevant to more complex situations (for which animations are often the preferred form of representation). Indeed, although it is critical for researchers to continue crafting well-controlled research studies that isolate variables one at a time, studies that address how designers use these variables in combination are also needed. Principles about the effective use of individual design elements such as cueing, color, user control, pacing and more have emerged from the literature and serve as a useful foundation upon which effective design is built. However, we also need to devise experimental approaches that address how these principles are interdependent and can (or cannot) be used in combination.

This rift in quality between the stimuli used in research and the animations coming out of client-driven projects undermines any hope of applying research-derived principles to real world design projects. There are likely to be many reasons why researchers are sometimes left with selecting poor animations as part of their studies (including lack of availability of high quality materials and/or prohibitive costs to develop one's own quality materials). Because sometimes the stimuli are obtained from existing (but disparate) sources, researchers will claim the authenticity and 'ecological validity' of these materials in that they represent materials widely available for educational purposes. But this reasoning does not justify the use of poor research stimuli since it potentially leads to conclusions born out of poorly designed starting materials. As a scientist, designer and animator having gained a relatively recent and fresh look into the visualization and multimedia design literature, I am occasionally left wondering whether we are studying the effects of 'bad' animation on learning. Are the observations and conclusions applicable to all media or do they primarily reflect the quality of the stimuli in the first place? Another criticism that we will need to address as a community if we are to bridge the cultural gap between designers and researchers is the belief that researchers sometimes 'discover' and claim design principles that designers have already understood and put into practice for years based on their personal intuition and experience of 'what works.' There may be some truth to this in select cases but, generally speaking, gaining experimental confirmation for such principles work is a positive endeavor that strengthens everyone's base of knowledge and ability to justify, especially in the context of a heated design meeting with a client, why some approaches work and others don't!

Even if we are able to overcome the barriers described above, we will still be left with the challenge of better communicating the research results to the designer community in ways that facilitate their immediate application in design. Theoretical frameworks for design can be inspiring and shift one's thinking productively, but as we have seen in most of this chapter, this information is still a long way from addressing the specific challenges that designers face every day. Therefore, dissemination is critical if the research community is serious about improving the design of instructional materials in the field. Simply put, it is most likely that a designer in the field of biomedical illustration and animation would have a copy of either 'The Digital Biomedical Illustration Handbook' (De La Flor, 2004), 'The Guild Handbook of Scientific Illustration' (Hodges, 2003), '100 Things Every Designer Should Know About People' (Weinschenk, 2011), or 'Visual Strategies—A Practical Guide to graphics for Scientists and Engineers' (Frankel & De Pace, 2012) on his or her shelves. These books contain a wealth of information and are very useful in terms of teaching the practical and technical aspects of the craft mostly through examples, but they do not replace the kind of research-based knowledge that is presented in books like the 'Cambridge Handbook of Multimedia Learning' (Mayer, 2014) or 'Learning with Animation' (Lowe, Schnotz, & Rasch, 2011) and others. As noted above, part of the reason may be that the research findings remain to be shared in a way that facilitates their application by designers. As we have discussed, there is also somewhat of a cultural divide between these communities whereby designers cherish and trust their experience and intuition (Johnson & Pierce, 2014) and some of the 'prescriptive' guidelines offered by the research community can feel too restrictive. So we need to strike a balance between powerful guidelines that designers should know versus more 'prescriptive' laws that squelch creative flexibility and visual design thinking. An example of such a 'bridge' can be found in the collection of one-page data visualization and design-oriented 'Points of View' articles published in Nature Methods (Evanko, 2013). These are short, well-crafted articles that discuss design principles in the specific context of how they should be applied to journal figure preparation. Another approach would be to develop visual examples of 'dos and dont's' that offer an immediate and intuitive appreciation of the design principles in action (something that we are attempting in the realm of molecular animation with a series of 'paired visualizations' each addressing a different concept in molecular animation) (Jantzen, Jenkinson, & McGill, 2015).

6.8 Conclusion

The design of dynamic visualizations and instructional multimedia has much to gain from the wealth of research studies in the fields of human perception, cognitive psychology, learning sciences as well as heuristic design experiences accumulated by designers during the course of practicing their craft. However, to gain a realistic view of how the design process unfolds in the context of working with clients, it is also important to be cognizant of the pressures and disruptive variables that arise in such environments. Some of these pressures can be strategically avoided through experience and planning in order to ensure optimal results—in this case, 'optimal' means a close match between intended learning objectives, client preferences and effective design. Other variables however are so unique to a given production environment, type of client, or specific subject matter that they can be difficult to apprehend and avoid impacting the design process.

Success integrating research-based knowledge with practical design craft will require a collaborative, 2-way communication effort. On the one hand, researchers would do well to integrate the myriad ways in which the client-designer interaction impacts design decisions. In this chapter we discussed a number of such design and project management variables that have a very real impact on the success of visualizations in fulfilling intended learning objectives. The temptation to investigate what is readily studied sometimes irrespective of its usefulness to designers (perhaps a result of the 'publish or perish' pressures of academia) should be considered when devising research studies. For example, the research community rarely tackles the complexity of variables typical of live projects, preferring instead the analysis of isolated variables. Granted that it may not be realistic to expect researchers to test every possible combination of design variables, especially since the efficacy of particular combinations is very likely to be tied to the specific nature of the content being depicted—what may work with one type of content may be useless with another type. Whenever possible, however, efforts should be made to devise experiments that incorporate design variables and questions that more closely resemble those found in complex real-world projects.

Also, the commercial reality of the 'bottom line' is not something that typically concerns most researchers. However, these commercial imperatives turn out to be at least as influential as design principles, especially when a particular design choice impacts production deadlines. Although researchers may acknowledge that these issues are real and disruptive, they may nevertheless contend that they are only indirectly relevant to the learning principles addressed in their studies. The rich menu of stylistic and technical choices that are discussed during projects and offered to clients, however, remind us that design choices are often made entirely within the context of financial- and deadline-driven compromises. In reality these decisions are intertwined and researchers would do well to consider this additional complexity if they want the results of their studies to be relevant and useful to designers.

Collectively we also need to evolve a realistic understanding of how research-based knowledge is not only disseminated to the designer community, but also how results are most effectively integrated into the creative process and how designers go about deploying them as they wrestle with the pressures of project development. How can research-based design strategies and guidelines become incorporated into the design process and support designers' interactions with clients? Although certain handbooks and other publications attempt to address the craft of scientific visualization from a professional and practical perspective, few if any integrate and effectively summarize how results from leading studies can guide effective design. Realistically this challenge may only be addressed through a collaborative approach between the research and design communities. Researchers could experience a more direct and fine-tuned appreciation for the complexities of visualization production. Designers could benefit from having more knowledge of relevant aspects of perceptual and cognitive psychology (i.e., how the human visual system works and its relationship to thinking/learning processes), areas that are not always included (or rigorously covered) in their training. Only then are the results and insights from empirical studies likely to be successfully called upon within the context of the creative problem-solving process that characterizes real-world projects.

As demonstrated through the variety of production examples in this chapter, the latest multimedia technologies offer incredible opportunities for the creation of highly diverse types of dynamic visuals and interactive media. More than ever, this variety of techniques empowers designers to explore and implement the design recommendations found in the research literature (such as described in studies like Lowe & Boucheix, 2012, and applied to user-paced interactives or ones that offer menu-based animation segments). However, the variety of such features and the software tools and techniques that accompany them, also challenge designers to remain proficient with a fast-paced and evolving technology toolbox. We observed that many other distractions can also come into play including the lure of seductive technology, the fallacy of naïve realism and primacy of aesthetics as a potentially disruptive force in design. Finally, we considered the importance of incorporating some of the new 'mixed multimedia' formats into our controlled research studies so that we may not only gain a better understanding of their pedagogical value but also assess the potential 'cost of interaction' inherent to their user interfaces. In addition, more innovative assessments could be integrated within instructional media in order to reap the benefits of what we know about the brain's cycles of information capture and retention. The graphicacy of learners using dynamic visualizations for science instruction may, in the end, be as critical as the care expended to design such materials. As such, graphicacy training and visual thinking strategies should increasingly be deployed in schools and perhaps even shared with clients. Those concepts are germane not only to our ability to offer strong S.T.E.M. (Science, Technology, Engineering and Mathematics) education programs, but also our efforts to communicate and leverage the power of design thinking to clients who commission dynamic visualizations.

Finally, a key factor that influences designers' interest and trust in principles derived from research-based knowledge is the quality of animations used in these studies. The overall difference in quality between the stimuli used in the research community and the visualizations that result from client-driven projects undermines the research community's desire to more directly impact the design of instructional multimedia. In the same way that some scientists have increasingly begun to join the ranks of designers and programmers and thereby bringing the unique benefits of a combined skillset to the scientific multimedia design process, this field would benefit from having professional designers carry out controlled research studies and, vice versa, have academic researchers become more closely involved and proficient in multimedia production. Although increased collaboration between researchers and designers is a natural direction to explore, it is one that still awaits tangible implementation. Few meetings and journals exist where such cross-fertilization can occur and blossom. This is exacerbated by the fact that these communities have very different professional reward mechanisms (publication is the 'currency' of academic researchers, while portfolios drive success in the world of design). At a time when we increasingly need to gain audience confidence that dynamic visualizations and instructional multimedia accomplish more than fleeting engagement, such collaboration could not only help stem the use of poor stimuli in research but also bring additional rigor to the design of visualizations 'in the trenches.'

Acknowledgments I wish to thank members of both my collaborative academic team at Harvard Medical School and the University of Toronto, as well as members of my Digizyme team who were involved in creating many of the examples shown in this chapter. Funding in support of some of the research activities was provided by the NSF and NASA.

References

Ainsworth, S. (2008). How do animations influence learning? In D. Robinson & G. Schraw (Eds.), *Current perspectives on cognition, learning, and instruction: Recent innovations in educational technology that facilitate student learning* (pp. 37–67). Charlotte, NC: Information Age Publishing.

Andrew, A. D., & Wickens, C. D. (2011). When users want what's NOT best for them. *Ergonomics in Design, 3*, 10–14.

Boucheix, J.-M. (2008). Young learners' control of technical animations. In R. K. Lowe & W. Schnotz (Eds.), *Learning with animation: Research implications for design* (pp. 208–234). New York: Cambridge University Press.

Burgstahler, S. E., & Cory, R. C. (2010). *Universal design in higher education*. Cambridge, MA: Harvard Education Press.

De Koning, B. B., Tabbers, H. K., Rikers, R. M. J. P., & Paas, F. (2009). Towards a framework for attention cueing in instructional animations: Guidelines for research and design. *Educational Psychology Review, 21*, 113–140.

De La Flor, M. (2004). *The digital biomedical illustration handbook*. Hingham: Charles River Media.

Evanko, D. (2013, July 30). *Data visualization: A view of every points of view column*. Nature Methods [Web log post]. Retrieved from http://blogs.nature.com/methagora/2013/07/data-visualization-points-of-view.html

Frankel, F. C., & De Pace, A. H. (2012). *Visual strategies—A practical guide to graphics for scientists and engineers*. New Haven, CT: Yale University Press.

Fry, R., & Kolb, D. (1979). Experiential learning theory and learning experiences in liberal arts education. *New Directions for Experiential Learning, 6*, 79–92.
Gremmler, T. (2014). *Creative education and dynamic media*. Hong Kong: City University of Hong Kong Press.
Hasler, B. S., Kersten, B., & Sweller, J. (2007). Learner control, cognitive load and instructional animation. *Applied Cognitive Psychology, 21*, 713–729.
Hattie, J. (2009). *Visible learning—A synthesis of over 800 meta-analyses relating to achievement*. New York: Routledge.
Hattie, J., & Yates, G. (2014). *Visible learning and the science of how we learn*. New York: Routledge.
Hodges, E. R. S. (2003). *The Guild handbook of scientific illustration*. Hoboken, NJ: Wiley.
Hoffman, D. D. (1998). *Visual intelligence—How we create what we see*. New York: W. W. Norton & Company Inc..
Hübscher-Younger, T., & Narayanan, N. H. (2008). Turning the tables: Investigating characteristics and efficacy of student-authored animations and multimedia representations. In R. K. Lowe & W. Schnotz (Eds.), *Learning with animation: Research implications for design* (pp. 235–259). New York: Cambridge University Press.
Jantzen, S., Jenkinson, J., & McGill, G. (2015, August 5). *Molecular visualization principles*. Retrieved from https://bmcresearch.utm.utoronto.ca/sciencevislab/index.php/portfolio/molecular-visualization-principles/
Jenkinson, J. (2017). The role of craft-based knowledge in the design of dynamic visualizations. In R. Lowe, & R. Ploetzner (Eds.), *Learning from dynamic visualization—Innovations in research and application*. Berlin: Springer (this volume).
Jenkinson, J., & McGill, G. (2012). Visualizing protein interactions and dynamics: Evolving a visual language for molecular animation. *CBE Life Sciences Education, 11*, 103–110.
Jenkinson, J., & McGill, G. (2013). Using 3D animation in biology education: Examining the effects of visual complexity in the representation of dynamic molecular events. *Journal of Biocommunication, 39*, 42–49.
Johnson, B., & Pierce, J. T. (2014). *Design school wisdom*. San Francisco: Chronicle Books.
Kaplan, A. (1964). *The conduct of inquiry: Methodology for behavioral science*. New Brunswick, NJ: Transaction Publishers.
Kirby, J. R. (2008). Mental representations, cognitive strategies, and individual differences. In R. K. Lowe & W. Schnotz (Eds.), *Learning with animation: Research implications for design* (pp. 165–180). New York: Cambridge University Press.
Kolb, D. (1976). *Learning style inventory: Technical manual*. Boston: McBer and Company.
Kosslyn, S. M. (1994). *Image and brain—The resolution of the imagery debate*. Cambridge, MA: MIT Press.
Kriz, S., & Hegarty, M. (2007). Top-down and bottom-up influences on learning from animations. *International Journal of Human-Computer Studies, 65*, 911–930.
Lowe, R. (2000). Visual literacy in science and technology education. *Connect—UNESCO International Science, Technology & Environmental Educational Newsletter, XXV*(2), 1–3.
Lowe, R. (2006). Animations: A key advance for open and distance learning? In M. Tulloch, S. Relf, & P. Uys (Eds.), *Breaking down boundaries: International experience in open, distance and flexible learning: Selected papers from the 2005 ODLAA Conference* (pp. 189–195). Bathurst: Charles Sturt University.
Lowe, R. K. (2008). Learning from animation: Where to look, when to look. In R. K. Lowe & W. Schnotz (Eds.), *Learning with animation: Research implications for design* (pp. 49–68). New York: Cambridge University Press.
Lowe, R. K., & Boucheix, J.-M. (2012). Dynamic diagrams: A composition alternative. In P. Cox, B. Plimmer, & P. Rogers (Eds.), *Diagrammatic representation and inference* (pp. 233–240). Berlin: Springer.

Lowe, R., Boucheix, J.-M., & Fillisch, B. (2017). Demonstration tasks for assessment. In R. Lowe & R. Ploetzner (Eds.), *Learning from dynamic visualization—Innovations in research and application*. Berlin: Springer (this volume).

Lowe, R. K., Jenkinson, J., & McGill, G. (2014, August). *Learning functional relations in complex animations*. Paper presented at the EARLI SIG 2 (Comprehension of Text and Graphics) Meeting, Rotterdam, Netherlands.

Lowe, R., Schnotz, W., & Rasch, T. (2011). Aligning affordances of graphics with learning task requirements. *Applied Cognitive Psychology, 25*, 452–459.

Marx, V. (2013). Data visualization: Ambiguity as a fellow traveler. *Nature Methods., 10*, 613–615.

Mayer, R. (2014). *The Cambridge handbook of multimedia learning* (2nd ed.). New York: Cambridge University Press.

Mayer, R. E., & Moreno, R. (2002). Animation as an aid to multimedia learning. *Educational Psychology Review, 14*, 87–99.

McGill, G. (2014). Crafting scientific visualization—Creative process and best practices. *Biozoom, 3*, 17–19.

McGill, G., Nowakowski, D., & Blacklow, S. C. (2017). Creating molecular visualizations: It's the journey and the destination that counts. Cell (submitted).

Metros, S. E. (2008). The educator's role in preparing visually literate learners. *Theory Into Practice, 47*, 102–109.

Munzner, T. (2014). *Visualization analysis and design*. Boca Raton, FL: CRC Press.

Pinker, S. (2014). *The sense of style: The thinking person's guide to writing in the 21st century*. New York: Penguin.

Plass, J. L., Homer, B. D., & Hayward, E. O. (2009). Design factors for educationally effective animations and simulations. *Journal of Computing Higher Education, 21*, 31–61.

Ploetzner, R., & Lowe, R. (2012). A systematic characterization of expository animations. *Computers in Human Behavior, 28*, 781–794.

Ploetzner, R., & Lowe, R. (2014). Simultaneously presented animations facilitate the learning of higher-order relationships. *Computers in Human Behavior, 34*, 12–22.

Ploetzner, R., & Lowe, R. (2017). Looking across instead of back and forth—How the simultaneous presentation of multiple animation episodes facilitates learning. In R. Lowe & R. Ploetzner (Eds.), *Learning from dynamic visualization—Innovations in research and application*. Berlin: Springer (this volume).

Rappolt-Schlichtmann, G., Daley, S. G., & Rose, L. T. (2012). *A research reader in universal design for learning*. Cambridge, MA: Harvard Education Press.

Roediger, H. L., Agarwal, P. K., Kang, S. H. K., & Marsh, E. J. (2010). Benefits of testing memory—Best practices and boundary conditions. In G. M. Davies & D. B. Wright (Eds.), *New frontiers in applied memory* (pp. 13–49). Brighton, UK: Psychology Press.

Roediger, H. L., & Karpicke, J. D. (2006). The power of testing memory: Basic research and implications for educational practice. *Perspectives on Psychological Science, 1*, 181–210.

Schnotz, W., & Rasch, T. (2008). Functions of animation in comprehension and learning. In R. K. Lowe & W. Schnotz (Eds.), *Learning with animation: Research implications for design* (pp. 92–113). New York: Cambridge University Press.

Schwan, S., & Papenmeier, F. (2017). Learning from animations: From 2d to 3d? In R. Lowe & R. Ploetzner (Eds.), *Learning from dynamic visualization—Innovations in research and application*. Berlin: Springer (this volume).

Smallman, H. S., & Cook, M. B. (2011). Naïve realism: Folk fallacies in the design and use of visual displays. *Topics in Cognitive Science, 3*, 579–608.

Strobelt, H., Oelke, D., Kwon, B. C., Schrek, T., & Pfister, H. (2015). Guidelines for effective use of text highlighting techniques. *IEEE Transactions on Visualizations and Computer Graphics, 22*, 489–498.

Sweller, J. (1988). Cognitive load during problem solving: Effects on learning. *Cognitive Science, 12*, 257–285.

Tufte, E. R. (1997). *Visual explanations*. Cheshire, CT: Graphics Press.
Tversky, B., Morrison, J. B., & Bétrancourt, M. (2002). Animation: Can it facilitate? *International Journal of Human Computer Studies, 57*, 247–262.
Ware, C. (2008). *Visual thinking for design*. Waltham, MA: Morgan Kaufman.
Ware, C. (2013). *Information visualization—Perception for design*. Waltham, MA: Morgan Kaufman.
Weinschenk, S. M. (2011). *100 things every designer should know about people*. Berkeley, CA: New Riders.
Yenawine, P. (2013). *Visual thinking strategies—Using art to deepen learning across school disciplines*. Cambridge, MA: Harvard Education Press.

Part II
Innovations in Assessment

In the context of research on animation and learning, assessment may be considered from three different perspectives. First, learner characteristics that are assumed to be relevant to learning from animation might be assessed. Second, the learners' understanding of the displayed animation might be assessed. Third, animations might be employed as diagnostic tools to assess learning from other external representations. This part includes three chapters, each chapter focusing on one of these perspectives on assessment.

With respect to the first perspective, assessment of learner characteristics has long been a central focus of educational measurement. Researchers use psychometric methods to investigate how learner characteristics and learning situations interact with respect to learning performance. Knowledge of such aptitude-treatment interactions can then be recruited to adapt learning situations to target learners' abilities (cf. Cronbach & Snow, 1977). Because animations are visuospatial representations, a learner characteristic that has frequently been investigated in the context of research on animation and learning is *spatial ability* (see Berney & Bétrancourt, 2017, this volume; Sanchez & Wiley, 2017, this volume). In this research, spatial ability is commonly assessed by tasks that require the learners to identify and mentally preserve visuospatial relations (e.g., the Figure Rotation Task; Cooper & Shepard, 1973) or to mentally manipulate visuospatial relations (e.g., the Paper Folding Task; Ekstrom, Frensch, Harman, & Dermen, 1976). Although higher spatial ability – as assessed by tasks such as those mentioned – can enhance learning from static graphics, it barely affects learning from dynamic graphics such as animations (cf. Hegarty & Kriz, 2008; Höffler, 2010; Höffler & Leutner, 2011). However, animations are not only visuospatial representations; they also provide direct representations of spatiotemporal information. Sanchez and Wiley (2017, this volume) therefore differentiate between two kinds of spatial ability: one that focusses on spatial relations within objects, and another that focusses on the change of spatial relations between multiple objects over time. Furthermore, the authors propose an innovative method that allows a learner's *dynamic spatial ability* to be assessed. Because the ability to process dynamic spatial relations between multiple objects closely corresponds with the processing demands during learning from animation as put forward in the Animation Processing Model (see Lowe & Boucheix, 2017, this volume), the proposed assessment method is especially promising with respect to research on the interaction between spatial ability and learning from animation.

Concerning the second perspective, text-based assessments of learning have dominated the history of education. This heavy reliance on verbal information both for presenting assessment tasks and for collecting learner responses was perhaps understandable when the spoken and written word were the major representations used for instruction. However, the recent proliferation of first static and then dynamic visualizations within educational resources poses new challenges for making valid assessments of student performance and educational effectiveness. Early research on learning from animation unquestioningly adopted the established verbal approaches to assessment, with written questions in both closed and open formats being the favored types of measurement. Unfortunately, the strict linear sequential structure of verbal representations can be ill-suited to capturing the visuospatial and spatiotemporal information portrayed. Too often, it is just very difficult to articulate in words aspects of the knowledge that has been acquired about systems that do not fit these structural attributes. Likely consequences of the poor match between the non-verbal form of representation used in dynamic portrayals and the verbal form of representation used for assessment include incomplete tapping of a learner's actual understanding and a distorted reflection of that understanding due to its necessary transformation from non-verbal to verbal format. An innovative alternative is to use graphics or even physical models rather than text as the primary means of assessment. Correspondingly, Lowe, Boucheix and Fillisch (2017, this volume) propose to use *manipulable models* for assessing the effectiveness of interventions intended to improve learning from animation. Such models resemble the material shown in an animation with respect to both appearance and potential behaviors. Learners can manipulate the models' component parts to demonstrate directly what has been learned from the animation. Thus, there is no need for learners to perform the often demanding and error-prone additional step of transforming their non-verbal understandings of the animation into a verbal representation in order to respond to the assessment task.

Although graphics and physical models are promising for assessing learning from animation, they are also applicable to other learning contexts in which visuospatial and spatiotemporal aspects need to be understood. For instance, in science learning inquiry practices such as making and structuring observations, collecting and analyzing data, as well as modeling dynamic systems have become increasingly important in recent years. Although text-based assessments may effectively measure declarative science knowledge, they commonly fail to provide evidence of science inquiry practices. Since technological advances have opened up new possibilities for developing graphics-based assessments of science learning, Davenport and Quellmalz (2017, this volume) conceptualize and empirically evaluate innovative forms of assessment that make use of animations and interactive simulations. Special emphasis is is given to alignment between the type of graphic used in the assessment and the learning objectives being targeted. Furthermore, assessments made on the basis of animations and simulations require the learners to understand the dynamic and interactive task formats in their own right. Therefore, the authors emphasize that the construction of dynamic and interactive assessment tasks needs to take into account the design principles that emerge from research on learning from animation and simulation (see Parts I and III of this volume).

References

Berney, S., & Bétrancourt, M. (2017). Learning three-dimensional anatomical structures with animation: Effects of orientation references and learners' spatial ability. In R. Lowe & R. Ploetzner (Eds.), *Learning from dynamic visualization – Innovations in research and application*. Berlin: Springer Publishers (this volume).

Cooper, L. A., & Shepard, R. N. (1973). Chronometric studies of the rotation of mental images. In W. G. Chase (Ed.), *Visual information processing* (pp. 75–176). New York: Academic Press.

Cronbach, L. J., & Snow, R. E. (1977). *Aptitudes and instructional methods – A handbook for research on interactions*. New York: Irvington Publishers.

Davenport, J. L., & Quellmalz, E. S. (2017). Assessing science inquiry and reasoning using dynamic visualizations and interactive simulations. In R. Lowe & R. Ploetzner (Eds.), *Learning from dynamic visualization – Innovations in research and application*. Berlin: Springer (this volume).

Ekstrom, R. B., Frensch, J. W., Harman, H. H., & Dermen, D. (1976). *Kit of factor-referenced cognitive tests*. Princeton, NJ: Educational Testing Service.

Hegarty, M., & Kriz, S. (2008). Effects of knowledge and spatial ability on learning from animation. In R. K. Lowe & W. Schnotz (Eds.), *Learning with animation: Research implications for design* (pp. 3–29). New York: Cambridge University Press.

Höffler, T. N. (2010). Spatial ability: Its influence on learning with visualizations – A meta-analytic review. *Educational Psychology Review, 22*, 245–269.

Höffler, T. N., & Leutner, D. (2011). The role of spatial ability in learning from instructional animations – Evidence for an ability-as-compensator hypothesis. *Computers in Human Behavior, 27*, 209–216.

Lowe, R., & Boucheix, J.-M. (2017). A composition approach to design of educational animations. In R. Lowe & R. Ploetzner (Eds.), *Learning from dynamic visualization – Innovations in research and application*. Berlin: Springer (this volume).

Lowe, R., Boucheix, J.-M., & Fillisch, B. (2017). Demonstration tasks for assessment. In R. Lowe & R. Ploetzner (Eds.), *Learning from dynamic visualization – Innovations in research and application*. Berlin: Springer (this volume).

Sanchez, C. A., & Wiley, J. (2017). Dynamic visuospatial ability and learning from dynamic visualizations. In R. Lowe & R. Ploetzner (Eds.), *Learning from dynamic visualization – Innovations in research and application*. Berlin: Springer (this volume).

Chapter 7
Dynamic Visuospatial Ability and Learning from Dynamic Visualizations

Christopher A. Sanchez and Jennifer Wiley

7.1 Introduction

Developing understanding of many phenomena in STEM areas is a complex cognitive activity that theoretically requires not only the accumulation of rote knowledge of individual domain concepts, but also the creation of internal dynamic visuospatial representations that capture the interaction and integration between those concepts across space and time (Friedman & Miyake, 2000; Hegarty, 1992; Hegarty et al., 2010; Rinck, 2005; Wiley & Sanchez, 2010). These mental representations, or mental models of dynamic visuospatial systems, likely provide access to some of the same information as the actual experience, although often created in the absence of actual perceptual input. One marked benefit of this kind of mental simulation is that it offers knowledge-seekers the opportunity to better appreciate relationships that are not readily apparent in linguistic form, essentially permitting learners to see patterns or interactions that are otherwise 'invisible'. Indeed, some of the most critical advances in scientific thinking have occurred due to the ability of individuals to spatially recreate or imagine scientific content (e.g., DNA, benzene ring, etc.; National Research Council (NRC), 2006), allowing for insight that would otherwise not be possible. This suggests it may be critical to present information to potential learners in such a way that maximizes the likelihood that they will be able to form coherent and appropriate visuospatial representations of the material while learning. From a motivational perspective, the presence of animations might also positively affect levels of motivation within students, in addition to the learning benefits suggested above. For example, it has been demonstrated previously that the inclusion

C.A. Sanchez (✉)
Oregon State University, Corvallis, OR 97331, USA
e-mail: christopher.sanchez@oregonstate.edu

J. Wiley
University of Illinois at Chicago, Chicago, IL 60607, USA
e-mail: jwiley@uic.edu

R. Lowe, R. Ploetzner (eds.), *Learning from Dynamic Visualization*,
DOI 10.1007/978-3-319-56204-9_7

of animations made it more likely that students would 'continue-on' with learning in a STEM content domain (Rieber, 1991). Based on intuitions such as these, it is common for instruction in STEM areas to include dynamic visualizations such as animations and videos in the hopes that they will help learners appreciate such dynamic relationships. Yet, some research on learning from dynamic visualizations has shown that they sometimes fail to produce this same facilitative benefit. Benefits of dynamic visualizations may depend to the nature of the material to be learned, as well as the way information is presented in the animations themselves. For example, a recent meta-analysis found that more 'decorational' animations (i.e., those that do not explicitly depict the representation to-be-learned) do not appear to demonstrate a benefit above static illustrations, and more importantly, also produce learning effects that are significantly smaller than animations that do explicitly demonstrate the target representation (Höffler & Leutner, 2007). Features of the learner are another factor that could determine whether benefits of dynamic visualization are seen (Wiley, Sanchez & Jaeger, 2014). The main purpose of this chapter is to explore a particular *aptitude-by-treatment interaction* that can help to explain when dynamic visualizations may be most likely to facilitate learning. The studies reported here assess *Multiple-Object Dynamic Spatial Ability (MODSA)*, a particular set of spatial skills involving integrating information from multiple objects over time and space, and discuss its relation to learning from dynamic visualizations.

7.2 Visualizations and Instructing STEM Topics

One common approach that has been taken to enhance learning of STEM topics, particularly topics that have a temporal or spatial component, has been to include explicit external visualizations to augment instruction. This approach involves the addition of visualizations to text to potentially provide a mechanism of external support to help the learner form their mental model of the STEM phenomena. For example, including appropriate static images has been shown to produce better learning of biology (Ainsworth & Th Loizou, 2003), physics (Chi, Bassok, Lewis, Reimann, & Glaser, 1989; Loftus & Harley, 2004), and mechanical devices (Hegarty, 1992; Mayer, 1989; Mayer & Gallini, 1990). Similarly, the addition of animations or videos has produced facilitation in learning meteorology, mechanical tasks, and computer programming tasks (Mayer & Moreno, 1998; Moreno & Mayer, 1999; Palmiter & Elkerton, 1993; Schnotz, Böckheler, & Grzondziel, 1999). The explanation given for such facilitative effects is quite simple: because these content areas all contain an explicit visuospatial component, providing relevant visuospatial information to learners in a pre-packaged form allows for the better development of understanding. Unfortunately, this assumption might prove to be overly simplistic.

Despite these successes, there are also numerous examples of failed attempts to enhance learning through the simple addition of visualizations, with some cases even leading to lower learning (Chanlin, 1998; Harp & Mayer, 1997; Rieber, Boyce, & Assad, 1990; Schnotz & Rasch, 2005; Westelinck, Valcke, Craene, & Kirschner,

2005; Wiley, 2003). Why is this the case, and how is the simple assumption described above flawed? While there are numerous potential explanations, there is some suggestion that the facilitative effects of visualizations is directly dependent on the interaction of such visual material with characteristics of the learner themselves (Geiger & Litwiller, 2005; Hannus & Hyönä, 1999; Sanchez & Wiley, 2006). In other words, an aptitude-by-treatment interaction, or individual differences in particular cognitive skills, might dictate the circumstances under which the use of visualizations is not only warranted, but also most effective. The general class of cognitive abilities that seem most relevant for understanding the 'how' and 'when' to use visualizations, and that are explored further in the following studies, are visuospatial aptitudes (see also Berney & Bétrancourt, 2017, this volume; Wagner & Schnotz, 2017, this volume).

7.3 Assessments of Visuospatial Aptitudes

A long history of psychometric research has established that the ability to represent and manipulate visuospatial relationships is directly tied to a set of discrete aptitudes that exist independent of such general cognitive factors as fluid intelligence or working memory capacity (WMC). Traditionally, these visuospatial abilities have been divided into two distinguishable but related sub-classes: those that evaluate the preservation of visuospatial relationships of an item, and those that examine how individuals can manipulate existing visuospatial relations to transform them into a set of novel new relations (Carroll, 1993; Cooper, 1975; Cooper & Shepard, 1973; Mumaw, Pellegrino, Kail, & Carter, 1984; Pellegrino & Hunt, 1991). The distinction between these sub-classes becomes more apparent when considering tasks that are frequently used to assess these different abilities. For example, visuospatial relations (VSR) are commonly evaluated with tasks that require the learner to mentally rotate or move the existing item in some way to make a subsequent judgment about whether a second item is the original item, or not. Prototypical VSR tasks are the Cube Comparisons task (French, Ekstrom, & Price, 1963) and Figure Rotation Task (Cooper & Shepard, 1973). On the other hand, visuospatial visualization (VSV) tasks instead require individuals to intake a given set of visuospatial relations, and then modify these relations in some constrained way into a new set of relations. The Paper Folding task (French et al., 1963) and Form-board task (French et al., 1963) are common examples of a VSV task. Example items of VSR and VSV tasks are available in Fig. 7.1.

Again, although distinguishable, there can be difficulties drawing strict boundaries between these different sub-classes of ability, and the tasks that measure them (Carroll, 1993; Just & Carpenter, 1985; Stumpf & Eliot, 1995). VSR and VSV tasks do tend to correlate at a moderate level (~.40), and also tend to cluster together in factor analytic solutions that also contain measures of verbal or reasoning ability (Kane et al., 2004). Perhaps a reason for the difficulty in fully segmenting these types of abilities from one another has to do with how the tasks that measure them

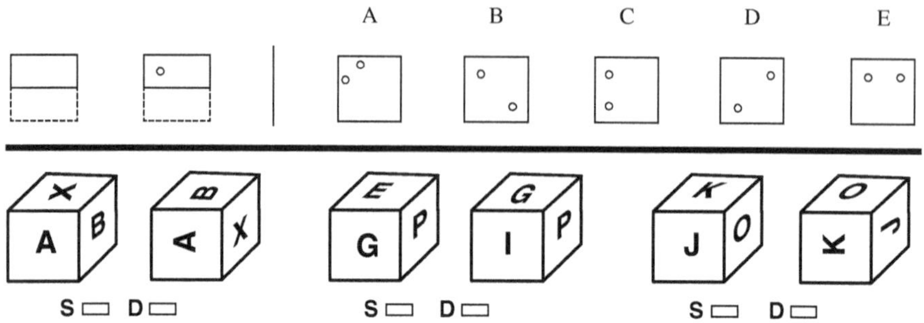

Fig. 7.1 Example items from the Paper Folding (*top*) and Cube Comparisons (*bottom*) tasks

are themselves constructed. For example, a common feature of nearly every VSR and VSV task is that they require the representation of relations within a single item, without the requirement to capture transitional changes over time, or relations outside of the referent itself. In other words, while these tasks require the manipulation and preservation of visuospatial relations, the nature of these relations is strictly self-referential. Thus, these tasks can be classified more broadly as measures of *within-object manipulation spatial ability (WOMSA)*, a description that is consistent with other frameworks of visuospatial processing that also emphasize the focus on intrinsic visuospatial processing required for these type of tasks (Newcombe & Shipley, 2012; Uttal et al., 2013).

Higher performance on WOMSA tasks has been shown to predict performance across a wide range of tasks that also contain a requirement to process visuospatial information. This includes tasks of mechanical reasoning (Boucheix & Schneider, 2009; Hegarty & Sims, 1994; Hegarty & Steinhoff, 1997), route learning (Sanchez & Branaghan, 2009), and even the comprehension of narrative texts about character movement in physical space (Bower & Morrow, 1990; De Beni, Pazzaglia, Gyselinck, & Meneghetti, 2005; Fincher-Kiefer, 2001; Fincher-Kiefer, & D'Agostino, 2004; Haenggi, Kintsch, & Gernsbacher, 1995; Meneghetti, De Beni, Pazzaglia, & Gyselinck, 2011). A recent meta-analysis also found that WOMSA tasks have been found to predict how well people learn from visualizations or illustrations, especially those that are non-dynamic in nature (Höffler, 2010). Related work has also suggested that the positive effect of learning from more *dynamic* visualizations or animations is also largest for lower WOMSA individuals (Höffler & Leutner, 2011; Mayer & Sims, 1994; Sanchez & Wiley, 2010). Thus, these WOMSAs appear critical not only for the formation of visuospatial knowledge derived from text, but also for the decomposition or understanding of explicit visuospatial referents that are used to instruct in these areas (i.e., visualizations and animations).

Given the above discussion, and the often visuospatial nature of STEM learning, it has been suggested that WOMSAs may also be critical for developing understanding of STEM topics (e.g., Halpern et al., 2007; Wu & Shah, 2004). However, studies exploring the relation between WOMSA and STEM learning have not provided a clear pattern of results. While a small number of studies have found that these WOMSAs do positively correlate with classroom performance in STEM topics such as organic chemistry and earth science (Black, 2005; Carter, LaRussa, & Bodner, 1987; Pribyl & Bodner, 1987; Sanchez, 2012; Sibley, 2005; Wu & Shah, 2004), there are also examples of WOMSAs failing to predict learning in other STEM domains like biology and physics (ChanLin, 2000; Koroghlanian & Klein, 2004). This lack of a consistent relationship between WOMSAs and STEM learning challenges the somewhat simple assumption that because many STEM topics have a visuospatial component, WOMSAs should also always be relevant for learning in STEM. An alternative explanation is that these WOMSAs, although likely relevant for STEM education, may not be *as relevant* for predicting learning dynamic concepts in dynamic STEM areas as visuospatial abilities that better capture the dynamic nature of most STEM topics.

7.4 Multiple Object Dynamic Spatial Ability

Tests of *Multiple-Object Dynamic Spatial Ability (MODSA)* focus on the change of spatial relations between multiple items, and also as they unfold over time. MODSAs (originally identified through the work of Hunt, Pellegrino and colleagues nearly two decades ago; Fischer, Hickey, Pellegrino, & Law, 1994; Hunt, Pellegrino, Frick, Farr, & Alderton, 1988; Law, Pellegrino, & Hunt, 1993), were proposed as distinguishable from traditional measures of WOMSA, and have been shown to be separable from not only typical assessments of WOMSA (Contreras, Colom, Hernandez, & Santacreu, 2003; D'Oliveira, 2004), but also measures of visuospatial perspective taking like the Guilford-Zimmerman task (Hunt et al., 1988). MODSA has also been shown to be independent of verbal intelligence (Jackson, Vernon, & Jackson, 1993) and education level (Contreras, Colom, Shih, Alava, & Santacreu, 2001), further confirming its validity as a novel and independent indicator of visuospatial processing.

As MODSA theoretically focuses on the processing of visuospatial relationships across multiple items, over time, it is natural for assessments of MODSA to exhibit this kind of dynamic focus. A typical example of a MODSA task is the Intercept task. In this task (described in more detail below), a target item moves across the screen, and participants must intercept this item with a second moving item (a missile) which they control the timing of release from the launchpad (cf. Fig. 7.2). To successfully achieve an interception, the learner must first represent visuospatial movement over time; effectively computing a relative velocity for both the target *and* the interceptor. It is this information that can then be used to calculate an intersection between the visuospatial items, and subsequently produce a valid release point for the interceptor (missile). While other measures of MODSA do exist, such as the Race task (Hunt et al., 1988), appropriate measures of MODSA all share this focus on relative velocity between multiple visuospatial items. Importantly, tasks which measure these factors in isolation (e.g., time/velocity or visuospatial change

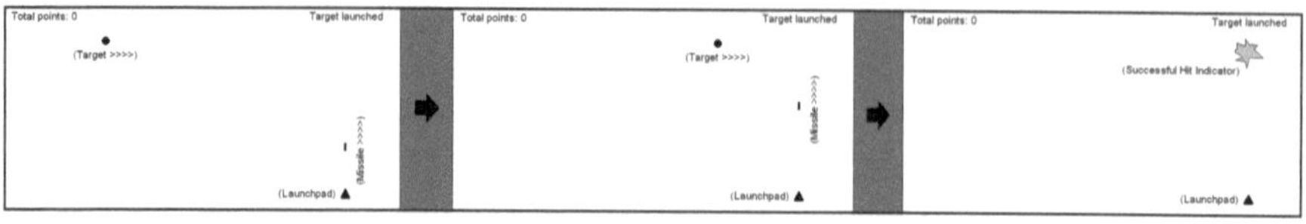

Fig. 7.2 Three progressive screenshots of the Intercept task. Labels in parenthesis do not appear in the actual task display

alone) often fail to correlate with validated measures of MODSA (Fischer et al., 1994). Thus, it appears that the integration of visuospatial and temporal information is the key element of effective MODSA assessments, and an effective measurement of MODSA must focus on both.

While the use of MODSA assessments to predict real-world performance has been less frequent than work on their WOMSA counterparts, there has been some demonstration that these dynamic abilities predict performance in tasks that require participants to integrate visuospatial and temporal information together. For example, there was found to be a positive relationship between MODSA and performance in an air-traffic controller task; a task that overtly involves representing multiple spatial objects that are moving and changing over time (Contreras et al., 2003; D'Oliveira, 2004). Importantly, MODSAs were also recently found to predict STEM learning about plate tectonics, and revealed an aptitude-by-treatment interaction between MODSA and the use of dynamic visualizations in instruction (Sanchez & Wiley, 2014). The details of this pivotal study are discussed next.

7.5 An ATI for MODSA and Science Learning

To evaluate the possible role of MODSAs in learning from dynamic visualizations, first it was necessary to select a topic that required the construction of a visuospatial mental model in order to represent key systemic and dynamic interactions. Plate tectonics was selected as the topic for the lesson as a fundamental tenet of understanding the theory of plate tectonics is the idea that the entire process is cyclical in nature, and progresses across multiple components, in multiple locations, and across time. This is consistent with research on learning plate tectonics that suggests that the main struggle of most learners in this area is to integrate the conceptual units into a coherent cyclical process (Smith & Bermea, 2012). Quite simply, the Earth is composed of a dense molten core, on top of which floats a hard rock crust, which is the surface we live on. Critically, this crust is not uniform. In areas that are unbroken, the crust is usually flat and free of deformation. However, there are also several breaks in the crust which lead to the topography (i.e., mountains, volcanoes, etc.) that make up the more interesting features on the Earth's crust. These breaks represent the intersection of different tectonic plates, and the subsequent deformations at these plate boundaries are a result of the different types of collisions at these points.

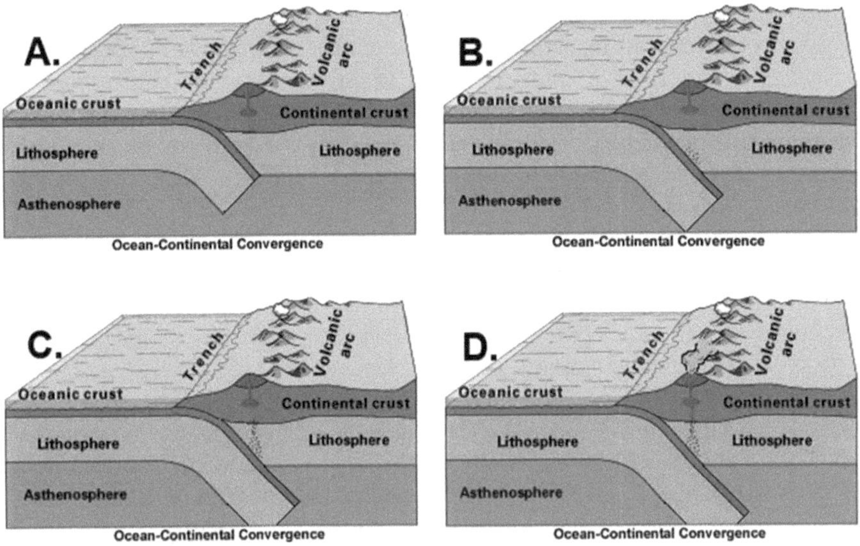

Fig. 7.3 Example visualizations used to instruct the process of subduction

For example, convergent boundaries can produce mountains or volcanoes, whereas divergent boundaries produce more slowly erupting volcanoes that underlie the formation of some islands (e.g., Hawaii) and sea-floor spreading. These plates interact at all because they are floating on top of the sea of liquid rock that makes up the core of the planet, which itself moves and circulates in convection currents within the innermost areas of the Earth.

As such, if a learner is presented with a purely textual description of the above phenomenon (i.e., not supplemented with any kind of visualizations), to successfully develop an understanding of plate tectonics, the learner would be required to not only mentally represent the spatial units themselves (e.g., plates), but also the interactive processes between these spatial units. Such a situation would likely place a very high demand on visuospatial resources given the concurrent need to both represent and integrate the conceptual material in the text. Further, misconceptions are not only possible at the level of basic representation of the concepts themselves, but also regarding how these units interact. In other words, learners may not only misunderstand the conceptual units themselves, but potentially compound this issue with further misunderstanding of the interaction between said units. Contrast this now with at text that is given a basic level of visuospatial support, in the form of static visualizations. A typical static visualization that might illustrate a portion of the above overall interaction is visible in Fig. 7.3d. This figure demonstrates the process of subduction, a specific type of plate collision where the ocean plate collides with (and is forced underneath) the continental plate. This process causes the ocean plate to not only grind apart against the continental plate, but in so doing produces a thick and viscous magma that traps gases, eventually leading to an explosive eruption from a volcano located at the plate boundary. While all of these discrete concepts are captured in the visualization in Fig. 7.3d, the interaction between these concepts is not necessarily prominently highlighted. Instead, as the change of relationships is not explicitly demonstrated, the independent concepts themselves take the forefront, forcing the learner to mentally 'fill-in-the-blanks' regarding how they interact. This 'filling-in' process is expected to be an effortful process, not only requiring preservation of spatial relations, but also integrating these changes over the event.

Now contrast this with a simple dynamic visualization, which would consist of an animated sequence of 4 frames (Fig. 7.3a–d). Note that the end frame is identical to the static visualization discussed above. Thus, while the visuospatial relations and concepts are ultimately consistent between these two genres of illustrations, what is fundamentally different is the conveyance of the process leading up to the final presented state. As is visible throughout Fig. 7.3a–d, the change across frames would receive the primary emphasis. The ocean plate is shown to move and subduct, while the magma slowly rises, fills magma chambers and eventually leads to an eruption. Again, the spatial concepts (e.g., ocean plate, subduction, etc.) are all present in both static and dynamic visualizations, however the dynamic visualization places a greater emphasis on the relationships between concepts, rather than just the concepts alone.

As is visible in Fig. 7.3, and is also hopefully apparent in the above discussion of the topic of plate tectonics, forming a well-developed and complete model of tectonic theory requires learners to not only understand visuospatial concepts in isolation, but also appreciate the interaction of these units over time, and any subsequent changes these interactions produce in the system. Thus, there appears to be a basic requirement in this domain to represent these conceptual relationships as the process unfolds, and this requirement should rely heavily on visuospatial abilities that deal with the representation and understanding of multiple relationships over time (e.g., MODSA). Further, as developing understanding in tectonic theory is an inherently dynamic process, one might also predict that MODSAs should predict unique variance in learning over and above any contributions of WOMSA or basic cognitive abilities such as working memory capacity.

To test for a possible ATI between MODSA and illustration condition, low-knowledge undergraduates ($N = 162$) from a large public university read a text about plate tectonics that contained either no visualizations, static visualizations or dynamic visualizations, and were then tested on their understanding of the content. The text itself was approximately 3500 words long (adapted from the Classrooms of the Future 'Exploring the Environment – Volcanoes & the Earth' module (Center for Educational Technologies, 1997; http://www.cotf.edu/ete/modules/volcanoes/volcano.html). Eight critical concepts underlying volcanic eruptions were identified within the text (Fig. 7.4). Given the nature of the material, it was expected that in order to truly understand the content area, learners would need to integrate these concepts with one another, and understand how they might fit together into a dynamic causal model of volcanic eruptions. For example, they must not only understand that plates move, but also that these collisions can lead to plate subduction, which in turn leads to the formation of magma. This magma then rises and

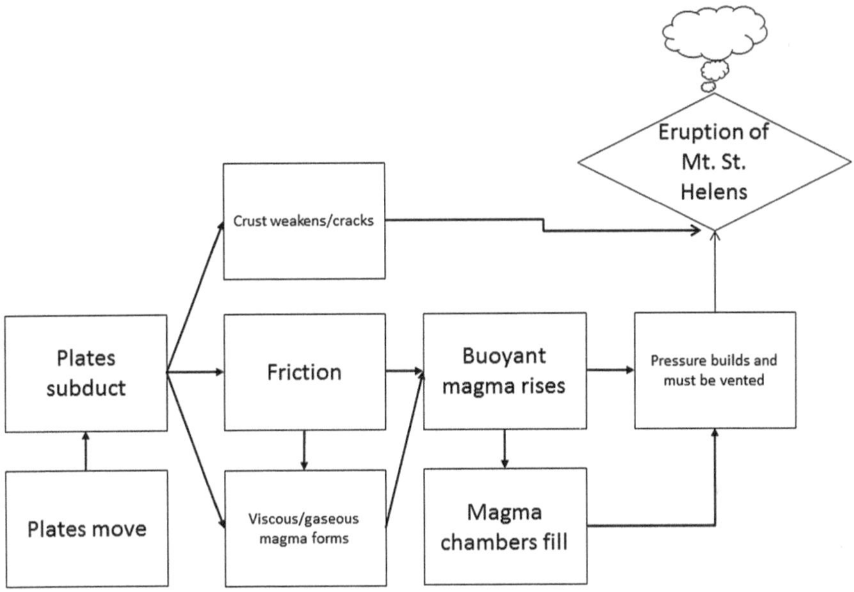

Fig. 7.4 Critical causal concepts within a model of plate tectonics

builds pressure within the crust, eventually culminating in an explosive volcanic eruption. Thus, this text does contain information of a very temporally dynamic and visuospatial cyclical nature, made up of the interaction of multiple visuospatial objects, rather than single items which only reference where understanding is localized within the object itself. As such, to form a more complete understanding of the content domain itself, there is an explicit demand to internally generate a dynamic representation between objects that is consistent with the actual external phenomenon.

Based on experimental condition, this lesson was further modified to contain different levels of external support for the need to mentally simulate visuospatial interactions. The first group was not given any diagrams or illustrations, while the second group read the same text instead illustrated with relevant static diagrams. Finally, the third group was given the same text as the first two groups, however their lesson contained animated versions of the static illustrations seen by the second group. All visualizations in the static and dynamic condition provided a visual analogue of the textual presentation, consistent with general interactions described in the text. These different visualizations do provide differing levels of explicit support for the representation of the visuospatial interaction between relevant concepts. For example, the non-illustrated condition offers no external support, while the static illustrated condition provides at least a visual representation of the operators and how they might be structured within a system. However, the interaction of these operators is not emphasized in these static illustrations. Dynamic illustrations (i.e., animations), on the other hand, not only highlight the visuospatial concepts themselves, but also provide an external representation of the interaction between these concepts. Thus, these animated visualizations provide maximal external support for learning the topic, highlighting not only 'what' but also the 'how' these various visuospatial components come together and interact.

To evaluate how well individuals learned in the different conditions, participants were asked to generate a written response to the question 'What caused Mt. St. Helen's to erupt?' Importantly, the instance of Mt. St. Helens was not explicitly mentioned in the text, so in order to answer this question participants would have to transfer the knowledge they learned from the lesson to this specific application. These essay responses were then evaluated for the presence of the eight critical concepts identified in the target text (cf. Fig. 7.4).

All participants were also assessed for their WOMSA and MODSA. MODSA was assessed using a version of the Intercept task (Hunt et al., 1988), with adjustments based on Law et al. (1993; Fig. 7.2). The appearance of the Intercept task is very similar to a simple video game. In this task, a small target moves across the screen (from left to right) at one of three potential preset speeds. Participants are required to release a second item that travels at a constant speed vertically, in an effort to intercept the horizontally moving target. In order to successfully hit the target, and subsequently earn a higher score in the task, the participant must launch their vertically traveling 'missile' so it reaches the point of intersection at the same time as the target. Thus, successful performance on this task involves representing not only where items are on the screen, but also where they will be after a certain amount of time, which can then be used to decide when to release the 'missile'. The Intercept task lasts approximately 15 minutes from start to finish, and previous iterations of this task have been shown to be not only reliable measures of MODSA (Spearman-Brown $r > .87$; Law et al., 1993), but also correlate positively with other valid measures of MODSA (e.g., Race task; Hunt et al., 1988).

WOMSA was measured with the Paper Folding task (VZ-2; French et al., 1963). In this task, participants were shown a series of 20 diagrams of an irregularly folded piece of paper, and asked to imagine a single hole being punched through the paper at an indicated point. Participants were then required to mentally unfold this piece of paper to decide between a set of alternatives. This task has been shown to be a reliable and valid indicator of WOMSA (Kane et al., 2004), and is traditionally considered a measure of VSV.

Participants were also evaluated for their working memory capacity (WMC; Kane et al., 2004) using two standard complex span tasks: Operation Span (OSpan), and Reading Span (RSpan). In each trial on these tasks, participants are first required to verify a given piece of information (i.e., the sum for a simple math equation in OSpan, or the grammaticality of a simple sentence in RSpan), then remember an unrelated target item (word for OSpan, and letter for RSpan) for a later test at the end of each set of trials. Set size is generally manipulated between two and five trials, and proactive interference increases throughout the tasks. Points are awarded for correct recall of the target items (words or letters). The scores for these two tasks were averaged together to form a composite working memory score, thereby reducing any variance unique to each corresponding WMC task (Conway et al., 2005).

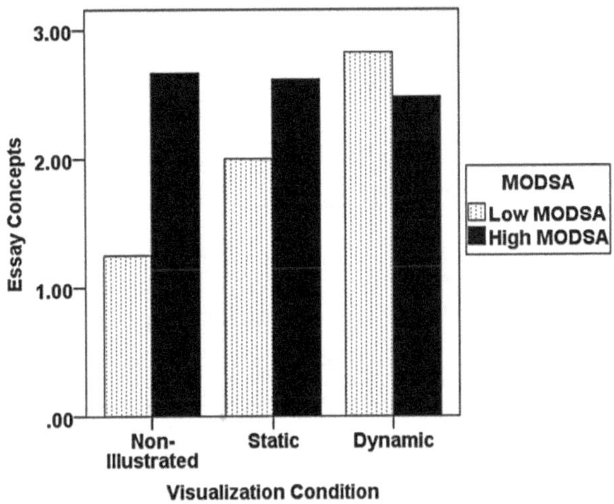

Fig. 7.5 Interaction between MODSA and illustration condition

Although the main purpose for including these assessments was to explore aptitude-by-treatment interactions with these individual differences, it is important to note that the three experimental conditions did not differ in WMC, WOMSA or MODSA scores. There were also no differences in the number of science courses taken across conditions. To examine the influences of ability and visualizations on learning, a set of hierarchical linear regressions was conducted on essay performance. The cognitive ability measures and prior coursework (number of classes taken) were entered into the first block of the analysis, followed by illustration condition in the second block. Illustration condition was decomposed into two dummy coded variables: the first capturing the presence of illustrations or not (illustrated dummy variable), and the second capturing whether the illustrations were dynamic or not (dynamic dummy variable). Finally, interaction terms between the illustration dummy variables and each ability variable were entered into the subsequent blocks of the analysis.

Results from the first block of this analysis showed that WMC and MODSA both predicted unique variance in learning about plate tectonics, but WOMSA, and number of previous science courses did not predict unique variance. Results from the second block showed that the visualization condition failed to explain any variance in essay performance. However, MODSA was found to significantly interact with the visualization condition, but only with the dynamic dummy variable (and not the illustrated dummy variable). These results suggest that MODSA is *less* related to learning content when a lesson contains dynamic visualizations, and the influence of MODSA does not depend on whether the lesson contains any visualizations or not. In other words, dynamic visualizations appear to compensate for lower MODSAs, leading to overall higher performance. But, when dynamic visualizations are not provided, then MODSA strongly predicted learning about plate tectonics. This pattern of results is evident in Fig. 7.5. Finally, both WMC and WOMSA

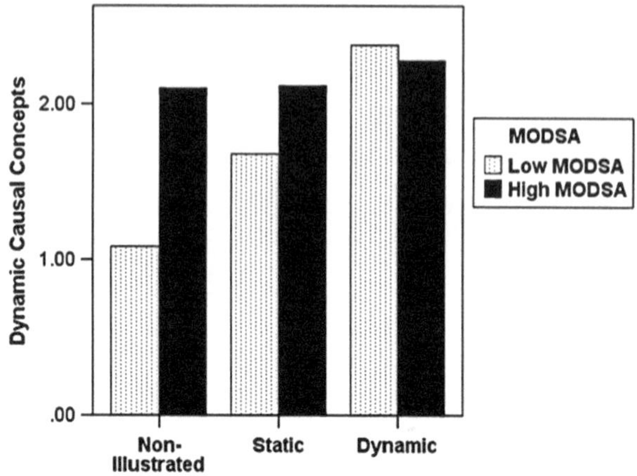

Fig. 7.6 Learning of dynamic concepts across different visualizations by MODSA

did not appear to interact with the visualization condition at any level. These results raise two interesting issues: first, WMC does not appear to influence the ability to use visualizations, dynamic or not, and second, WOMSAs did not account for any unique variance either in the ability to use visualizations, or in learning in the content domain (as evidenced by the lack of an initial significant main effect above).

A second follow-up analysis examined a subset of concepts from Fig. 7.6 that are more explicitly dynamic in nature. These five concepts were: (1) plates move, (2) plates converge, (3) heated magma rises, (4) magma chambers fill, and (5) pressure builds and is released. In contrast with the remaining three concepts that lack dynamic aspects, these five dynamic concepts represent changes in the conceptual system over time. The three non-dynamic concepts appear to be more connected to outcomes of these dynamic processes (e.g., magma forms because plates converge), than being processes in and of themselves. Learning of these dynamic concepts was then compared across the different visualization conditions, for high and low MODSA learners (defined by a median split on MODSA performance), and is visible in Fig. 7.6. Here main effects for visualization condition, MODSA, and a significant interaction were found. As is visible in Fig. 7.6, dynamic visualizations provided the greatest opportunity for learning these dynamic concepts ($F(2, 156) = 7.50$, $p < .01$), significantly more so than both the non-illustrated and static illustration conditions as evidenced by post-hoc comparisons ($p < .05$). Higher MODSA also again predicted better learning of dynamic concepts ($F(1, 156) = 8.35$, $p < .01$). Most importantly, there was also a significant interaction between MODSA group and visualization condition ($F(2, 156) = 4.25$, $p < .05$), just as was observed in the overall analysis. While there was little change in performance in the different visualization conditions for high MODSA learners, low MODSA learners learned the dynamic concepts best in the dynamic visualization condition. This further supports the suggestion that dynamic visualizations make the learning of these dynamic concepts more accessible to all individuals, and not solely for those that are high in MODSA.

Taken together with the above regression results, this final analysis provides a more complete picture on the role of MODSA in learning, and the interaction between MODSA and providing dynamic visualizations. To begin, it appears that MODSA generally facilitates learning about plate tectonics, especially for those concepts that themselves are dynamic in nature. This facilitation was observed over and above measures of general ability and WOMSA. Second, and directly relevant for the focus of this chapter, this study demonstrated a significant aptitude-by-treatment interaction between MODSA and visualization type, suggesting that dynamic visualizations can compensate for lower MODSA scores, and essentially eliminate the observed difference between low and high MODSA individuals on learning. By making the implicit requirements for comprehension of the domain explicit through dynamic visualizations, learning was improved specifically among individuals who might be less likely or able to engage in dynamic mental simulation on their own. Dynamic visualizations were most useful for those individuals who were lower in a particular spatial aptitude (MODSA) and were neither beneficial (nor detrimental) for those individuals who were already high on this ability. The benefit of dynamic visualizations was therefore localized to a specific group of individuals who were most likely to benefit from this kind of external support. This result is consistent with the 'ability-as-compensator' hypothesis originally proposed by Mayer and Sims (1994).

7.6 Specificity of Benefits for Dynamic Visualizations and MODSA

A parallel study using a different subject matter helps to highlight when MODSA and dynamic visualizations will specifically benefit learning. As in the plate tectonics study, a second group of undergraduates (N = 119) read a similar length text (~3500 words) about the Irish Potato Famine (adapted from Wiley, 2001) in order to understand the causes of the drastic change in population that occurred between 1841 and 1851. Again, this text was either not illustrated (n = 40), or instead illustrated with static (n = 40) or dynamic (n = 39) visualizations that portrayed changes in agricultural products and their diversity, death rates, and other economic indicators such as rent costs by county (cf. Fig. 7.7). Like the plate tectonics text, eight *a priori* concepts were identified in this text that represented a thorough understanding of population changes in Ireland. Critically, although this Irish Potato Famine text does reference visuospatial locations (e.g., towns or counties on a map), the causal concepts themselves are not inherently based in dynamic spatial relations between entities. Thus, while the content does contain a small discrete component of visuospatial information, this topic seems less likely to require the construction of a runnable visuospatial mental model in order to represent key systemic and

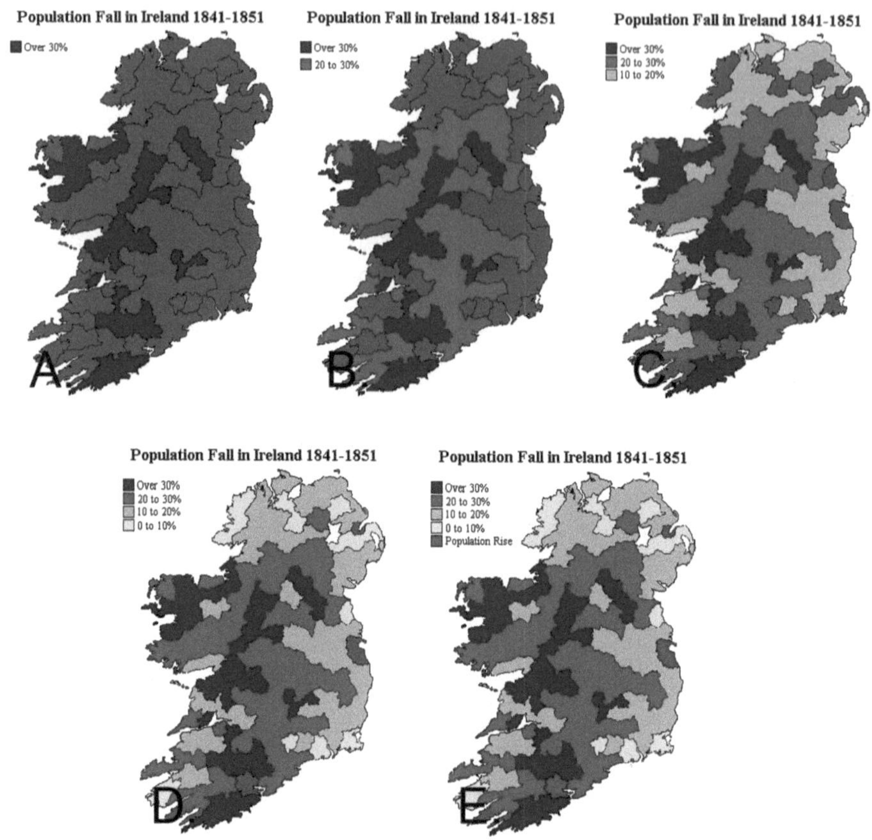

Fig. 7.7 Sample visualization from Irish Potato Famine text

dynamic interactions compared to a topic such as plate tectonics. Participants were again assessed for their WMC, WOMSA and MODSA. Of interest was whether the same pattern of relationships would be observed here as demonstrated previously with the plate tectonic content, or whether the interaction between DSA and visualizations on learning might depend on the subject matter.

To examine the influences of ability and visualizations on learning about the Potato Famine, a set of hierarchical linear regressions was again conducted on essay performance. Results from the first block of analysis ($R^2 = .10$, $F(3, 116) = 4.10$, $p < .01$) indicated that the only significant predictor of learning was WOMSA ($\beta = .30$, $p < .01$). WMC ($\beta = .02$, $p > .05$) and MODSA ($\beta = -.01$, $p > .05$) did not contribute unique variance for learning about the Irish Potato Famine. In the second block, no differences were seen in learning due to visualization condition ($R^2\Delta = .01$, $p > .05$). Both visualization condition dummy variables also failed to significantly predict performance (both p-values > .05), and failed to interact with any of the cognitive ability variables in the later blocks (all $R^2\Delta < .025$, $p > .05$). Thus, no interaction between MODSA and visualization condition was seen for this content.

As a whole, the results of this follow-up experiment allow for a number of important observations. First, advantages due to individual differences in MODSA were not found on a topic that did not seem to require the creation of a runnable visuospatial mental model. Because there is no inherently dynamic component to be understood from this text, there was little need to invoke MODSA to form understanding. This helps to rule out alternative explanations for MODSA effects on learning as being due to more general differences in ability, since it does not always relate to superior learning. Second, dynamic visualizations also do not always lead to improved understanding. This helps to rule out alternative explanations for dynamic visualizations as being necessarily more interesting or engaging to students.

Third, the only ability measure that was uniquely related to learning for this topic was WOMSA. Although the reasons for this observed relation are less clear than the observed relation between MODSA and learning about plate tectonics, one speculative interpretation is that as the information in the potato famine text does reference several spatial locations (i.e., different counties/towns of Ireland), it is possible that the processing of these simple spatial orientations was required to contextualize the rest of the factual information contained with the text. This is somewhat consistent with previous findings regarding WOMSA abilities being related to following character movements within narrative texts (Bower & Morrow, 1990; De Beni et al., 2005; Fincher-Kiefer, 2001; Fincher-Kiefer, & D'Agostino, 2004; Haenggi et al., 1995; Meneghetti et al., 2011). Further, because learners were also presented with visualizations in two of the conditions, it is possible that WOMSA might have been needed to help readers to decode these diagrams, and therefore it resulted in an overall relationship with learning. As a simple test of this potential explanation, a final hierarchical regression was conducted that examined only WOMSA and the visualization dummy variable that evaluated whether the text was illustrated or not. Results indicated that while there was still only a main effect of WOMSA ($\beta = .30$, $p < .01$) and not the presence of visualizations ($\beta = -.11$, $p > .05$) in the first block ($R^2 = .11$, $F(2, 117) = 7.14$, $p < .01$), WOMSA did interact with the presence of illustrations in the second block ($R^2\Delta = .03$, $p < .05$; $\beta = .74$). Again, this suggests that WOMSA was necessary for the decoding of the visualizations, both dynamic and not, and this relationship could underlie the main effect found in the overall analysis. This effect should be interpreted cautiously, however, because when explored in the full model, with all variables, this pattern did not reach statistical reliability. The failure to observe this interaction in the overall model is likely a result of intercorrelations between WOMSA and the other ability measures in this study. For example, when considering WOMSA alone, a portion of general ability variance that is usually shared with WMC (evidenced by the typically observed intercorrelation between WOMSA and WMC in this study, $r = .45$, $p < .01$) could be attributed inappropriately to WOMSA; thus producing an overestimate of the connection to learning, based on variance that is not specific to WOMSA (Jaeger, Jarosz, & Wiley, 2014). Obviously, when both WMC and WOMSA are present in the model, such overlapping variance would not be attributed to either factor, which although has the positive side effect of providing a more clear estimation of the effect, also potentially obscures smaller effects. Regardless of this WOMSA explanation, the observed patterns portrayed here at the very least provide an additional perspective on when MODSA and the use of dynamic visualizations are likely to impact learning.

7.7 Conclusions, Caveats, and Future Directions

The current chapter sought to explore the relationship between MODSAs and learning dynamic STEM topics through dynamic visualizations, and the potential for an aptitude-by-treatment interaction. Results from a study investigating the influence of MODSA on learning about plate tectonics showed not only that MODSA is relevant for predicting learning in dynamic domains, but also that MODSA significantly predicted the utility of dynamic visualizations used for instruction. While dynamic visualizations failed to lead to significant improvements in performance over non-illustrated or statically illustrated text when considered alone, an aptitude-by-treatment interaction revealed that the presence of dynamic visualizations specifically benefitted lower MODSA individuals. Further, these dynamic visualizations helped facilitate the learning of dynamic domain concepts more-so than the other visualization conditions, and specifically for lower MODSA individuals. This suggests that such visualizations allowed these lower ability individuals to better encode and learn such dynamic information; information that might have otherwise not been accessible to them. Essentially, these dynamic visualizations were most beneficial for those that likely struggle to mentally visualize such information themselves. This finding is consistent with other results suggesting that dynamic visualizations differentially impact high and low ability individuals (Höffler & Leutner, 2011; Mayer & Sims, 1994; Schnotz & Rasch, 2005), and help to clarify when facilitation may be found by adding relevant visualizations to learning environments (ChanLin, 2000; Craig, Gholson, & Driscoll, 2002; Mayer & Moreno, 2002; Rieber, 1990).

Importantly, the influence of MODSA on learning was also observed above and beyond the influence of WMC and WOMSA. This suggests that high MODSA enabled understanding independent of higher general ability or other less relevant visuospatial abilities, further validating its consideration as an independent factor worth assessing when designing visualizations for learning (cf. Lowe & Boucheix, 2017, this volume). The results of a second study further support the distinction made above that this set of dynamic abilities is only invoked when there is an explicit demand for such processing made by the content area, and not invoked in situations that are less dynamically visuospatial (e.g., Irish Potato Famine). Encouragingly, the results of the plate tectonic study also suggest that this explicit demand can also be alleviated through the use of quality dynamic visualizations, thus allowing all learners to better access this kind of dynamic content information. It must be noted, however, that the caliber of dynamic visualizations does vary significantly across educational settings and applications. Note that a given visualization could be considered less-than-ideal for numerous reasons such as: being awkwardly constructed thus causing a focus on less relevant relationships (Fischer, Lowe & Schwan, 2008; Lowe, 2003), unrelated to the instructional content (e.g., decorational; Höffler & Leutner, 2007), or even being too complex despite being relevant (Lowe, 2004), to name a few. In these situations, MODSA might also play an additional role, namely the ability to decipher and extract information that is contained within a less-than-ideal visualization. For example, learners sometimes segment complex visualizations into smaller meaningful units when attempting to learn (Lowe, 2004). The unfortunate by-product of this type of segmentation is a reduced ability to integrate across segments. Higher MODSA might permit learners to maintain and integrate these isolated units, due to their enhanced ability to integrate temporal and visuospatial elements. Thus, it is possible that MODSA is not only useful for building internal dynamic mental representations, but also breaking down external dynamic representations (cf. Lowe & Boucheix, 2017, this volume). Future work is necessary to validate whether this is in fact the case.

Given that the relationship between WOMSA and learning through visualizations has been somewhat well established by previous research (Hays, 1996; Hegarty & Sims, 1994; Hegarty & Steinhoff, 1997; Höffler & Leutner, 2011; Koroghlanian & Klein, 2004; Mayer & Sims, 1994), it may seem curious that no role was seen for WOMSA in the plate tectonic study. When MODSA and WMC were taken into account, WOMSA failed to predict any unique variance in learning, and also failed to interact with visualizations in any way to predict how well learners understood plate tectonics. A tentative explanation is that by assessing all three aptitudes (WOMSA, MODSA and WMC) in this work, the independent role of each could be seen more clearly. Because MODSA and WOMSA are generally correlated, it is entirely possible that overlapping variance usually attributed to WOMSA was instead attributed to MODSA here, as it is again most relevant for learning within a dynamic domain, and also from dynamic visualizations, thus leaving little unique variance to be accounted for by WOMSA. When this content domain demand is removed, however, as was the case in the Irish Potato Famine study, MODSA then appears to take a back seat to WOMSA, and the relationship between WOMSA and learning from visualizations returns consistent with other research.

These results thus offer some insight from an individual differences perspective into why dynamic visualizations may sometimes fail to benefit learning. The results suggest that dynamic visualizations are most likely to facilitate learning under a specific set of conditions: when the topic and subject matter requires dynamic simulation for comprehension, and when the reader lacks MODSA. Although in these studies no harm was seen from providing dynamic visualizations in other conditions, there is some evidence from other work that suggests that there may be cases where animations can cause detriments in performance (Tversky, Morrison, & Bétrancourt, 2002). One class of concerns comes from studies on *seductive details* in which interesting illustrations or animations could cause readers to devote less attention to processing the ideas from the text (Harp & Mayer, 1997; Sanchez & Wiley, 2006; Wiley, Ash, Sanchez, & Jaeger, 2011). Another class of concerns arises from the subjective sense of fluency that readers may perceive after viewing a diagram or animation. Although visualizations can be a powerful tool for conveying a system of relations, they have also been shown to cause *illusions of comprehension* in which readers report having understood concepts better than they actually have (Jaeger & Wiley, 2014; Serra & Dunlosky, 2010; Wiley, 2003). For both of these reasons, further research that can help delineate the specific conditions under which dynamic visualizations are actually effective at improving learning is critical.

In conclusion, these studies have highlighted the benefits of assessing individual differences in learner characteristics when instructing in a visuospatial domain, and more specifically, while using dynamic visualizations. By incorporating an assessment of MODSA, educators will be able to more accurately tailor or scaffold the presentation of visual information so that it best meets the needs of the target population of learners. This research suggests that dynamic visualizations are most useful under constrained circumstances, such as when required by both the content domain *and* the needs of the learner themselves.

Acknowledgements Portions of this work were supported by the APA dissertation research award to the first author.

References

Ainsworth, S., & Th Loizou, A. (2003). The effects of self-explaining when learning with text or diagrams. *Cognitive Science, 27*, 669–681.

Berney, S., & Bétrancourt, M. (2017). Learning three-dimensional anatomical structures with animation: Effects of orientation references and learners' spatial ability. In R. Lowe & R. Ploetzner (Eds.), *Learning from dynamic visualization – Innovations in research and application*. Berlin: Springer (this volume).

Black, A. A. (2005). Spatial ability and earth science conceptual understanding. *Journal of Geoscience Education, 53*, 402–414.

Boucheix, J. M., & Schneider, E. (2009). Static and animated presentations in learning dynamic mechanical systems. *Learning and Instruction, 19*, 112–127.

Bower, G. H., & Morrow, D. G. (1990). Mental models in narrative comprehension. *Science, 247*, 44–49.

Carroll, J. B. (1993). *Human cognitive abilities: A survey of factor analytic studies*. New York: Cambridge University Press.

Carter, C. S., LaRussa, M. A., & Bodner, G. M. (1987). A study of two measures of spatial ability as predictors of success in different levels of general chemistry. *Journal of Research in Science Teaching, 24*, 645–657.

Center for Educational Technologies. (1997). Volcanoes and the Earth. In Exploring the environment. Retrieved from http://www.cotf.edu/ete/modules/volcanoes/volcano.html.

ChanLin, L. J. (1998). Animation to teach students of different knowledge levels. *Journal of Instructional Psychology, 25*, 166–175.

ChanLin, L. J. (2000). Attributes of animation for learning scientific knowledge. *Journal of Instructional Psychology, 27*, 228–238.

Chi, M. T., Bassok, M., Lewis, M. W., Reimann, P., & Glaser, R. (1989). Self-explanations: How students study and use examples in learning to solve problems. *Cognitive Science, 13*, 145–182.

Contreras, M. J., Colom, R., Hernandez, J. M., & Santacreu, J. (2003). Is static spatial performance distinguishable from dynamic spatial performance? A latent-variable analysis. *The Journal of General Psychology, 130*, 277–288.

Contreras, M. J., Colom, R., Shih, P. C., Alava, M. J., & Santacreu, J. (2001). Dynamic spatial performance: Sex and educational differences. *Personality and Individual Differences, 30*, 117–126.

Conway, A. R. A., Kane, M. J., Bunting, M. F., Hambrick, D. Z., Wilhelm, O., & Engle, R. W. (2005). Working memory span tasks: A methodological review and user's guide. *Psychonomic Bulletin and Review, 12*, 769–786.

Cooper, L. A. (1975). Mental rotation of random two-dimensional shapes. *Cognitive Psychology, 7*, 20–43.

Cooper, L. A., & Shepard, R. N. (1973). Chronometric studies of the rotation of mental images. In W. G. Chase (Ed.), *Visual information processing* (pp. 75–176). New York: Academic Press.

Craig, S. D., Gholson, B., & Driscoll, D. M. (2002). Animated pedagogical agents in multimedia educational environments: Effects of agent properties, picture features and redundancy. *Journal of Educational Psychology, 94*, 428–434.

De Beni, R., Pazzaglia, F., Gyselinck, V., & Meneghetti, C. (2005). Visuospatial working memory and mental representation of spatial descriptions. *European Journal of Cognitive Psychology, 17*, 77–95.

D'Oliveira, T. C. (2004). Dynamic spatial ability: An exploratory analysis and a confirmatory study. *International Journal of Aviation Psychology, 14*, 19–38.

Fincher-Kiefer, R. (2001). Perceptual components of situation models. *Memory and Cognition, 29*, 336–343.

Fincher-Kiefer, R., & D'Agostino, P. R. (2004). The role of visuospatial resources in generating predictive and bridging inferences. *Discourse Processes, 37*, 205–224.

Fischer, S. C., Hickey, D. T., Pellegrino, J. W., & Law, D. J. (1994). Strategic processing in dynamic spatial reasoning tasks. *Learning and Individual Differences, 6*, 65–105.

Fischer, S., Lowe, R. K., & Schwan, S. (2008). Effects of presentation speed of a dynamic visualization on the understanding of a mechanical system. *Applied Cognitive Psychology, 22*, 1126–1141.

French, J. W., Ekstrom, R. B., & Price, L. A. (1963). *Kit of reference tests for cognitive factors*. Princeton, NJ: Educational Testing Service.

Friedman, N. P., & Miyake, A. (2000). Differential roles for spatial and verbal working memory in the comprehension of spatial descriptions. *Journal of Experimental Psychology: General, 129*, 61–83.

Geiger, J. F., & Litwiller, R. M. (2005). Spatial working memory and gender differences in science. *Journal of Instructional Psychology, 32*, 49–58.

Haenggi, D., Kintsch, W., & Gernsbacher, M. A. (1995). Spatial situation models and text comprehension. *Discourse Processes, 19*, 173–199.

Halpern, D. F., Benbow, C. P., Geary, D. C., Gur, R. C., Hyde, J. S., & Gernsbacher, M. A. (2007). The science of sex differences in mathematics. *Psychological Science in the Public Interest, 8*, 1–51.

Hannus, M., & Hyönä, J. (1999). Utilization of illustrations during learning of science textbook passages among low-and high-ability children. *Contemporary Educational Psychology, 24*, 95–123.

Harp, S. F., & Mayer, R. E. (1997). The role of interest in learning from scientific text and illustrations: On the distinction between emotional interest and cognitive interest. *Journal of Educational Psychology, 89*, 92–102.

Hays, T. A. (1996). Spatial abilities and the effects of computer animation on short-term and long-term comprehension. *Journal of Educational Computing Research, 14*, 139–155.

Hegarty, M. (1992). Mental animation: Inferring motion from static diagrams of mechanical systems. *Journal of Experimental Psychology: Learning, Memory and Cognition, 18*, 1084–1102.

Hegarty, M., & Sims, V. K. (1994). Individual differences in mental animation during mechanical reasoning. *Memory & Cognition, 22*, 411–430.
Hegarty, M., & Steinhoff, K. (1997). Individual differences in use of diagrams as memory in mechanical reasoning. *Learning and Individual Differences, 9*, 19–44.
Hegarty, M., Crookes, R. D., Dara-Abrams, D., & Shipley, T. F. (2010). Do all science disciplines rely on spatial abilities? Preliminary evidence from self-report questionnaires. In *Spatial Cognition VII* (pp. 85–94). Berlin: Springer.
Höffler, T. N. (2010). Spatial ability: Its influence on learning with visualizations – A meta-analytic review. *Educational Psychology Review, 22*, 245–269.
Höffler, T. N., & Leutner, D. (2007). Instructional animation versus static pictures: A meta-analysis. *Learning and Instruction, 17*, 722–738.
Höffler, T. N., & Leutner, D. (2011). The role of spatial ability in learning from instructional animations – Evidence for an ability-as-compensator hypothesis. *Computers in Human Behavior, 27*, 209–216.
Hunt, E., Pellegrino, J. W., Frick, R. W., Farr, S. A., & Alderton, D. L. (1988). The ability to reason about movement in the visual field. *Intelligence, 12*, 77–100.
Jackson, D. N., Vernon, P. A., & Jackson, D. N. (1993). Dynamic spatial performance and general intelligence. *Intelligence, 17*, 451–460.
Jaeger, A. J., & Wiley, J. (2014). Do illustrations help or harm metacomprehension accuracy? *Learning and Instruction, 34*, 58–73.
Jaeger, A. J., Jarosz, A.F., & Wiley, J. (2014, November). Know when to hold em, know when to fold em: WMC and spatial reasoning. Poster presented at the 2014 Annual Meeting of the Psychonomic Society, Long Beach.
Just, M. A., & Carpenter, P. A. (1985). Cognitive coordinate systems: Accounts of mental rotation and individual differences in spatial ability. *Psychological Review, 92*, 137–172.
Kane, M. J., Hambrick, D. Z., Tuholski, S. W., Wilhelm, O., Payne, T. W., & Engle, R. W. (2004). The generality of working memory capacity: A latent variable approach to verbal and visuospatial memory span and reasoning. *Journal of Experimental Psychology: General, 133*, 189–217.
Koroghlanian, C., & Klein, J. D. (2004). The effect of audio and animation in multimedia instruction. *Journal of Educational Multimedia and Hypermedia, 13*, 23–45.
Law, D. J., Pellegrino, J. W., & Hunt, E. B. (1993). Comparing the tortoise and the hare: Gender difference and experience in dynamic spatial reasoning tasks. *Psychological Science, 4*, 35–40.
Loftus, G. R., & Harley, E. M. (2004). How different spatial-frequency components contribute to visual information acquisition. *Journal of Experimental Psychology. Human Perception and Performance, 30*, 104–118.
Lowe, R. K. (2003). Animation and learning: Selective processing of information in dynamic graphics. *Learning and Instruction, 13*, 157–176.
Lowe, R. (2004). Interrogation of a dynamic visualization during learning. *Learning and Instruction, 14*, 257–274.
Lowe, R., & Boucheix, J.-M. (2017). A composition approach to design of educational animations. In R. Lowe & R. Ploetzner (Eds.), *Learning from dynamic visualization – Innovations in research and application*. Berlin: Springer (this volume).
Mayer, R. E. (1989). Systematic thinking fostered by illustrations in scientific text. *Journal of Educational Psychology, 81*, 240–246.
Mayer, R. E., & Gallini, J. K. (1990). When is an illustration worth ten thousand words? *Journal of Educational Psychology, 82*, 715–726.
Mayer, R. E., & Moreno, R. (1998). A split-attention effect in multimedia learning: Evidence for dual processing systems in working memory. *Journal of Educational Psychology, 90*, 312–320.
Mayer, R. E., & Moreno, R. (2002). Aids to computer-based multimedia learning. *Learning and Instruction, 12*, 107–119.

Mayer, R. E., & Sims, V. K. (1994). For whom is a picture worth a thousand words? Extension of a dual-coding theory of multimedia learning. *Journal of Educational Psychology, 86*, 389–401.

Meneghetti, C., De Beni, R., Pazzaglia, F., & Gyselinck, V. (2011). The role of visuo-spatial abilities in recall of spatial descriptions: A mediation model. *Learning and Individual Differences, 21*, 719–723.

Moreno, R., & Mayer, R. E. (1999). Cognitive principles of multimedia learning: The role of modality and contiguity. *Journal of Educational Psychology, 91*, 358–368.

Mumaw, R. J., Pellegrino, J. W., Kail, R. V., & Carter, P. (1984). Different slopes for different folks: Process analysis of spatial aptitude. *Memory and Cognition, 12*, 515–521.

National Research Council. (2006). *Learning to think spatially*. Washington, DC: The National Academies Press.

Newcombe, N. S., & Shipley, T. F. (2012). Thinking about spatial thinking: New typology, new assessments. In J. S. Gero (Ed.), *Studying visual and spatial reasoning for design creativity* (pp. 1–18). New York: Springer.

Palmiter, S., & Elkerton, J. (1993). Animated demonstrations for learning procedural computer-based tasks. *Human-Computer Interaction, 8*, 193–216.

Pellegrino, J. W., & Hunt, E. B. (1991). Cognitive models for understanding and assessing spatial abilities. In H. A. H. Rowe (Ed.), *Intelligence: Reconceptualization and measurement* (pp. 203–225). Mahwah, NJ: Lawrence Erlbaum Associates.

Pribyl, J. R., & Bodner, G. M. (1987). Spatial ability and its role in organic chemistry: A study of four organic courses. *Journal of Research in Science Teaching, 24*, 229–240.

Rieber, L. P. (1990). Using computer animated graphics with science instruction in children. *Journal of Educational Psychology, 82*, 135–140.

Rieber, L. P. (1991). Animation, incidental learning, and continuing motivation. *Journal of Educational Psychology, 83*, 318–328.

Rieber, L. P., Boyce, M. J., & Assad, C. (1990). The effects of computer animation on adult learning and retrieval tasks. *Journal of Computer-Based Instruction, 17*, 46–52.

Rinck, M. (2005). Spatial situation models. In P. Shah & A. Miyake (Eds.), *The Cambridge handbook of visuospatial thinking* (pp. 334–382). New York: Cambridge University Press.

Sanchez, C. A. (2012). Enhancing visuospatial performance through video game training to increase learning in visuospatial science domains. *Psychonomic Bulletin and Review, 19*, 58–65.

Sanchez, C. A., & Branaghan, R. J. (2009). The interaction of map resolution and spatial abilities on map learning. *International Journal of Human-Computer Studies, 67*, 475–481.

Sanchez, C. A., & Wiley, J. (2006). An examination of the seductive details effect in terms of working memory capacity. *Memory & Cognition, 34*, 344–355.

Sanchez, C. A., & Wiley, J. (2010). Sex differences in science learning: Closing the gap through animations. *Learning and Individual Differences, 20*, 271–275.

Sanchez, C. A., & Wiley, J. (2014). The role of dynamic spatial ability in geoscience text comprehension. *Learning and Instruction, 31*, 33–45.

Schnotz, W., & Rasch, T. (2005). Enabling, facilitating, and inhibiting effects of animation in multimedia learning: Why reduction of cognitive load can have negative results on learning. *Educational Technology Research and Development, 53*, 1042–1629.

Schnotz, W., Böckheler, J., & Grzondziel, H. (1999). Individual and co-operative learning with interactive animated pictures. *European Journal of Psychology of Education, 14*, 245–265.

Serra, M. J., & Dunlosky, J. (2010). Metacomprehension judgments reflect the belief that diagrams improve learning from text. *Memory, 18*, 698–711.

Sibley, D. F. (2005). Visual abilities and misconceptions about plate tectonics. *Journal of Geoscience Education, 53*, 471–477.

Smith, G. A., & Bermea, S. B. (2012). Using students' sketches to recognize alternative conceptions about plate tectonics persisting from prior instruction. *Journal of Geoscience Education, 60*, 350–359.

Stumpf, H., & Eliot, J. (1995). Gender-related differences in spatial ability and the k factor of general spatial ability in a population of academically talented students. *Personality and Individual Differences, 19*, 33–45.

Tversky, B., Morrison, J. B., & Bétrancourt, M. (2002). Animation: Can it facilitate? *International Journal of Human-Computer Studies, 57*, 247–262.

Uttal, D. H., Meadow, N. G., Tipton, E., Hand, L. L., Alden, A. R., Warren, C., et al. (2013). The malleability of spatial skills: A meta-analysis of training studies. *Psychological Bulletin, 139*, 352–402.

Wagner, I., & Schnotz, W. (2017). Learning from static and dynamic visualizations: What kind of questions should we ask? In R. Lowe & R. Ploetzner (Eds.), *Learning from dynamic visualization – Innovations in research and application*. Berlin: Springer (this volume).

Westelinck, K., Valcke, M., De Craene, B., & Kirschner, P. (2005). Multimedia learning in social sciences: Limitations of external graphical representations. *Computers in Human Behavior, 21*, 555–573.

Wiley, J. (2001). Supporting understanding through task and browser design. In L. R. Gleitman & A. K. Joshi (Eds.), *Proceedings of the twenty-third annual conference of the Cognitive Science Society* (pp. 1136–1143). Hillsdale, NJ: Erlbaum.

Wiley, J. (2003). Cognitive and educational implications of visually-rich media: Images and imagination. In M. Hocks & M. Kendrick (Eds.), *Eloquent images: Word and image in the age of new media* (pp. 201–218). Cambridge, MA: MIT Press.

Wiley, J., Ash, I. K., Sanchez, C. A., & Jaeger, A. (2011). Clarifying readers' goals for learning from expository science texts. In M. McCrudden, J. Magliano, & G. Schraw (Eds.), *Text relevance and learning from text* (pp. 353–374). Greenwich, CT: Information Age Publishing.

Wiley, J., & Sanchez, C. A. (2010). Constraints on learning from expository science texts. In N. L. Stein & S. Raudenbush (Eds.), *Developmental cognitive science goes to school* (pp. 45–58). New York: Routledge.

Wiley, J., Sanchez, C. A., & Jaeger, A. J. (2014). The individual differences in working memory capacity principle for multimedia learning. In R. Mayer (Ed.), *The Cambridge handbook of multimedia learning* (2nd ed., pp. 598–619). New York: Cambridge University Press.

Wu, H., & Shah, P. (2004). Exploring visuospatial thinking in chemistry learning. *Science Education, 88*, 465–492.

Chapter 8
Demonstration Tasks for Assessment

Richard Lowe, Jean-Michel Boucheix, and Benjamin Fillisch

8.1 Introduction

Until quite recently, the textbooks that dominated the educational landscape typically contained few (if any) illustrations. These resources relied almost exclusively on text-based presentation of information because pictures were so much more expensive to include than the printed word (Houghton & Willows, 1987; Lowe & Schnotz, 2014). Today this is no longer the case because computer technology has fundamentally changed the way graphics can be generated, manipulated and distributed (Lowe, 2017). The mainstream electronic learning resources that have emerged from this technological revolution are far more reliant on pictorial approaches for presenting to-be-learned subject matter than were conventional textbooks. This change has been accompanied by a reduction in the traditional dominance of text-based representations within education. The ascendency of educational pictures is particularly notable in the case of animated explanations in which text (if used at all) may even play a subsidiary role to that of the animation. Despite this considerable shift in how information is *presented* to learners, there has been relatively little change in how learning from the newer types of educational resources is *assessed*. In contrast to the dynamic pictorial representations that abound in these resources, measurement of the

R. Lowe (✉)
Curtin University, Perth, Australia

University of Burgundy, Dijon, France
e-mail: r.k.lowe@curtin.edu.au

J.-M. Boucheix
University of Burgundy, Dijon, France
e-mail: Jean-Michel.Boucheix@u-bourgogne.fr

B. Fillisch
University of Education, Freiburg, Germany
e-mail: benjamin.fillisch@ph-freiburg.de

R. Lowe, R. Ploetzner (eds.), *Learning from Dynamic Visualization*,
DOI 10.1007/978-3-319-56204-9_8

educational outcomes from animations are still largely made using static, text-based assessment tools. Responding to such assessments therefore requires learners to comprehend a task expressed in a representational format very different from that used in the learning material. Under these circumstances, is it understanding of the learning material that is being assessed or is it understanding of the assessment task? The representational switch that both understanding and responding to the assessment requires may be demanding and the outcomes flawed. This major disconnect between presentation and assessment carries the danger that the measurements obtained may not provide an optimal indication of the learner's understanding of the animation's subject matter. With respect to conducting research on learning from animation in particular, it would appear preferable to employ assessment approaches that involve responses more closely related to nature of the stimulus instructional materials used. The present chapter considers the potential of assessing learning from animations by using manipulable models that allow learners to provide dynamic physical demonstrations of the understandings they have acquired from these resources.

8.2 Assessment of Dynamic Visualizations

An educational animation presents to-be-learned subject matter via an array of graphic entities that change in various ways during its time course. Learning about the subject matter depicted by an animation therefore requires internalization of visuospatial and spatiotemporal information that is relevant to the nominated learning task. For the purposes of this chapter, we assume the goal of assessment is to measure the quality of the mental model that the learner has constructed from this internalized information. Ideally, the way such measurements are made should be consistent with the inherently visual, dynamic nature of animations in order that the assessment's potential effectiveness is not compromised. This is important not only to help assessors match learner responses with target features of the animated presentation as directly as possible, but also to maximize the likelihood that those responses are a valid reflection of the learner's mental model.

A review of the literature on learning from animations and other forms of dynamic visualization indicates that the assessment tools used by researchers rarely match the way these representations present their subject matter (cf. Davenport & Quellmalz, 2017, this volume). In most cases, conventional text-based measures are used to assess learning effectiveness (see Bernay & Bétrancourt, 2009; Höffler & Leutner, 2007; Ploetzner & Lowe, 2012; Tversky, Morrison, & Bétrancourt, 2002). The bulk of such assessments have tended to require written answers or to involve multiple choice tests (e.g., de Koning, Tabbers, Rikers, & Pass, 2007, 2010a, 2010b; Khacharem, Zoudji, & Kalyuga, 2015; Mayer & Anderson, 1992; Mayer & Moreno, 2002; Scheiter, Gergets, Huk, Imhof, & Kammerer, 2009). Although this verbally-oriented approach is undoubtedly justifiable when the animation is accompanied by spoken or written text as part of a multimedia presentation (Mayer, 2014; Schnotz, 2002), it is perhaps less appropriate when animations have no such accompaniment.

As discussed later in this chapter, non-text alternatives to these conventional types of assessment have occasionally been used to measure learning from animations, but this is comparatively rare and tends not to be the case for declarative, explanatory types of knowledge.

In this section, we argue that the text-based nature of conventional assessment approaches can be ill-suited to capturing high quality information about learners' mental models of complex dynamic systems. It is important to avoid possible distorting effects of inter-representational mediation by fostering as far as possible direct, one-to-one mapping between the studied animation and the type of representation being used to assess learning. A major reason for text's lack of suitability are its limitations as a means of expressing the type of information presented in an animation. These limitations stem from the fundamental nature of animation versus text as representations. Schnotz (2001) makes an important distinction between *depictive* representations (such as animations) and *descriptive* representations (such as written text). The very different sign systems employed in these two types of representation have profound consequences for learner processing of pictorial and text-based information. Descriptive representations provide their information via linear strings of units made from a finite set of abstract symbols that are arranged according to agreed syntactic and formatting rules. Neither the symbols nor their arrangement resemble the referent subject matter they represent. By contrast, with depictive representations there is typically a strong, direct relationship between the appearance and configuration of the subject matter and how those attributes are portrayed. Unlike descriptive representations, the analog relationships between a depictive representation and its referent means that the matching of their corresponding aspects is usually relatively straightforward. Such differences in representational characteristics mean that it can be problematic to express information originally presented via one of these types of representation in terms of the other. As discussed below, conversion of a depictive representation into a descriptive representation involves substantial changes such as a reduction in dimensionality (from two-dimensions to effectively one-dimension) and a considerable distortion of temporal organization. Not only can it be a challenging task for the individual who must make the conversion, but the converted information can also be an inadequate representation of what was presented in the original depiction.

In order to capture a depiction's visuospatial and spatiotemporal information about the referent subject matter as a descriptive representation, it must be subjected to extensive transformation with respect to both the entities involved and their relationships. Rather than being identifiable by virtue of its distinctive appearance (as would be the case with a depictive representation), each entity must be assigned a verbal label. Further, the actual spatiotemporal arrangements and changes that are present in the referent subject matter must be re-cast as a linear sequence made up of labels that identify the referent entities, together with various other word classes that provide information about their relationships across space and time. These strings of words are organized not in a way that parallels the subject matter's real-life spatiotemporal relationships but rather according to conventionalized textual syntactics. Descriptive representations must use a range of prepositions to indicate how entities are arranged in space (above, beside, to the left of, inside, etc.).

However, this is a relatively crude way of specifying spatial configurations that is ill-suited to capturing the subtlety and complexity often present in the referent subject matter (which is of course a strong justification for the coupling of text with pictures for multimedia approaches in education). Suitable words simply do not exist for providing sufficiently precise and manageable accounts of such sophisticated information. Descriptive representations are possibly even more problematic when it comes to capturing temporal information. The linearization requirement involved in generating descriptive representations severely restricts their capacity to handle aspects such as the simultaneous occurrence of multiple events (see also Lowe & Boucheix, 2017, this volume). Dynamic aspects of animation that involve simultaneity cannot be represented directly in the way they actually take place but instead must be transmuted into a serial form that sequentially nominates the set of events concerned and specifies that they co-occur. Even when the simultaneous events involved are relatively few and straightforward, it may be challenging to generate and interpret such linearized accounts. The situation can be far worse when the simultaneity is extensive and involves complexities such as the staggered overlap of events. In sum, words are a very blunt instrument for dealing with the types of visuospatial and spatiotemporal information that are found in animations.

The characteristics of descriptive representations outlined above have important implications for assessment of animations. In order to respond to conventional ways of assessing learning from animation, learners must be able to convert information from the mental model that they have constructed during study of the depictive animation into an adequate descriptive representation. In the case of conventional multiple choice tests, appropriate interpretation of each item's text-based stem and distractors must take place if the chosen answer is properly to reflect what has been learned. In the case of questions requiring extended written answers, learners must be able to *generate* a suitable text rather than choose amongst existing alternative text responses. Various problems can arise in both these cases that may compromise their effectiveness. For example, even at the most basic level, multiple choice tests may result in flawed assessments merely because a learner does not recall the names of the entities being referred to in the stem or distractors. The learner could in fact have developed a mental model of the subject matter from the animation that is entirely satisfactory in terms of the referent's structure and operation but simply does not include labels for the constituent entities. Although the learner understands what was shown in the animation, this understanding is not captured by the multiple choice test.

Using learner-generated written answers to assess learning from animation may be even more problematic than employing multiple choice tests because it requires production rather than recognition. The research literature characterizes writing as a complex and highly demanding task (Bazerman, 2008; Bereiter & Scardamalia, 1987; Flower, 1979). When the topic to be written about involves capturing in text the types of visuospatial and spatiotemporal information referred to above, the processing costs for learners are likely to be particularly high. Expecting a learner to take an understanding that was built from an animation's analogical representation of the subject matter and convert it into a satisfactory text-based indication of that understanding is a very big ask. Learners may know the required information but may not be able to frame it as an adequate textual utterance – they simply cannot express their understanding in words. Observations made during our own work on learning from complex animations (e.g., of a traditional piano mechanism) are consistent with this possibility (Boucheix & Lowe, 2010; Boucheix, Lowe, Putri, & Groff, 2013; Boucheix, Lowe, & Soirat, 2006; Lowe & Boucheix, 2011; Lowe & Boucheix, 2016). Exit interviews with participants suggested that they frequently understood far more than was indicated by their written explanations. For example, whereas the formal assessments products often lacked clear, explicit connections between the written episodes, such deficiencies in coherence were far less evident in the exit interviews. There were also numerous cases where participants reported having deliberately produced minimal accounts of what they had learned from the animation because they found it so difficult to do it justice in a written explanation.

The coding of written explanations resulting from learners' study of complex animations can pose further challenges for researchers in this area. Our own experience is that on occasion, these learner accounts may be not only somewhat incomplete, imprecise, and lacking coherence, but also sometimes ambiguous. For example, in the piano animation studies mentioned above, participants may not correctly identify the piano mechanism's components correctly, either because they did not recall their names or because these labels were confused. This meant that some aspects of the explanations were contentious with respect to coding, such matters having to be resolved between coders by considering the broader context within the explanatory thread. In a recent study (Lowe & Boucheix, 2016), we attempted to ameliorate this difficulty by giving each of the components in the animation its own distinctive color so that it could be identified and referred to by the learner without having to know its proper name. While this did seem to remove some of the ambiguity in participants' written accounts, issues with lack of completeness and coherence remained. Our judgement after having examined many written explanations of animated presentations is that at best, they offer a somewhat impoverished measure of what has been learned from the animation.

8.3 Considering Assessment Alternatives

Given the limitations of conventional ways of assessing learning from animations flagged above, consideration of other options is warranted. As already noted, the underlying problem with these assessments is the lack of consistency between how animations present the to-be-learned information and how the resulting learning is measured. This inherent representational mismatch could possibly be prevented if assessment approaches gave priority the visual and dynamic character of animations (cf. Hegarty, Mayer, Kriz, & Keehner, 2005). There are clues to what such approaches might be like in how people behave in natural settings when trying to explain things that have important visuospatial and/or spatiotemporal aspects. A well know case is the fellow diner who recruits the tableware for use as props to aid his or her explanation and complements manipulations of those props through space and time with supportive gestures. In effect, the explainer uses these props (tokens for the referent entities), manipulations and gestures to create an informal dynamic model of the referent subject matter (cf. Jamalian, Giardino, & Tversky, 2013; Tversky, Heiser, Lee, & Daniel, 2009; Tversky, 2011). Demonstrations using this model are typically accompanied by verbal clarifications and elaborations that specify matters such as relevant properties of the entities, the actual sequencing of events, and the cause-effect relationships involved. It is important to note that in this dinner explanation example, words are an ancillary rather than (as in conventional assessments) a central part of the explanation. As characterized by Kang, Tversky and Black (2015), when explanations of this type succeed, they are "… a symphony of gesture, language, and props" (p. 1).

We suggest that a somewhat similar type of demonstration approach could provide more effective tools for assessing learning from animation than do conventional text-based approaches. Although such demonstration assessments would draw on the main three elements of the dinner table explanation example presented above (manipulable model, gesture, and verbal accompaniment), they would need to adopt a more rigorous, systematic approach to the design of the materials, task and regime used. An important potential advantage of using demonstrations to assess learning from animations is that they offer the learner a far more natural and unmediated way of representing of what was depicted in an animation than is possible with a written account. Without the need to allocate precious processing resources to composing a word-based account, learners can devote all their efforts to generating a rich multimedia explanation that is better suited than text to presenting visuospatial and spatiotemporal information.

8.4 Demonstrations for Assessing Learning

Although demonstrations have long been an important mainstream tool for instruction, they are less widely used for assessment purposes. Nevertheless, they have been adopted for assessing student learning in certain specific circumstances. In some cases, the demonstrations are observed and reported by the students being assessed, but not performed by them (e.g., Ramsey, Walczyk, Deese, & Eddy, 2000). However, in other cases, it is the student who performs the demonstration with the quality of that performance being the basis for assessment. Such performance-based assessment is widely used in areas such as medical education that aim to develop high levels of psychomotor skill and underpinning technical knowledge (e.g., Mandel, Lentz, & Goff, 2000; Van Hove, Tuijthof, Verdaasdonk, Stassen, & Dankelman, 2010). In the present chapter, we focus on demonstration tasks that are performed by the learner but for which the aim is not to assess the performance as an end in itself. Rather, we are interested in the demonstration as an indicator of how well the learner understands the subject matter that has been portrayed in an educational animation. For the purposes of this chapter, we characterize the goal of assessment as being to determine the quality of the mental model that a learner develops from the animation during its study. Our concern is with the potential of demonstration tasks to provide more useful information about the appropriateness, accuracy and comprehensiveness of learners' mental models of the to-be-learned subject matter (see also Lowe & Boucheix, 2017, this volume). In contrast to the medical education example referred to above, we are not primarily concerned with the quality of the demonstrated performance *per se* but with insights that demonstrations may give into learners' internal representations of the externally represented content. Indeed, as will be discussed later, the performance given by the learner during assessment demonstrations may actually seem to have relatively poor fidelity with respect to the temporal changes depicted in the animation. This is especially likely for demonstrations based on more complex animations in which many events take place simultaneously because of the inability of humans to mimic all the activity of subject matter as it actually occurs. However, we shall see that this limitation on the capacity of learners to provide demonstrations that are strictly faithful to what is shown in the animation does not necessarily rule the approach out as a potentially valuable assessment tool.

Despite the preponderance of traditional text-based assessment approaches in research, there has been some very limited use of demonstrations to assess learning from animations. One type of research in this area has involved studies in which participants were required to learn a linear series of steps constituting a procedure (such as in knot tying or origami; e.g., Marcus, Cleary, Wong, & Ayres, 2013; Schwan & Riempp, 2004; Wong, Leahy, Marcus, & Sweller, 2012). In such cases, assessments were scored with respect to the presence and sequencing of the steps exhibited in the participant demonstrations. This is very different from situations in which the goal is for learners to understand the subject matter rather than merely reproduce a series of actions. However, there has been some research in which manipulable models have been employed for assessing spatial processing (e.g., Adams, Stull, & Hegarty, 2014) and as the basis for subject matter understanding, in particular the quality of the mental model that learners develop while studying an animation (Boucheix, Lowe, Breyer, & Ploetzner, 2015; Fillisch & Ploetzner, 2015; Lowe & Boucheix, 2011; Lowe, Jenkinson, & McGill, 2014; Lowe & Schnotz, 2007). Before examining selected instances of where this latter approach has been used, we will consider a simple hypothetical example to raise some key issues regarding the use of demonstrations for assessment.

8.5 Demonstration Task Example

In this section, we use the example of a Scotch Yoke to illustrate how a demonstration task might be employed to assess learning from animation and the challenges involved in designing a suitable manipulable model. The Scotch Yoke is a mechanical device for converting between rotary and linear motion. It consists of a sliding

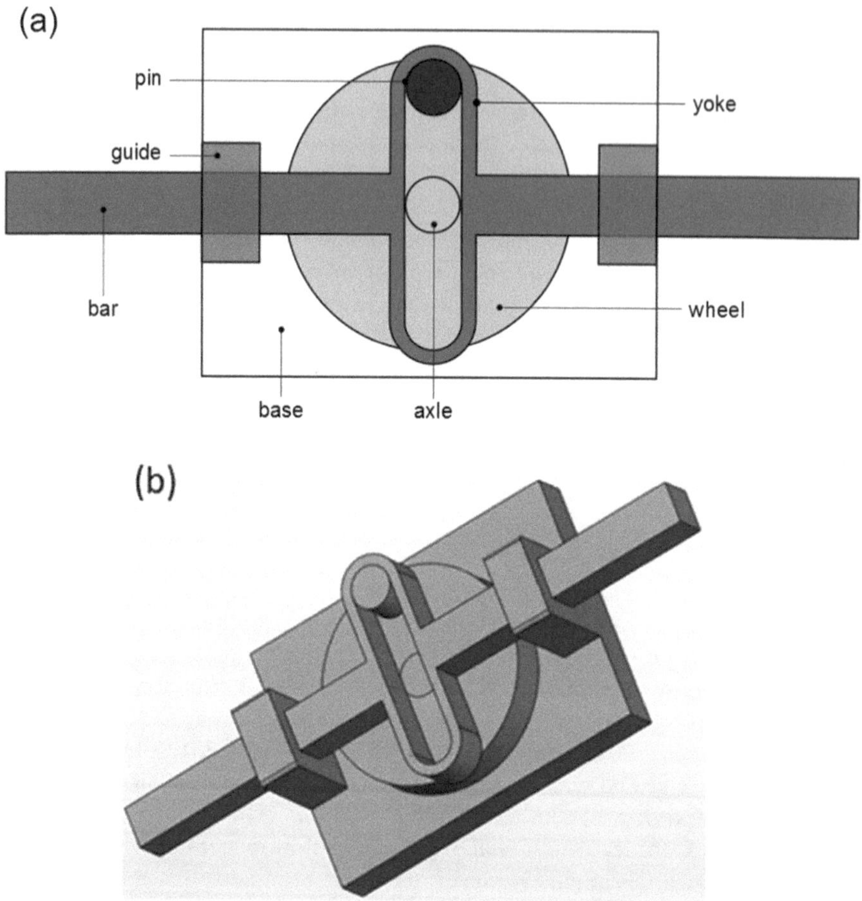

Fig. 8.1 (**a**) Parts of a Scotch Yoke and (**b**) 3D view of a Scotch Yoke

bar with a slot in the center (the 'yoke') which engages with a pin fixed to the edge of a wheel (Fig. 8.1). As the wheel rotates around its axle, the pin moves inside the slot, so pushing the bar back and forth within its guides (Fig. 8.2). Suppose that learners are asked to study an animation of a Scotch Yoke in order to understand how this mechanism performs its motion conversion function. To do this, a learner must appreciate the subtle visuospatial and spatiotemporal relationships that exist between its various parts.

If a Scotch Yoke model was to be constructed as it is shown in Fig. 8.1b, there is no doubt that it could (in principle at least) be used by a learner to provide an explanatory demonstration of how this device works. To be of most value for assessment of mental model quality, such a demonstration would not merely require the learner to reproduce the activity shown in the animation. Rather, the learner's task would be to explain how the various entities and relationships comprising the

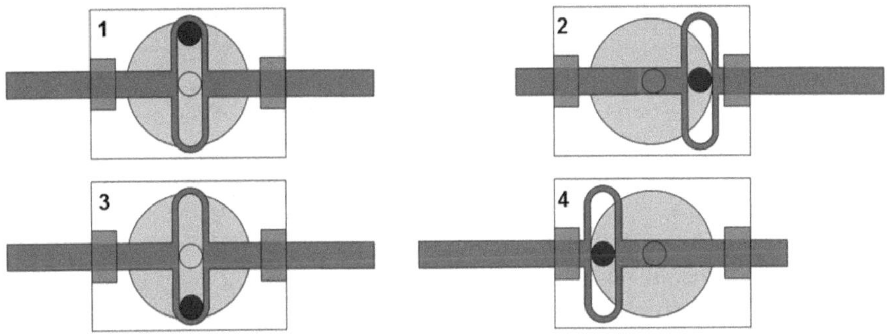

Fig. 8.2 Summary of the Scotch Yoke operational cycle

mechanism are responsible for producing that activity. To do this, the learners themselves would have to generate the activity (rather than that activity being largely self-generated by the manipulable model due to the physical interaction of its parts). In practice, a fully veridical model of the type depicted in Fig. 8.1b is likely to be unsuitable as an assessment tool because it is far too constrained in terms of its movement. The constraints it contains (such as exist in the pin-yoke combination and the guides-bar combination) mean that the model can 'automatically' perform a number of the movements correctly by itself without the need for the learner to do more than supply the basic motive force. For example, if the learner just turned the wheel (without manipulating any other parts of the mechanism), the movement of the pin inside the yoke would cause the yoke and its attached bar to move to-and-fro. Further, the guides would ensure that the bar's movement was executed in the required linear fashion. In this case it is the model (rather than the learner) that is determining how the mechanism's entities interact and the nature of the resulting movements. It does not therefore provide a reliable indication that the learner actually understands the individual patterns of component behavior that are involved in the Scotch Yoke's operation. It also circumscribes opportunities for the learner to explain how the entities and relationships present are able to produce the required conversions between rotary and linear motion.

For a manipulable model of a Scotch Yoke to provide a satisfactory assessment tool, such built-in constraints would need to be removed so that the learner alone determines how functionally crucial parts of the model behave during a demonstration task. The three-dimensional faithfulness of the model to an actual Scotch Yoke is at the heart of the problems identified here. Because the pin's vertical surface engages with the corresponding surface of the yoke's slot, it necessarily drives the yoke when the wheel rotates. One way to remove this constraint would be for the model to represent the pin as a two dimensional (flat) entity rather than as a solid 3D object. As such, it could no longer physically drive the yoke. Rather, the necessary coordinated movement of the yoke and pin would then have to be performed by the learner alone (so providing evidence of whether or not the learner actually understood the dynamic relationship between these entities). Similarly, instead of the guides being three-dimensional so that they ensured movement of the bar occurred in a straight line, they could be incorporated in the model as two-dimensional representations that provided no such constraint. This would require the learner to be solely responsible for producing the proper linear movement of the bar.

Although the intention behind removing these constraints is to obtain better insights into a learner's understanding, there are also possible negative consequences of loading so much responsibility for executing these movements onto the learner. There are clearly practical limits on how comprehensively the learner can execute all the dynamic changes depicted in the animation as they actually occur. Even for this relatively simple subject matter, there is a degree of simultaneity in the movements involved that may be impossible for learners to reproduce properly when they only have two hands with which to perform the demonstration. For example, when one hand is being used to rotate the wheel of the Scotch Yoke and the other is fully involved in moving the yoke in coordination with the pin, it is unlikely that the bar's linear movement within the bounds of the guides could be accurately maintained at the same time. It would be similarly problematic if the learner attempted to concentrate the demonstration on the to-and-fro movement of the bar – it would be difficult to simultaneously ensure that the movement of the yoke and pin were properly coordinated. Apart from learners experiencing frustration while attempting to synchronize these related movements, their inadequate efforts could be interpreted by the assessor as a lack of understanding of how the device works. Careful consideration is therefore required in both the design of manipulable models for use in assessment and in the development of procedures for eliciting demonstrations from learners. In the next section, we explore practical issues impinging on the design and use of demonstration materials by considering examples of manipulable models developed for use in a variety of experiments on learning from animation.

8.6 Experimental Materials

This section focuses on the development of manipulable models to be used as substrates for learner demonstrations. Although the examples given have a number of common characteristics that reflect the general types of design considerations flagged above, each individual manipulable model posed its own idiosyncratic challenges because of differences in the subject matter concerned.

Lowe and Schnotz (2007) investigated learners' extraction of macro versus micro level information from high and low speed versions of an animation depicting a five-ball Newton's cradle device. The manipulable model used in this study was very simple, consisting only of a line of five equally-sized small coins set on a horizontal surface (cf. Fig. 8.3). These coins represented the five balls shown in the animation but differed from the depicted situation in a number of respects. In contrast to an actual Newton's Cradle system, these token 'balls' were two-dimensional, were not connected to a supporting frame by strings, and did not hang in a freely

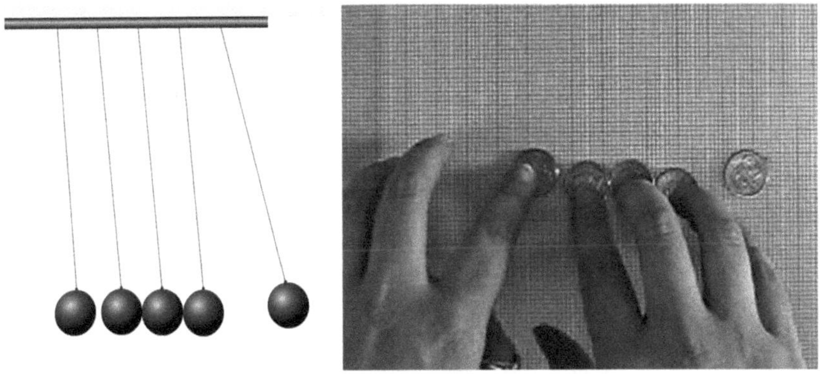

Fig. 8.3 Newton's Cradle animation frame and demonstration via coin-based manipulable model

suspended manner. Initially, the intention was to provide a manipulable model with a closer resemblance to the real device but trials showed this idea not to be feasible in practice. A pilot Newton's Cradle model with a high degree of fidelity in which the ball tokens hung vertically was found to be very awkward for participants to manipulate effectively. Further, it allowed gravitationally-induced ball token movements to occur that obviated the need for participants to self-generate all aspects of the dynamics on the basis of their own understanding of the animation. These drawbacks were eliminated in the coin-based model that was finally used in the study. This substantially simplified representation of the subject matter and minimized manipulative challenges whilst maximizing the need for the participant (rather than aspects of the model) to determine the presence and nature of ball token movements. Because of the small size of the coins and their close juxtaposition in the model, participants were able to use individual fingers to help them demonstrate finer-grained aspects of the movements involved. Demonstrations were video recorded from above with the coins resting on a paper grid to assist in quantifying subsequent analysis. To ensure that participants were adept at performing a range of movements required to demonstrate the Newton's Cradle behavior, they were pre-trained in execution of these possibilities. Requested accompanying think aloud productions facilitated the analysis of micro and micro aspects of the protocols in terms of configurations and movements (speed, distance, and acceleration). Despite the model's extreme simplicity, the coin-based demonstrations proved most effective for probing participants' understandings of the subtle changes that the Newton's Cradle underwent across time (Lowe & Boucheix, 2010).

A considerably more complex manipulable model was developed for research on learning from an animation depicting a traditional upright piano mechanism (Lowe & Boucheix, 2011). Whereas the active entities of a Newton's Cradle are five identical balls suspended by strings, the seven main entities shown in the piano animation are very different from each other in appearance, size, and movement possibilities (cf. Fig. 8.4).

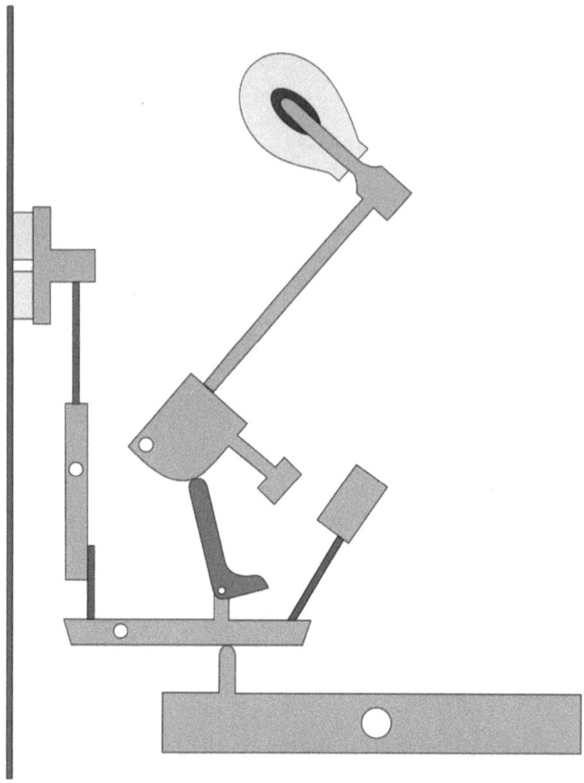

Fig. 8.4 Frame from an animation of upright piano mechanism operation

However, an important commonality in these two devices is the central role that the effect of gravity on entity behavior plays in their respective operations. Experience with developing the Newton's Cradle manipulable model indicated the need to remove this influence from the piano model as well. This was again done for the piano model by rotating the normal vertical orientation of the mechanism to the horizontal so that the participant was required to self-generate any movements of the mechanism's components. Pilot work with prototype piano models showed that video recording manipulations of the mechanism from above (as had been done with the Newton's Cradle coin model) was unsatisfactory. This was because the participant's hands obscured too much of the model's structure. The final model was therefore constructed from transparent plastic sheeting and rested on a glass-topped table to allow participants' manipulations of its components to be video recorded from below during their demonstrations. A real upright piano mechanism is largely made up from various pivoted components that act as levers for transferring motion from the piano keys to entities responsible for producing the musical notes. The plastic replica parts of the model were configured on a transparent base fitted with

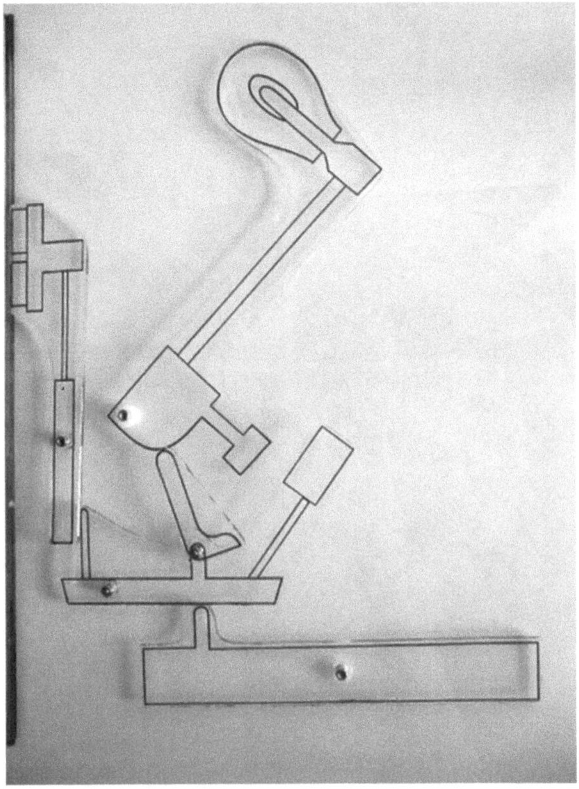

Fig. 8.5 Manipulable model of upright piano mechanism

pivots that allowed them to be manipulated in order to replicate operational events depicted in the animation (Fig. 8.5).

One of the challenges with this piano example is that proper operation of the mechanism involves several simultaneous actions occurring along two coordinated but distributed causal chains (see Lowe & Boucheix, 2017, this volume). It is physically impossible with two hands to execute all the required actions of the moveable components involved as they should actually take place. However, it is possible to cover the whole operational process less directly by breaking it down piecemeal fashion into smaller subgroups that can be demonstrated successively rather than all at once. For these partial demonstrations to provide accurate information about the piano mechanism's overall functioning, the participant's physical manipulations of the model must be accompanied by verbal and gestural framing that specify the true temporal relationships involved. In the Lowe and Boucheix (2011) study referred to above, this required that participants be trained (using a model of different, non-related content) to provide such explanatory elaborations. Due to the complexity of the piano mechanism's operational cycle, participants' initial demonstrations with the manipulable model tended to be incomplete, fragmented, and error-prone. It took repeated viewings of the animations and interleaved demonstration trials for

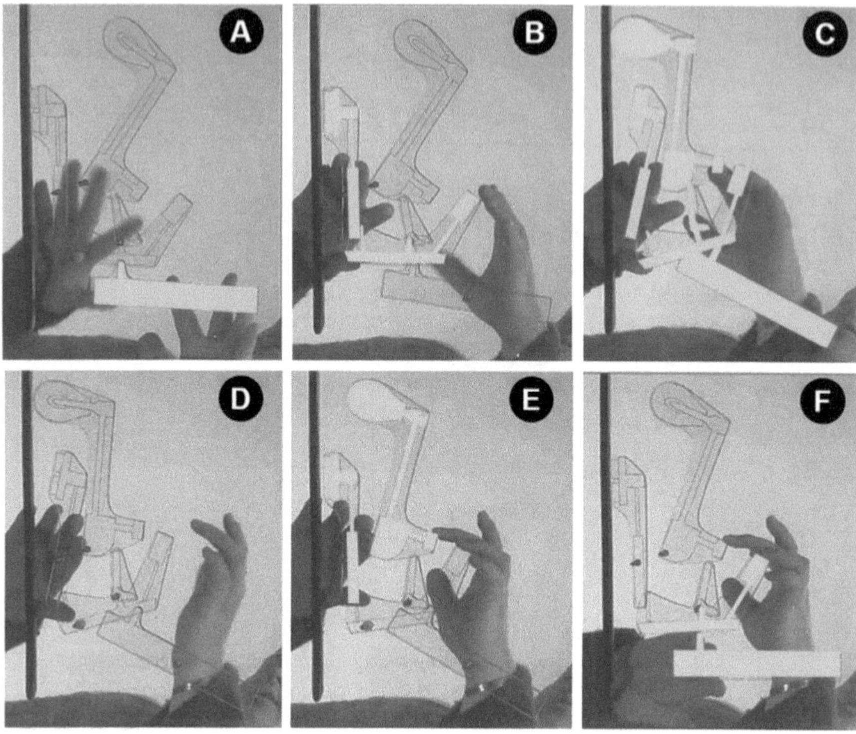

Fig. 8.6 Participant's manipulation of piano model across successive trials

them to produce more satisfactory accounts of what was happening. However, this was not considered a disadvantage because an important aspect of the study was to investigate process aspects of learning from animation and the multiple demonstrations provided important revelations about how extraction of information from animation proceeds over time (cf. Fig. 8.6).

Further challenges involved in designing and administering demonstration-based assessments are highlighted in our next example that targets learning from a four-stroke engine animation (Fillisch & Ploetzner, 2015). Like the previous piano example, this device consists of a set of very different and spatially distributed components that interact in various ways to make distinctive contributions to the device's overall functioning (cf. Fig. 8.7).

The engine animation also has a number of sites widely dispersed across the display area at which multiple functionally-important events take place simultaneously or in rapid succession, with some of these being very short lived. Such characteristics pose challenges not only for learners trying to understand how the device operates, but also for researchers wishing to develop effective demonstration-based assessments to measure learning from this animation. With respect to the Fillisch and Ploetzner study (2015), these assessment challenges were made even more demanding by the fact that three related but very different aspects of learning were

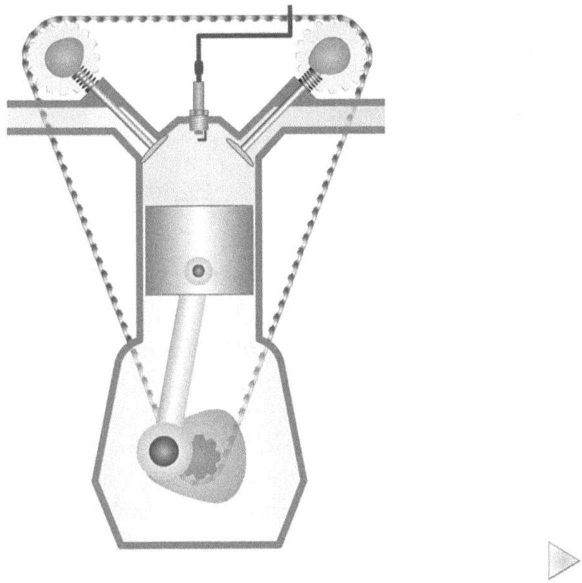

Fig. 8.7 Frame from four-stroke engine animation

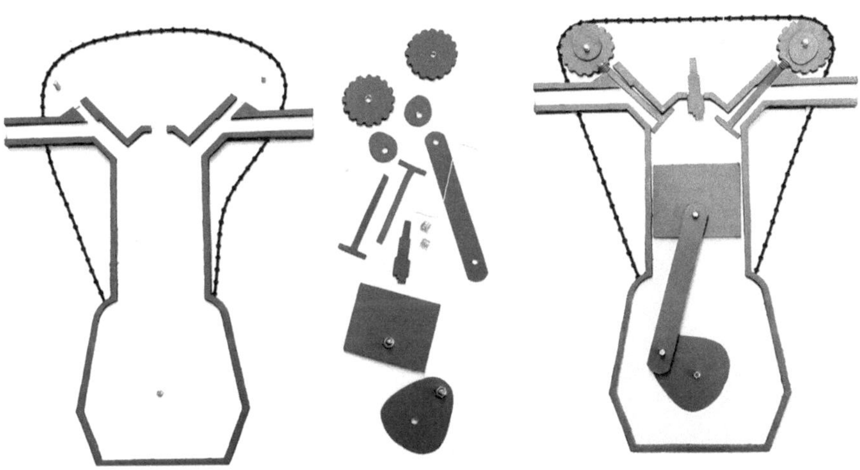

Fig. 8.8 Engine model before assembly (*left*) and correctly assembled (*right*)

targeted: (i) knowledge of the engine's structure (ii) knowledge of how each of the engine's individual components behaved in its own right, and (iii) understanding of how interactions of the engine's various components contributed to its overall operation. The manipulable model developed for use in this demonstration-based assessment was constructed from cardboard, plastic, and metal. Although most of these components made from these materials were rigid, two exceptions were the timing belt (flexible) and the valve springs (flexible and elastic). The horizontal baseboard upon which the engine components were to be positioned contained holes for mounting these items using paper fasteners. Figure 8.8 shows the arrangement of

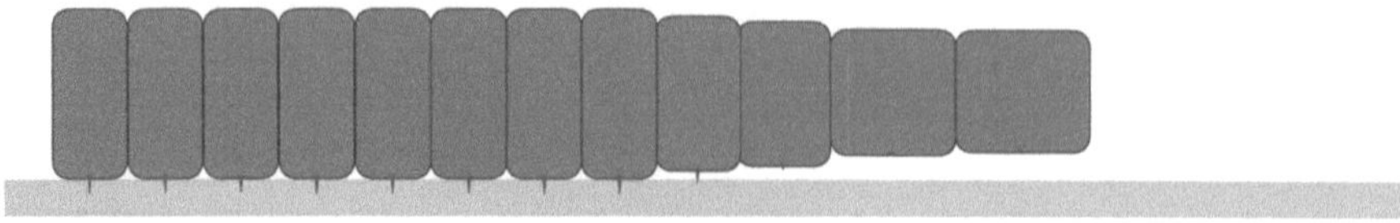

Fig. 8.9 Frame from an animation showing the principle of worm locomotion. Note protracted setae on narrower segments and elongation of other segments

the separate engine components before assembly and after they have been correctly assembled.

In the first step of the assessment procedure, knowledge of the engine's structure was measured by presenting the set of its separated components (see Fig. 8.8 left) and asking the participant to assemble them into the arrangement displayed in the animation. For the second step in which knowledge of individual component behaviors was measured, participants were allowed only to point to individual components in the model they had previously assembled, but not to manipulate them. This gestural signaling was to be accompanied by verbal explanations of the behaviors that each component would exhibit. The justification for this 'hands-off' approach to the second assessment step was that it would prevent participants from finding out about component behaviors as a consequence of their physical interactions with the model (rather than relying only on what had learned from the animation). The final step once more involved the participant in 'hands-on' interaction with the assembled model. This task required the model's parts to be manipulated in order to demonstrate how their different individual movement patterns were combined and coordinated during the engine's operation. Participants were required to provide verbal explanations to accompany their demonstrations. All steps in the procedure were video recorded for analysis. Data collected from these three stages of the assessment were analyzed using event unit based coding schemes (Lowe & Boucheix, 2008).

In the previous examples, the subject matter whose behavior is to be demonstrated involved a purely mechanical system. The entities depicted in the animations were nearly all rigid objects that are shown to translate through space but undergo no intrinsic changes in their form. However, there are many types of to-be-learned subject matter (especially in the biological sciences) in which such transformations are commonplace and are central to understanding the phenomena involved. Animal locomotion is a case in point. Our next example explores the additional challenge of designing a manipulable model that represents subject matter in which the entities involved do not merely translate (as in the Newton's Cradle, Piano mechanism and engine examples), but also undergo transformations. Consider an animation showing a simplified portrayal of the mechanism by which an earthworm moves along that has been used in recent empirical investigations (cf. Fig. 8.9). As a soft-bodied invertebrate, its locomotion is based not on muscles acting on rigid skeletal structures but rather on changes in its flexible, elastic, segmented body. In order to move forward, the earthworm executes a set of coordinated elongations and contractions

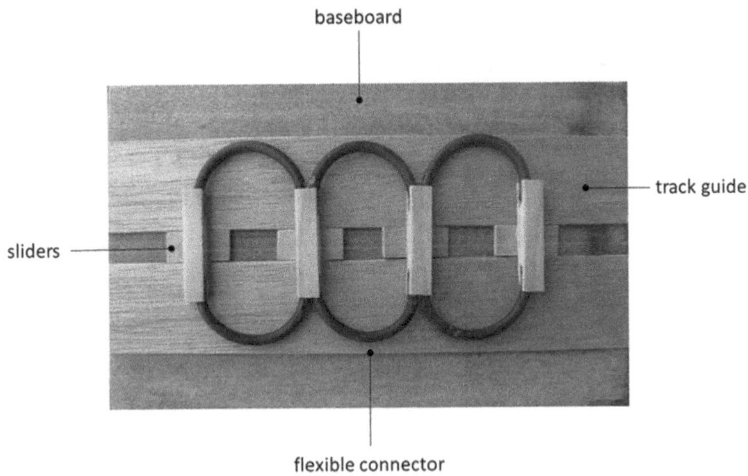

Fig. 8.10 Prototype model for demonstrating worm locomotion

of its individual body segments that are performed in concert with retraction and protraction of tiny spikes ('setae') located on those segments. When protracted into the soil, these setae anchor their respective, contracted segments and allow the non-anchored segments ahead of them to elongate in order to move that leading part of the body forward. The net effect of the alternating anchoring and release of body segments via the setae combined with the contraction and elongation of the segments is that the worm's whole body progresses forwards.

Figure 8.10 shows a prototype for a manipulable worm model developed for use in assessing learning from the worm locomotion animation. It consists of three 'segments' made from lengths of flexible rubber cording connected via wooden sliders that are free to move along a guide track. This arrangement allows both translation of the segments (to enable forward movement of 'the worm') and their transformation (between contracted and elongated) to be demonstrated.

Participants can change the shape of segments by moving pairs of adjacent sliders closer together (contraction) or farther apart (elongation; cf. Fig. 8.11). They can also use their fingers to represent the setae in order to demonstrate the role these anchoring entities play in the overall locomotion process.

Compared with the previous examples discussed, the design of this manipulable model has considerably less fidelity to what is shown in the animation. The most striking difference between the model and the animation is that it has only three (rather than twelve) segments. However, it was also decided to omit setae from the model. One reason for using fewer segments is that, unlike the other types of subject matter, the goal of the animation is to teach the *principle* of worm locomotion. Analysis shows that it is possible to explain this principle with just three segments. Another reason is that it makes two-handed manipulation of the model far more tractable than it would be if all twelve segments were provided. The decision to omit setae was made to avoid complications in construction and operation of the model.

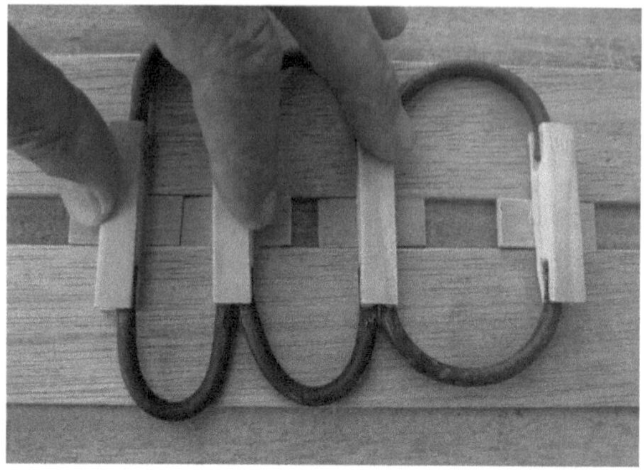

Fig. 8.11 Using fingers to demonstrate segment transformation

Instead of trying to incorporate physical components that could be moved in and out of the segments (difficult to implement in practice), participant fingers were used to represent the setae. However, it can be difficult to simultaneously demonstrate both the behaviors of the setae and the coordinated complementary behaviors of the segments. A practical solution to this problem is to demonstrate these aspects on separate occasions but indicate verbally that they actually occur at the same time. In its combination of physical props, gestures and verbal framing, this approach is consistent with the dinner party explanation example given above.

Although the examples dealt with so far have involved subject matter that is actually three-dimensional, the animations depicting them were all two dimensional. Accordingly, the corresponding manipulable models were designed as essentially two dimensional representations. However, the dynamics of some other types of subject matter are very difficult or impossible to explain without also considering the third dimension. Incorporating all three dimensions into a manipulable model introduces some additional challenges with respect to both the design of the model and the implications for its use in demonstration. Like the previous earthworm example, the final manipulable model we consider also involves biological subject matter and locomotion – how an ant walks. A frame from a cued version of an ant locomotion animation is shown in Fig. 8.12 (see McGill, 2017, this volume).

In contrast with the previous earthworm example, the ant's body undergoes no visible transformations (due to its rigid exoskeleton). However, its segmented legs do perform complex and subtle translational motions during locomotion. These individual motions are both tightly coordinated and hierarchically organized into overarching patterns of movement. For a proper understanding of ant locomotion, a learner must develop a sophisticated internal representation that is multilayered with respect to both space and time. In order to assess the quality of the mental model the learner develops from an ant locomotion animation, the demonstration materials and procedure must be capable of capturing the extent to which this level of sophistication is present.

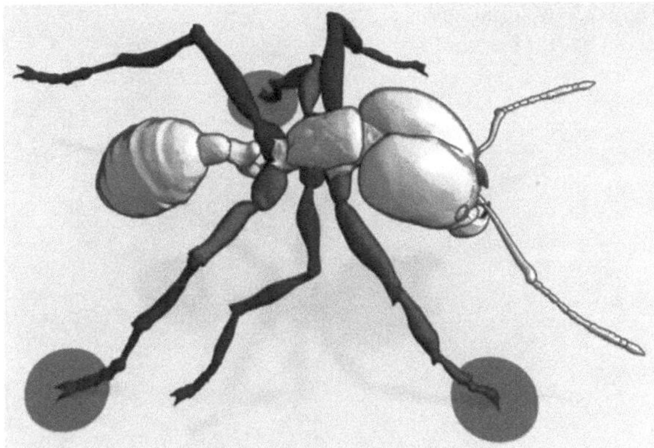

Fig. 8.12 Frame from ant locomotion animation (cued version)

Figure 8.13 shows the 3D manipulable model used to investigate learning from animations of ant locomotion (Lowe et al. 2014). The model was designed as a reasonably faithful representation of the ant's jointed legs with respect to the movements that participants can perform with them. However, other aspects of the ant's body that are essentially irrelevant to locomotion (such as the head and abdomen) were omitted. Further, the shapes of the thorax and legs are greatly simplified since they are of no real consequence in this context. The fact that this model needed to be 3D in order to permit proper demonstration of ant locomotion raised design issues not encountered with the previous 2D examples discussed above. In particular, there were the practical challenges of (i) minimizing constraints on how the legs could be moved so learner demonstrations mainly reflected knowledge acquired from the animation, and yet (ii) maximizing stability of the interim configurations generated by the participants during the course of their demonstrations so that it was physically feasible for them to perform manipulations of the model that were suitably informative. Balancing these two competing aspects was addressed by (i) using a small support beneath the thorax section to keep it sufficiently raised for the legs to adopt a 'natural' elevated position relative to the walking surface (rather than being incorrectly splayed out flat across the surface), and (ii) attaching small beads of sticky putty to the ends of the legs in contact with the floor so that once the participant had placed each leg in position, its configuration would be maintained after it was released. Both of these provisions were essentially preventive measures take to avoid disruptive effects of gravity on the demonstration. As would be expected, the presence of these features required participants to receive preliminary training about how to manipulate the model in order to make the best of their demonstrations. Figure 8.14 is a frame from a video of this manipulation in progress. The animation shows that in reality, an ant's six legs all act together, each with its own distinctive and complex pattern of motion. As with the earlier examples, it would be physically impossible for a participant to demonstrate this array of simultaneous movements as they actually occur. Consequently, participants were instructed to partition their demonstrations into tractable chunks and complement them with verbal and gestural indications of the true temporal structure shown in the animation.

Fig. 8.13 Manipulable model ant for locomotion demonstration

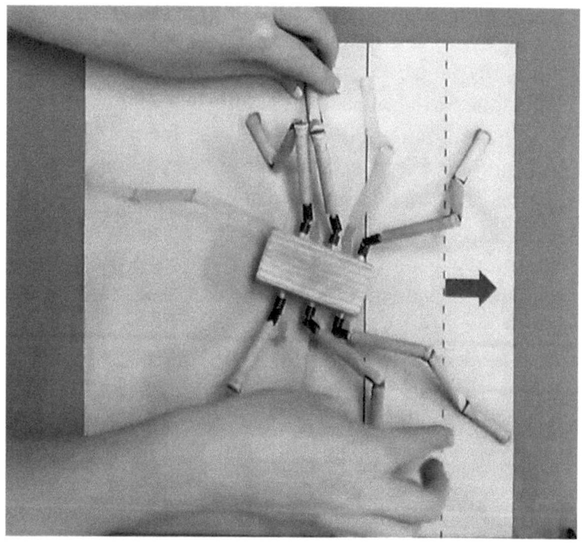

Fig. 8.14 Manipulation of ant model during locomotion demonstration

8.7 Conclusion

This chapter suggests that appropriate, well-designed demonstrations are potentially superior to conventional assessments for assessing learning from dynamic visualizations, particularly with respect to the measurement of mental model quality. However, the potential of such demonstrations is not limited to assessing learning outcomes. They may also provide valuable insights into how learners process animations that can in turn indicate ways to improve the educational effectiveness of these materials. We take a broad view of the nature of demonstrations that encompasses the model with which the demonstration is executed, the learner gestures accompanying that execution, and the verbalizations used to frame or explain the demonstrated activity. To work well as tools for assessment, demonstrations should be designed to ensure these aspects are highly complementary and mutually supporting so that the best possible evidence is obtained about learner understandings.

Learner manipulation of a model that represents the target subject matter is at the heart of the demonstrations for assessment approach. For this reason, the design features of such manipulable models and the protocols for using them can play a crucial role in the ultimate effectiveness of demonstrations as tools for assessment. These materials and procedures need to ensure that the dynamics exhibited during a demonstration are deliberately generated by the learner (rather than being mere artefacts of contingencies present in the model). To do this, any physical constraints that allow the 'model to do the work', or aspects that give hints to learners about what the dynamics should be need to be minimized. In other words, no components of the model should be able to self-operate in the correct way – the proper behavior of each component must occur as a direct result of the learner's interaction with that component and not as a secondary result of manipulating some other component. Perhaps counter-intuitively, this requires limits to be placed on the fidelity of the manipulable model to the referent subject matter. Absolute fidelity can mean that the model is over-constrained so that learner manipulations readily produce the correct behavior without actually requiring any understanding. Limiting a model's fidelity can also help make the learner's task of manipulating the model better targeted and more tractable. Potential manipulative difficulties should be minimized wherever possible. A model that is a simplified, abstract version of the referent encourages focus on the essentials and makes it easier to manipulate. The more complex the dynamics, the simpler the model needs to be so that the necessary actions can be performed with just two hands and ten fingers.

The presence of multiple events that occur simultaneously is an inherent feature of many complex animations. A key issue in developing suitable demonstration regimes for such animations is how to deal with this simultaneity so that learners can properly demonstrate their knowledge of the referent subject matter's spatio-temporal structure. In cases where the subject matter dynamics includes even a modest degree of simultaneity, it is often simply not physically possible for the learner to produce a veridical demonstration that portrays the events as they actually occur. If the number of things that need to happen together exceeds the learner's manipulative capacity to execute them concurrently, an alternative must be found. One way to do this is to allow the learner to demonstrate these events sequentially rather than simultaneously. Instead of requiring a continuous demonstration, we can plan for learners to produce discontinuous, temporally separated sub-episodes that are tractable to execute and have them frame these with appropriate verbal and gestural elaborations that make the true temporality evident to assessors. In this way, learners can make clear which aspects of the dynamics actually take place sequentially and which occur simultaneously but are being demonstrated sequentially because of manipulative limitations.

The design of suitable manipulable models and accompanying demonstration regimes can be challenging without some systematic way of characterizing aspects such as simultaneity versus sequentiality. We have found event unit analysis (Lowe & Boucheix, 2017, this volume) to be useful for this purpose because its comprehensive mapping of the behaviors of a system's components across the time course reveals instances of coincidence amongst multiple event units that may be problematic for learners to demonstrate. Once these instances have been identified, a demonstration regime can be devised that allows learners to re-cast the simultaneous events into manageable sequential episodes, with the actual simultaneity being indicated gesturally and verbally. An event unit analysis is also useful in the initial design of a manipulable model for identifying and avoiding possibilities for unwanted self-operation of model components during a demonstration. For example, it can show which movements in a fully faithful model of the subject matter would be caused by gravity or secondary inter-component contact interactions (rather than by direct actions of the demonstrator). Such revelations can help the designer of a manipulable model decide which constraints to keep or remove. Further, an event unit analysis can help fine tune the model's design so that it is well aligned with the instructional goal being targeted in the assessment. For example, the design of a model intended to assess learners' understanding of an abstract principle (such as the principle of worm locomotion) could be very different from that intended to assess a real-life instance based on that principle (how an actual worm moves). The gathering and interpretation of evidence for the quality of a learner's mental model presents another daunting challenge for researchers. Although video records can very useful as a source of data about learners' demonstrations, some ingenuity in the recording set-up may be required (such as videoing from below or using high speed recording) in order to capture crucial aspects of model manipulations that may otherwise be missed. The coding of collected data presents further challenges, particularly when these data involve combinations of manipulations, gestures and verbalizations. Here again, event unit analysis appears to provide a systematic and defensible basis for developing credible coding schedules.

Much of what has been summarized so far in this section with respect to the design and use of demonstrations for assessment appears to be relatively generalizable. However, this is not to suggest that a 'one size fits all' approach is appropriate when very different types of learning are to be assessed. For example, trying to use the same model to assess both the structure of a device and its operation is likely to be problematic. If learners were required to assemble a model of the target subject matter from its components by themselves in order to demonstrate a knowledge of its structure, errors present in their assembled model would inevitably compromise their subsequent attempts to demonstrate of its operation. Further, the very act of assembling the components may in itself give the learner hints about what types of dynamics are likely to be possible in the finished assembly. Conversely of course, using a pre-assembled model to measure operational knowledge would sabotage any subsequent attempt to measure knowledge of structure. Instead, it is probably wiser to design different models to assess different aspects of learning, such as ensuring that a model for assessing structure is not in fact operational and that a model for assessing operations is structurally correct. Similarly, the demonstration regimes that are devised for use with these different types of models need to be carefully tailored to ensure optimal assessment outcomes. Although the use of demonstrations for assessing learning from dynamic visualizations has the potential to provide a superior alternative to conventional assessment approaches, it is clear that much research into design and implementation will be needed for this potential to be fulfilled.

References

Adams, D. M., Stull, A. T., & Hegarty, M. (2014). Effects of mental and manual rotation training on mental and manual rotation performance. *Spatial Cognition and Computation, 14*, 169–198.

Bazerman, C. (Ed.). (2008). *Handbook of writing research*. Hillsdale, NJ: Lawrence Erlbaum.

Bereiter, C., & Scardamalia, M. (1987). *The psychology of written composition*. Hillsdale, NJ: Lawrence Erlbaum.

Berney, S., & Bétrancourt, M. (2009, August). *When and why does animation enhance learning? A meta-analysis*. Paper presented at the 13th European Conference for Research on Learning and Instruction, Amsterdam, Netherlands.

Boucheix, J.-M., & Lowe, R. K. (2010). An eye tracking comparison of external pointing cues and internal continuous cues in learning with complex animations. *Learning and Instruction, 20*, 123–135.

Boucheix, J.-M., Lowe, R. K., Breyer, B., & Ploetzner, R. (2015, August). *How self-generated drawing may impact learning from animation*. Paper presented at the 16th European Conference for Research on Learning and Instruction, Limassol, Cyprus.

Boucheix, J.-M., Lowe, R. K., Putri, D. K., & Groff, J. (2013). Cueing animations: Dynamic signaling aids information extraction and comprehension. *Learning and Instruction, 25*, 71–84.

Boucheix, J.-M., Lowe, R. K., & Soirat, A. (2006, August). *Online processing of a complex technical animation: Eye tracking investigation during verbal description*. Paper presented at the Comprehension of Text and Graphics Conference, Nottingham, UK.

Davenport, J. L., & Quellmalz, E. S. (2017). Assessing science inquiry and reasoning using dynamic visualizations and interactive simulations. In R. Lowe & R. Ploetzner (Eds.), *Learning from dynamic visualization—Innovations in research and application*. Berlin: Springer (this volume).

De Koning, B. B., Tabbers, H. K., Rikers, R. M. J. P., & Paas, F. (2007). Attention cueing as a means to enhance learning from an animation. *Applied Cognitive Psychology, 21*, 731–746.

De Koning, B. B., Tabbers, H. K., Rikers, R. M. J. P., & Paas, F. (2010a). Learning by generating vs. receiving instructional explanations: Two approaches to enhance attention cueing in animations. *Computers & Education, 55*, 681–691.

De Koning, B. B., Tabbers, H. K., Rikers, R. M. J. P., & Paas, F. (2010b). Attention guidance in learning from a complex animation: Seeing is understanding. *Learning and Instruction, 20*, 111–122.

Fillisch, B., & Ploetzner, R. (2015, August). *The role of support in drawing for learning from complex animation*. Paper presented at the 16th European Conference for Research on Learning and Instruction, Limassol, Cyprus.

Flower, L. S. (1979). Writer-based prose: A cognitive basis for problems in writing. *College English, 41*, 19–37.

Hegarty, M., Mayer, S., Kriz, S., & Keehner, M. (2005). The role of gestures in mental animation. *Spatial Cognition and Computation, 5*, 333–356.

Höffler, T. N., & Leutner, D. (2007). Instructional animation versus static pictures: A meta-analysis. *Learning and Instruction, 17*, 722–738.

Houghton, H. A., & Willows, D. M. (1987). *The Psychology of illustrations: Vol 2: Instructional issues*. New York: Springer.

Jamalian, A., Giardino, V., & Tversky, B. (2013). *Gestures for thinking*. Paper presented at the 35th annual meeting of the Cognitive Science Society, Berlin, Germany.

Kang, S., Tversky, B., & Black, J. B. (2015). Coordinating gesture, word, and diagram: Explanations for experts and novices. *Spatial Cognition and Computation, 15*, 1–26.

Khacharem, A., Zoudji, B., & Kalyuga, S. (2015). Perceiving versus inferring movements to understand dynamic events: The influence of content complexity. *Psychology of Sport and Exercise, 19*, 70–75.

Lowe, R.K. (2017). Designing static and animated diagrams for modern learning materials. In A. Black, S. Lund, & S. Walker (Eds.), *Handbook of information design* (pp. 361–376). London: Gower.

Lowe, R. K., & Boucheix, J.-M. (2008). Learning from animated diagrams: How are mental models built? In G. Stapleton, J. Howse, & J. Lee (Eds.), *Diagrammatic representation and inference* (pp. 266–281). Berlin: Springer.

Lowe, R. K., & Boucheix, J.-M. (2010). Manipulatable models for investigating processing of dynamic diagrams. In A. K. Goel, M. Jamnik, & N. H. Narayanan (Eds.), *Diagrammatic representation and inference* (pp. 319–321). Berlin: Springer.

Lowe, R. K., & Boucheix, J.-M. (2011). Cueing complex animation: Does direction of attention foster learning processes? *Learning and Instruction, 21*, 650–663.

Lowe, R. K., & Boucheix, J.-M. (2016). Principled animation design improves comprehension of complex dynamics. *Learning and Instruction, 45*, 72–84.

Lowe, R. K., & Boucheix, J.-M. (2017). A composition approach to design of educational animations. In R. Lowe & R. Ploetzner (Eds.), *Learning from dynamic visualization—Innovations in research and application*. Berlin: Springer (this volume).

Lowe, R. K., Jenkinson, J., & McGill, G. (2014, August). Learning functional relations in complex animations. Paper presented at the EARLI SIG 2 (Comprehension of Text and Graphics) Meeting, Rotterdam, Netherlands.

Lowe, R. K., & Schnotz, W. (2007, August). *Animations and temporal manipulations: Supporting comprehension of complex dynamic information*. Paper presented at the 12th European Conference for Research on Learning and Instruction, Budapest, Hungary.

Lowe, R. K., & Schnotz, W. (2014). Animation principles in multimedia learning. In R. E. Mayer (Ed.), *The Cambridge handbook of multimedia learning* (2nd ed., pp. 513–546). New York: Cambridge University Press.

Mandel, L. P., Lentz, G. M., & Goff, B. A. (2000). Teaching and evaluating surgical skills. *Obstetrics and Gynecology, 95*, 783–785.

Marcus, N., Cleary, B., Wong, A., & Ayres, P. (2013). Should hand actions be observed when learning hand motor skills from instructional animations? *Computers in Human Behavior, 29*, 2172–2178.

Mayer, R. E. (Ed.). (2014). *The Cambridge handbook of multimedia learning* (2nd ed.). New York: Cambridge University Press.

Mayer, R. E., & Anderson, R. B. (1992). The instructive animation: Helping students build connections between words and pictures in multimedia learning. *Journal of Educational Psychology, 84*, 444–452.

Mayer, R. E., & Moreno, R. (2002). Animation as an aid to multimedia learning. *Educational Psychology Review, 14*, 87–99.

McGill, G. (2017). Designing instructional science visualizations in the trenches: where research meets production reality. In R. Lowe & R. Ploetzner (Eds.), *Learning from dynamic visualization – Innovations in research and application.* Berlin: Springer (this volume).

Ploetzner, R., & Lowe, R. (2012). A systematic characterization of expository animations. *Computers in Human Behavior, 28*, 781–794.

Ramsey, L. L., Walczyk, J., Deese, W. C., & Eddy, D. (2000). Using demonstration assessments to improve learning. *Journal of Chemical Education, 77*, 1511–1516.

Scheiter, K., Gerjets, P., Huk, T., Imhof, B., & Kammerer, Y. (2009). The effects of realism in learning with dynamic visualizations. *Learning and Instruction, 19*, 481–494.

Schnotz, W. (2001). Sign systems, technologies, and the acquisition of knowledge. In J.-F. Rouet, J. J. Levonen, & A. Biardeau (Eds.), *Multimedia learning: Cognitive and instructional issues* (pp. 9–29). London: Pergamon.

Schnotz, W. (2002). Towards an integrated view of learning from text and visual displays. *Educational Psychology Review, 14*, 101–120.

Schwan, S., & Riempp, R. (2004). The cognitive benefits of interactive videos: learning to tie nautical knots. *Learning and Instruction, 14*, 293–305.

Tversky, B. (2011). Visualizing thought. *Topics in Cognitive Science, 3*, 499–535.

Tversky, B., Heiser, J., Lee, P., & Daniel, M.-P. (2009). Explanations in gesture, diagram, and word. In K. R. Coventry, T. Tenbrink, & J. A. Bateman (Eds.), *Spatial language and dialogue* (pp. 119–131). Oxford: Oxford University Press.

Tversky, B., Morrison, J. B., & Bétrancourt, M. (2002). Animation: Can it facilitate? *International Journal of Human Computer Studies, 57*, 247–262.

Van Hove, P. D., Tuijthof, G. J. M., Verdaasdonk, E. G. G., Stassen, L. P. S., & Dankelman, J. (2010). Objective assessment of technical surgical skills. *British Journal of Surgery, 97*, 972–987.

Wong, A., Leahy, W., Marcus, N., & Sweller, J. (2012). Cognitive load theory, the transient information effect and e-learning. *Learning and Instruction, 22*, 449–457.

Chapter 9
Assessing Science Inquiry and Reasoning Using Dynamic Visualizations and Interactive Simulations

Jodi L. Davenport and Edys S. Quellmalz

9.1 Introduction

Though the majority of this book focuses on *learning* from dynamic visualizations, the aim of the current chapter is to explore how dynamic visualizations and simulations can be effectively used as tools for *assessing* what students have learned. Science learning requires students to understand, reason about, and use inquiry to explore complex, dynamic systems. How can we know if students have attained proficiency? What kinds of evidence can dynamic and interactive displays gather about students' understandings of dynamic phenomena exhibited by science systems, as well as students' abilities to use these understandings to conduct science investigations?

In traditional multiple choice test items, students demonstrate mastery by selecting correct responses from a set of text-based alternatives. These static items are adequate for providing evidence that students have obtained declarative knowledge, such as scientific facts or definitions. However, this type of test cannot provide direct evidence for other skills deemed important by science educators around the world, such as the ability to extract information from dynamic presentations or carry out inquiry in realistic scientific contexts (Mullis & Martin, 2013; NGSS Lead States [NGSS], 2013; Organization for Economic Cooperation and Development [OECD], 2009).

With computers becoming increasingly ubiquitous, digital learning environments can use embedded ongoing assessment of learning progress to inform provision of immediate feedback and scaffolding *during* instruction. In addition,

J.L. Davenport (✉)
WestEd, Oakland, CA, USA
e-mail: jdavenp@wested.org

E.S. Quellmalz
WestEd, Redwood City, CA, USA
e-mail: equellm@wested.org

R. Lowe, R. Ploetzner (eds.), *Learning from Dynamic Visualization*,
DOI 10.1007/978-3-319-56204-9_9

accountability testing to report the status of learning proficiency can move beyond traditional static testing to include dynamic, interactive, and simulation-based tasks that potentially capture richer information about students' abilities to use science knowledge and engage in inquiry practices.

Yet, these innovative assessments raise questions about validity. Cognitive psychology and multimedia learning research suggest that visualizations can place substantial demands on attention and memory, and individuals may not always interpret the displays or response requirements as intended. Many test developers who are highly experienced in creating static items are less experienced with dynamic tasks and items and may be unaware of relevant multimedia learning research. Deeper study of the affordances and limitations of technology-enhanced assessment designs can shed light on the potential benefits of different formats and modalities for measuring understanding of complex science system dynamics and use of inquiry practices.

In this chapter, we describe the range of science content knowledge and practice skills we seek to measure, outline research-based design principles for the design of innovative, interactive, technology-enhanced assessments, and review a test creation process that is intended to capitalize on the unique features of dynamic assessments and mitigate their potential limitations. Next, we provide examples of how the stimulus and response features of dynamic visualizations allow us to assess a range of scientific practices. Finally, we provide findings from an efficacy study that support the claim that carefully-designed, dynamic and interactive assessments can provide better measures of student proficiency with complex scientific inquiry and reasoning than static assessments. Throughout the chapter, we give concrete examples of the principles and processes in the context of two science assessment platforms, SimScientists and ChemVLab+.

9.2 What Science Knowledge and Practice Skills Do We Want to Assess?

Internationally, science education standards and frameworks, e.g., Trends in International Mathematics and Science Study (TIMSS), the Programme for International Student Assessment (PISA), and the United States' National Assessment of Educational Progress (NAEP), state goals of teaching and assessing scientific reasoning skills, integrated knowledge of science systems, and the ability to conduct scientific inquiry (Mullis & Martin, 2013; National Assessment Governing Board [NAGB], 2010; OECD, 2009).

In the domain of science, knowledge and practice skills are strongly intertwined. Proficiency in science requires students to move beyond recall of scientific facts to apply knowledge to generate explanations and carry out and interpret the results of investigations. Further, science assessment requires students to show what they can do in the context of a content area. Schematically, many science systems share a similar structure. A science system can be described by explaining how components of the system interact in proscribed ways to produce emergent system behaviors (cf. Schweingruber, Keller, & Quinn, 2012). Science systems can be thought of as iteratively nested levels of components, interactions, and emergent behaviors. These levels (components, interactions, emergent behaviors) describe a system at a specific grain size. The components of a science system are the parts that interact in rule-governed ways that are described by the interactions level. For example, in chemistry, components could be atoms that interact to form bonds according to rules of chemistry. In ecosystems, the components could be animals that engage in predator-prey relationships. In the human body system, components could be individual organs that interact to form the circulatory system. Taken at the emergent behavior level, the aggregate of these components working together yields phenomena such as the properties of matter, population dynamics, or the overall health of a person.

9.2.1 Science Practice Skills from the NAEP Science Framework

How can we operationalize the integrated science knowledge and skills to assess? Because assessments require students to take actions to demonstrate their knowledge, we can describe the to-be-assessed learning goals of science as *science practices* that require students to engage in different types of cognitive processing. The NAEP Science Framework, from the United States' test of national science proficiency, specifies three science practices that increase in complexity and processing demands (NAGB, 2010). For the rest of the chapter, we will use these three types of science practices, described by the NAEP framework, as the basis for our discussion of science assessment: (1) *identifying science principles*, (2) *using science principles,* and (3) *conducting scientific inquiry*. Below we outline the three science practices, their cognitive demands, and the possible advantages of using dynamic interactive assessments rather than traditional static test items to elicit evidence of science proficiency. As acknowledged by the authors of the NAEP framework, the practices build on each other and are somewhat interdependent (e.g., to effectively carryout inquiry, you first need to identify the relevant science principles). The goal for assessing these science practice skills is to ensure the processing demands of the tasks being used for assessment are aligned with the targeted science practice.

9.2.2 Identifying Science Principles

The first science practice described in the NAEP framework is dubbed *identifying science principles*. Science requires students to learn facts about the physical world and the principles that explain causal relationships between parts of systems. The authors operationalize *identifying principles* as the primary level of scientific understanding that includes the abilities to state or recognize correct principles and to describe, measure, or classify observations.

In the science system context, *identifying principles* involves engaging with a single level of a science system; that is, either the components, interactions, or the emergent properties rather than connecting across the levels of a science system. For instance, in an ecosystem context, *identifying principles* requires students to recall information about categorizing components (e.g., classifying animals as predators or prey), stating the rules for how the components interact (e.g., predators eat prey), or recognizing different types of emergent properties (e.g., recognizing a population crash on a graph).

Although other processing demands are required, assessing the ability to *identify principles* is most strongly associated with the cognitive demand to recall or recognize factual or declarative knowledge. To summarize the complexity of the *identifying principles* science practice for assessment developers, the framework authors refer to this demand as "knowing that" (NAGB, 2010). Tasks assessing whether students can identify principles most commonly involve prompts that require students to recall or recognize correct principles, labels, or categories of objects. In many cases, static items with multiple choice or short answer responses are able to provide evidence of mastery of this type of knowledge. Because task formats that include animations or interactive simulations are much more costly to create and potentially introduce task-irrelevant processing demands, static task formats are likely well-suited to assessing the science practice of identifying science principles.

9.2.3 Using Science Principles

The next science practice outlined in the NAEP Science Framework, *using science principles*, requires students to move beyond simply storing and retrieving facts from long-term memory to applying the knowledge in novel ways to make predictions, explain observations, and transfer knowledge of principles to different settings (NAGB, 2010). The NAEP framework describes using science principles as requiring schematic knowledge – "knowing why." In addition to processing demands associated with attention, perception, and retrieval, the cognitive demand of schematic knowledge requires students to both retrieve information from memory and apply that knowledge to novel contexts and stimuli.

In the context of a science system, *using science principles* involves explaining or describing relationships between the component, interaction, and emergent levels of a science system. Using science principles would involve students creating explanations for how the components interact and how the interactions yield different emergent phenomena. Similarly, students proficient in *using science principles* are able to make predictions that integrate their knowledge of science principles with observations of phenomena. For example, students might be asked to explain how interactions between predator and prey affect changes in emergent system population levels.

To allow a student to demonstrate mastery of *using science principles*, the assessment environment must present students with a context, situation, or real world problem to which the knowledge can be applied. Tasks that include animations or dynamic visualizations of science phenomena can provide a format for students to demonstrate ability to make predictions based on observations, explain patterns, and apply knowledge in a range of contexts. Although animations may place greater processing demands on students and are much costlier than text descriptions, they have the distinct advantage of being able to provide more valid portrayals of the science phenomena "in action." Therefore, the demands for processing the information portrayed in an animation of a dynamic visualization are aligned with the processing demands required by the to-be-assessed science practice of *using science principles* to make predictions, analyze observations, and explain patterns as the principle plays out in multiple contexts.

9.2.4 Conducting Science Inquiry

Finally, the NAEP Science Framework describes *conducting science inquiry* as the ability to design an investigation, collect data using appropriate tools and techniques, analyze data, and make inferences. In addition to the processing demands described above, the practice of *conducting science inquiry* requires both procedural knowledge ("knowing how") and strategic knowledge ("knowing when and where to apply knowledge"; NAGB, 2010). The cognitive demands of *conducting inquiry* require students to understand how to frame a scientific argument and carry out procedures that generate the appropriate observations to answer an empirical question. A distinctive feature of *conducting inquiry* is that the practice requires data to be generated, not merely observed; that is, designing an experiment has processing and performance demands very different from recognizing a good design when you see one. In the science systems context, using inquiry requires both an understanding of how all three levels (components, interactions, emergent behaviors) of the science systems work together as well as strategies for generating appropriate data to support claims about dynamic system phenomena.

Scientists frequently use simulations and modeling tools to dynamically represent spatial, temporal, and causal processes in science systems and permit active, virtual investigations of phenomena that are too big or small, fast or slow, or dangerous to be conducted in hands-on labs (de Jong, 2006; Lehrer, Schauble, Strom, & Pligge, 2001; Quellmalz & Pellegrino, 2009; Stewart, Cartier, & Passmore, 2005). As technology becomes increasingly accessible in educational settings, simulations and models allow students to demonstrate their ability to *conduct inquiry* by designing and carrying out investigations. The interactive nature of simulation environments allows students to see the results of their manipulations, perform iterative trials, and make inferences based on the data students themselves generate, rather than canned data provided by test developers.

9.2.5 Contexts for Assessing Science Practice Skills

Due to pragmatic constraints of large-scale, performance-based testing and available technology, most assessments have historically been paper-and-pencil and have items that are scored for a single correct response per item. These primarily multiple-choice forms of testing rely on recognition of correct responses and fail to elicit evidence that students possess the full range of desired science competencies (cf. Pellegrino, 2013; Quellmalz & Pellegrino, 2009). In the past decade, findings from trials of innovative assessments suggest that dynamic and interactive items are able to elicit a broader range of skills than traditional assessments. The United States' 2009 National Assessment of Educational Progress (NAEP) included interactive computer tasks to measure science knowledge, and the results showed that proficiency across science practice skills was uneven. Although students were generally successful in making low-level observations from data, most students performed poorly on complex assessment tasks involving multiple variables or strategic decision-making (National Center for Education Statistics [NCES], 2012).

In the discussion below, we provide examples of assessing these three types of science practices in the context of two simulation-based assessment systems, SimScientists (simscientists.org) and ChemVLab+ (chemvlab.org). The SimScientists system provides simulation-based formative and summative assessments for a range of middle school science concepts in life, physical, and the Earth sciences (Quellmalz, Timms, & Buckley, 2010; Quellmalz, Timms, Silberglitt, & Buckley, 2011). The ChemVLab+ system provides a virtual chemistry lab and interactive items to be used as formative assessments for high school chemistry (Davenport et al., 2014a; Davenport, Rafferty, Yaron, Karabinos, & Timms, 2014b; Davenport, Rafferty, Timms, Yaron, & Karabinos, 2012a; Davenport, Timms, Yaron, & Karabinos, 2012b). Both systems have reporting features that allow students and teachers to view the results of the assessments.

9.3 Design Principles and Process

Although animations and simulations can provide potentially powerful environments for assessing a range of science practices, they must be carefully designed to be effective. As others have discussed at length (cf. Lowe & Boucheix, 2017, this volume; Lowe & Schnotz, 2008; Mayer, 2014), animations and simulations can have far greater processing demands than static displays, such as paper-and-pencil items. Thus the goal for assessment is to ensure the processing demands of the displays are aligned with the processing demands of the to-be-assessed science practices. Cognitive and multimedia learning research provide guidance regarding the identification and creation of a scientifically appropriate context, the alignment between assessment tasks and learning objectives, and the minimization of extraneous, construct irrelevant cognitive processing. In this section we frame recommendations for assessment developers as design principles for integrating dynamic simulations into science assessments and review a process to guide the development of these innovative items. We focus on four main recommendations: (1) create representations that are scientifically appropriate for the population being assessed, (2) ensure alignment between tasks and the cognitive demands of the learning objectives, (3) minimize extraneous processing, and (4) validate the evidence model specifying how responses will be scored and reported. An iterative design process is used to ensure that the principles are effectively enacted in the final version of the assessment.

9.3.1 Select Scientifically Appropriate Representations

Animations and simulations can provide rich environments that allow students to demonstrate complex, interconnected knowledge. However, selecting appropriate representations requires designers to simultaneously consider the target populations' experience with the scientific context and the role of the dynamic visualization or simulation.

Research in science education has shown that authentic and contextualized practice using simulations promotes learning in classrooms contexts (Adams et al., 2008; Buckley, Gobert, Horwitz, & O'Dwyer, 2010; Cuadros, Leinhardt, & Yaron, 2007; Horwitz, Gobert, Buckley, & O'Dwyer, 2010; Krajcik, Marx, Blumenfeld, Soloway, & Fishman, 2000; Schwartz & Heiser, 2006; Zacharia, 2007). Learning that is connected with intellectual and practical use is more memorable, more readily accessible, and thus more available to transfer to new settings (cf. Bransford, Brown, & Cocking, 2000).

If students are expected to *learn* conceptual and inquiry processes within authentic contexts, it follows that assessments should also provide comparable settings when *assessing* what students know and can do. In addition, designers need to be sensitive to the prior knowledge of students, because this knowledge influences how representations are perceived (e.g., Hegarty & Just, 1993; Petre & Green, 1993; Goldman, 2003; Lowe & Boucheix, 2017, this volume; Lowe & Schnotz, 2008; Mayer, 2014). Test designers should be careful to ensure that all representations are scientifically appropriate, grade level appropriate, and appropriate in their role as either stimuli representing the science phenomena or as response options for demonstrating knowledge and skills.

9.3.2 Ensure Alignment Between Tasks and Cognitive Demands of Learning Objectives

Despite their tendency for greater complexity and higher processing demands, we suggest that dynamic visualizations and simulations have considerable potential as tools for effective assessment. However, as demonstrated by research on learning from animations and multimedia, the potential of these representations is not necessarily fulfilled in practice. In assessment contexts, dynamic visualizations and interactive presentations require students to process both what they are seeing (i.e., the representation of the scientific phenomenon as an animation or simulation) and what they are being asked to do (i.e., the assessment prompt). If the processing demands of understanding the representation and interpreting the prompt are not clearly aligned with the to-be-assessed science practice, the test items will not be valid indicators of student proficiency. For example, if a task depicts a visualization of the scientific phenomena with more details than necessary for the student to respond, or if the response format requires complicated maneuvers, the student may not attend to the relevant features of the scientific representation or may not understand how to produce the response. Thus, a student's response to the prompt may not be a valid indicator of his or her science proficiency, but may instead reflect his or her difficulty processing the item.

The evidence-centered assessment design framework offers an approach for increasing the probability that student responses provide accurate indicators of proficiency for to-be-assessed skills (Mislevy, Almond, & Lukas, 2003). For assessment of science learning, the aim is to develop tasks that elicit evidence of the specified science content knowledge and practice skills. The evidence-centered assessment design framework structures an argument by linking claims about student proficiency with evidence elicited by principled tasks to support inferences about those student proficiencies. Essential components of the evidence-centered design framework are the student model, the task model, and the evidence model.

The student model provides a detailed specification of the knowledge and skills that are targeted by the assessment. These specifications focus the design of tasks and items on the essential stimuli and response formats necessary for demonstrating the targeted knowledge and skills. In the study we describe in the final section of the chapter, our student model specified science practice skills (e.g., identifying principles, using principles, and using inquiry) in the context of science content knowledge described by science system model levels (e.g., components, interactions, emergent processes) For instance, in the ecosystems context, *identifying principles* may be more specifically stated as "identify the role of components of an ecosystem as consumers, producers or decomposers". Similarly, *conducting inquiry* could be specified as "design an experiment to test the impact of an increasing shrimp population on the algae population".

The task model outlines the features of tasks that will elicit evidence of student proficiency related to the student model. Creating the task model requires the developer to align the processing demands of the task stimuli and responses with the targeted knowledge and skills. The developer must ensure that fidelity of the task's representations of the scientific phenomena and the use of multimedia are appropriate for the targeted knowledge. The developer must also ensure that the task processing demands and response formats are appropriate for eliciting evidence that the student can recognize and actively apply the targeted knowledge and inquiry skills.

Finally, the evidence model specifies exactly how the student response will be used to estimate proficiency. For instance, on a multiple-choice test, an item would have one correct response (e.g., "c"), and we would assume student proficiency was higher if the student selected the correct response than if the student selected an incorrect response. However, online, dynamic, and interactive environments enable students to demonstrate a range of correct strategies rather than confining them to a single correct response. For instance, an effective experimental design may require students to hold one variable constant while another is varied. The computer would capture the "observable variables," that is the specific values the student selected for each trial. The evidence model would then specify that proficiency was high if different values were selected for the manipulated variable but the same value was maintained for the control variable. This logic may be programmed into the assessment for automated scoring. Rather than having a single correct approach to designing an experiment, the online format allows students to demonstrate proficiency though they may select different values in their experiments or choose to run trials in a different order.

9.3.3 Minimize Extraneous Processing Demands

Although dynamic visualizations and simulations have features that make them well-suited for science learning and assessment, the additional information they present may also distract or overwhelm students. Multimedia learning researchers have examined the effects of pictorial and verbal stimuli in static, dynamic, and interactive formats, as well as the effects of active versus passive learning enabled by degrees of learner control (Clark & Mayer, 2011; Mayer, 2005; Lowe & Schnotz, 2008). Mayer (2014) and Clark and Mayer (2011) summarize multimedia research and offer principles for multimedia design.

Many multimedia design principles address how to focus students' attention and minimize extraneous processing as students engage with learning materials. Research suggests attention should be specifically guided by making the most important information salient and omitting irrelevant representations (cf. Betrancourt, 2005; Clark & Mayer, 2011; De Koning & Jarodzka, 2017, this volume; Mayer & Fiorella, 2014). The use of visual cues such as text consistency, color, and arrows can help students map between representations and gain a deeper conceptual understanding (cf. Ainsworth, 2008; Boucheix & Lowe, 2010; Kriz & Hegarty, 2007; Larkin & Simon, 1987; Lowe & Schnotz, 2008; Pedone, Hummel, & Holyoak, 2001).

Research specific to animations and interactive environments finds that the temporal nature of dynamic displays can increase perceptual and cognitive demands on learners (cf. Lowe & Boucheix, 2017, this volume; Lowe & Schnotz, 2014). Although more research is required to determine what types of supports are best suited to which learners and content, some research suggests it may be helpful to: (1) align the task requirements with the presentation format, (2) allow learners to replay animations, (3) signal upcoming changes, and (4) ensure that the fidelity of the display is appropriate for the task. Task-format alignment suggests that dynamic or interactive features should be included only when they are required for the task. Although not always effective (e.g., Boucheix, 2008; Lowe, 2008), allowing dynamic presentations to be paused, rewound, and replayed offers students multiple opportunities to extract information from the display (cf. Lowe & Schnotz, 2008; Schwartz & Heiser, 2006). Signaling complex animations by giving cues such as "there will be three steps" and directly instructing students to reason through the components of systems increases student comprehension (Hegarty, 2004; Schwartz & Black, 1999; Tversky, Heiser, Lozano, MacKenzie, & Morrison, 2008). Mayer and Johnson (2008) found that redundancy of text in multimedia presentations may be beneficial when on-screen text is short, highlights the key action described in the narration, and appears next to the portion of the graphic that it describes to highlight salient features of a multimedia presentation. Finally, the fidelity principle suggests that the complexity of a simulation should be appropriate for the learner outcomes. Rather than realistically portraying every detail of a system, it is more important to ensure that the most relevant parts of the system are easily discernible (cf. Lee, Plass, & Homer, 2006; van Merrienboer & Kester, 2005). Lowe and Boucheix (2017, this volume) suggest that learners' inability to extract key aspects of presented information in an animation may be due to fundamental inadequacies of the animations' design. They recommend that a principled approach be taken in the initial design of the animations such that key information presented in the animation is progressively built up into a task-appropriate mental model of the referent in order to avoid the need for post hoc remediation.

9.3.4 Validate the Evidence Model

Evidence-centered design requires making inferences that tasks will elicit desired student knowledge. Because processing demands may be hard to fully predict, assessments that include dynamic visualizations and interactive simulations require rounds of revisions to produce effective assessments. These revisions should be data-driven. Expert reviews, think-alouds, and feasibility studies can be used to ensure the validity of the items. Expert reviews can validate factors such as scientific accuracy, grade-level appropriateness, and alignment with standards. Student think-alouds and follow-up interviews can provide data about student problem solving and allow developers to verify that the items elicit the expected knowledge and skills and minimize extraneous processing. Finally, classroom feasibility tests ensure the assessments function in authentic settings using existing equipment.

9.4 Why Use Dynamic Visualizations and Interactive Simulations for Science Assessment?

As suggested by research in this volume and elsewhere, carefully-designed animations and interactive simulations have the potential to display the complexity and dynamics of science systems. In a number of science contexts, simulations have been shown to support the development of deeper understanding and problem-solving skills in areas such as genetics, environmental science, and physics (cf. Adams et al., 2008; Buckley et al., 2010; Horwitz et al., 2010; Krajcik et al., 2000; Schwartz & Heiser, 2006; Zacharia, 2007). For instance, students using an aquatic ecosystem simulation or a collective simulation of multiple human body systems were able to demonstrate causal connections among the levels of these systems (Hmelo-Silver et al., 2008; Ioannidou et al., 2010; Slotta & Chi, 2006; Vattam et al., 2011). Using a computer-based simulation tool that allowed students to create, test, and revise models helped students develop more robust and transferrable modeling skills than worksheet-based instruction (Papaevripidou, Constantinou, & Zacharia, 2007).

A key difference between dynamic visualizations and simulations for learning versus assessing is that the design of *learning* environments focuses on the stimulus features (e.g., what is being presented to the student), whereas the design of *assessment* environments requires thoughtful consideration of the evidence of learning that can be elicited by both the stimulus (i.e., the item prompt that presents the question or task), and the response (i.e., a means of capturing student behavior related to the prompt). Dynamic and simulation-based assessments enable new kinds of prompts as well as different types of responses. As part of the prompt, dynamic visualizations can provide an explicit spatiotemporal representation of the complex scientific phenomena to be understood and so relieve students from having to mentally animate models of science systems. As part of a response, dynamic

Table 9.1 Levels of student/interface interactivity with stimulus and response features and examples

Level of interactivity	Stimulus features	Stimulus example	Response features	Response examples
Static	All information is available at the outset	Text, still images, other visual representations	Responses do not change what is on screen	Selected responses (multiple choice, true/false) or constructed responses (text, drawings)
Active	Not all information available at the outset, but same information unfolds over time	Animations	May move items spatially or change place in animation	Drag and drop, user control of animation pace and direction
Interactive	Information present is contingent on student responses	Simulation environment	Student actions alter subsequent available information	Set parameters, collect data, create visualizations

visualizations and simulations allow students to manipulate change processes (see also Lowe, Boucheix, & Fillisch, 2017, this volume). As we move from static text and images to animations and simulations, the level of possible interactivity between the user and the display on the screen changes. Quellmalz et al. (2013) provides a framework that categorizes levels of interactivity as static, active, or interactive. Table 9.1 shows a summary of how the stimulus and response features of the differing levels of interactivity as the students engage with an environment to complete an assessment task. The level of interactivity relates to what information is available at the outset to students and how students may interact to change and manipulate what information is provided. As shown in the table below, the level of interactivity may not be the same for the stimulus and response of a single item. For instance, an active prompt, such as an animation, could have a static response if students use multiple-choice to describe their observations. Here, we briefly describe the three levels of interactivity and their common stimulus and response features.

9.4.1 Static

The level of interactivity is static when the information does not change over time and students are not able to alter the information that is presented. A static item stimulus, or prompt, can include text or other visual representations; however, all students will see the same information and the displayed information remains constant over time. Responses are considered static if the responses do not alter the prompt. Static responses can be either selected or constructed responses. Selected responses require students to recognize correct response options presented and to indicate their selection with radio buttons, check boxes, or dropdown menus. Constructed response formats require additional cognitive processing as students must generate rather than recognize the correct answer. Formats for constructed responses can range from writing words to drawing arrows, diagrams, lines, tables or charts.

9.4.2 Active

The level of interactivity is active when the information available changes over time, such as in the case of animations or other dynamic visualizations. Animations are considered particularly useful for providing visualizations of dynamic phenomena that are not easily observable in real space and time scales such as plate tectonics and animal movement (Betrancourt, 2005; Kühl, Scheiter, Gerjets, & Edelmann, 2011). Thus, active stimulus features are ideally suited to provide opportunities for students to make observations in realistic science contexts. An additional type of active stimulus is in the form of tools that can minimize errors due to perceptual limitations. For instance, in the SimScientists project, items that require students to retrieve information from graphs include a Data Inspector, a tool that allows students to drag a point through a graph and view the exact values of the graph at different points. This tool minimizes possible error due to trouble reading precise values from the axis. As shown in Fig. 9.1, students can drag the gray triangle and the system will produce the value of the graph at that point and populate the table. This level of interactivity is active rather than interactive because all students see the same values at the same points.

Responses are active when a student is able to actively manipulate the display or prompt. When the prompt is an animation, students may pause or scroll to a desired frame to indicate understanding (e.g., "freeze the animation when a reaction occurs." Online formats offer increased options for active responses with drag-and-drop functionalities that allow students to label or categorize by moving objects around the screen. In this way active responses can minimize the need to process and produce verbal information and allow students to more directly demonstrate spatial understanding.

9.4.3 Interactive

The highest level of interactivity is the interactive format. In contrast to animations, interactive simulations provide environments that allow learners to manipulate parameters as they generate hypotheses, test them, and see the outcomes, therefore taking advantage of technological capabilities well-suited to conducting scientific

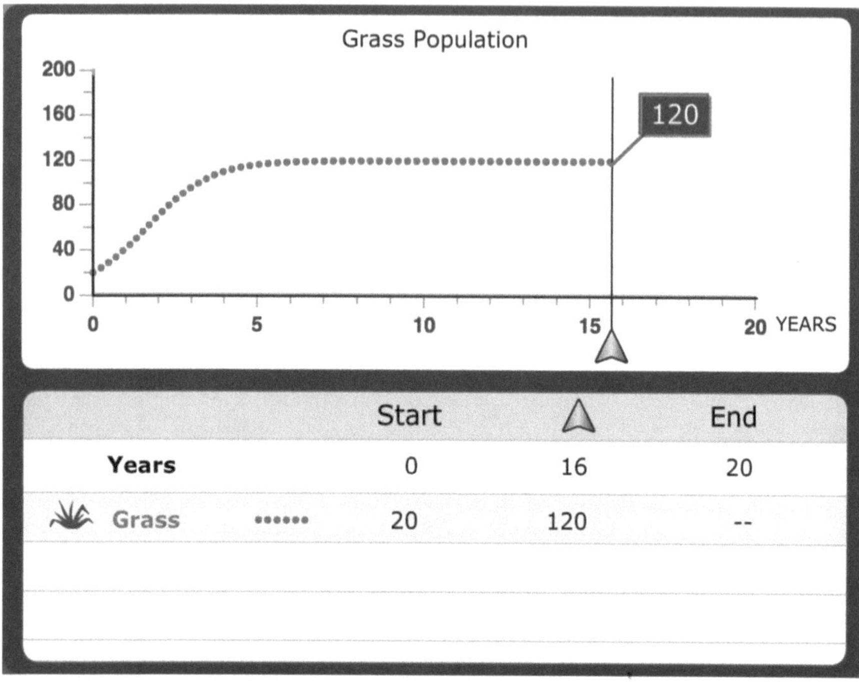

Fig. 9.1 Example of an active stimulus tool, the data inspector, displaying the y-axis value

inquiry. A stimulus is interactive if the system is responsive to student input and the feedback is contingent to the options a student selects or produces. Simulation environments can range from virtual laboratories to dynamically generated graphs to platforms that allow students to construct their own models. Responses are interactive when students must engage with a simulation to actively create new information, such as designing and running trials for an experiment. For example, Rieber, Tzeng, and Tribble (2004) created an environment that provided graphical feedback during a simulation on laws of motion. Plass, Homer, and Hayward (2009) provided students with environments in which they could actively manipulate contents of a visualization, not just the timing and pacing. The manipulations students make by changing variables provide evidence of students' knowledge of the variables' likely effects and of understanding of the design of investigations. Students' interpretations and explanations of the data generated provide evidence of science knowledge and investigation practices.

9.4.4 Examples of Tasks that Elicit Science Practice Skills

How do differing levels of interactivity provide evidence for different science practice skills? We provide some illustrative examples of how active and interactive formats can be leveraged to measure proficiency from two science assessment platforms: SimScientists (simscientists.org) and ChemVLab+ (chemvlab.org; Davenport et al., 2012a, b, 2014a, b; Quellmalz et al., 2011). Although there is some overlap between the processing demands across the science practices, in general, higher levels of interactivity are more appropriate for eliciting evidence of more cognitively demanding tasks. Carefully-designed static items are able to tap both declarative and schematic knowledge when they require students to recall facts and generate explanations of science systems. However, static items are less well-suited to providing evidence that students are able to make predictions based on observations of dynamic systems or carryout multistep inquiry. The SimScientists system provides formative and summative simulation-based science assessments for middle school students in the life, earth, and physical sciences. The ChemVLab+ system provides formative assessment for key concepts in high school chemistry. In each example, we discuss features designed to minimize extraneous processing irrelevant to eliciting evidence of the knowledge and skills to be assessed (cf. Table 9.2).

Active Items Active items are well suited to displaying dynamic science systems and are able to provide students with response formats that allow them to demonstrate their ability to apply knowledge of science principles to accomplish a task. For instance, a core understanding in chemistry is that temperature is a measure of the average kinetic energy of particles in a system. Figure 9.2 provides an example from the ChemVLab+ system that requires students to observe the speed of particles and use those observations to order the systems from lowest to highest temperature. We fostered highly targeted processing by allowing students to sequence boxes representing the systems directly, rather than having to map to an additional label (e.g., selecting the correct order with a multiple choice prompt).

Testing the ability to make observations from dynamic systems is another beneficial use of active assessment items. For instance, to determine predator-prey relationships, students must understand that the act of eating determines which species is predator and which is prey. A common misconception is that larger organisms always eat smaller organisms (Gallegos, Jerezano, & Flores, 1994; Reiner & Eilam, 2001). Providing an animation of an ecosystem in action allows us to evaluate whether students are able to apply science principles to draw conclusions about the role of different organisms. In the SimScientists project, we provided an animation that depicts a mountain lake (see Fig. 9.3). In this system, the large fish is an herbivore, so students must observe carefully to avoid categorizing it as a predator of the smaller organisms. We used recommendations from the multimedia learning literature to incorporate features that should facilitate relevant processing. These features include visual mapping of the shapes of the organisms next to the labels in the legend and highlighting that appears on both the organism and label when students mouse over either. Further, because animation research has found that user control

Table 9.2 Response types aligned with cognitive demands and science practice skills

Format	Type of response	Cognitive demands	Science practice	Evidence elicited
Static	Selected responses (multiple choice, true/false, check boxes)	Declarative/schematic	Identifying or using principles	Recognize principles/facts, recognize correct conceptions or explanations in the context of misconceptions
	Constructed responses (written responses, drawings)	Declarative/schematic	Identifying or using principles	Reveal whether students can recall facts or generate explanations
Active	Drag and drop	Declarative/schematic	Identifying or using principles	Reveal ability to label, sort, categorize (can be verbal or non-verbal)
	Scroll/move slider	Declarative/schematic	Identifying or using principles	Reveal ability to categorize observations
Interactive	Can actively design experiments and collect data	Procedural/strategic	Using inquiry	Reveal whether students can design (as opposed to recognize) or plan a correct design
	Carry out complex procedures involving multiple steps	Procedural/strategic	Using inquiry	Reveal whether students select appropriate tools and carryout procedures correctly
	Opportunity to correct errors after running simulations	Strategic	Using technological design	Reveal metacognitive skills, where students evaluate whether a result was expected
	Can design their own simulation	Strategic	Using technological design	Reveal student conceptions about complex systems

may be efficacious, students are given the opportunity to play and pause the animation as they wish and are able to use the scrub bar to view specific parts of the animation.

Interactive Assessment Interactive assessment formats provide students with the opportunity to design and carryout experiments rather than merely recognize pre-

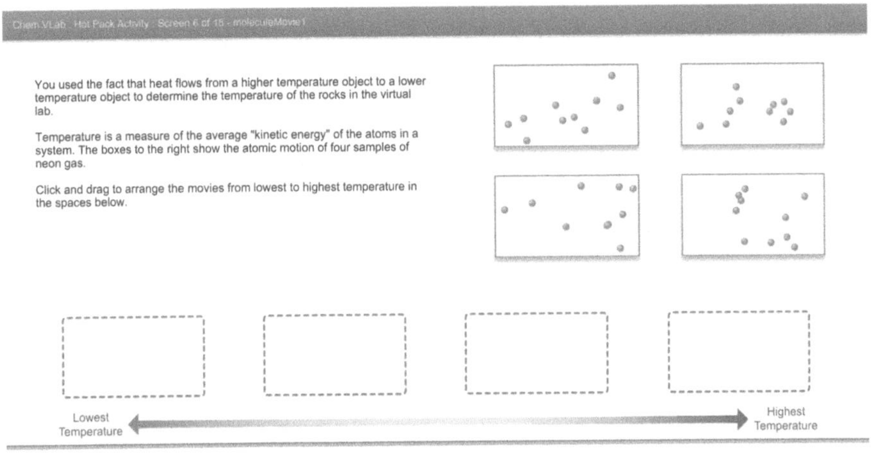

Fig. 9.2 Students sequence boxes showing moving particles (*top right*) by dragging them to the spaces below

Fig. 9.3 Students observe a lake ecosystem with the aim of determining predator-prey relationships

determined designs as correct or otherwise. At the heart of the ChemVLab+ activities is a virtual chemistry lab that allows students to select appropriate glassware, choose then mix chemicals, and see the results of their experiments through various types of instrumentation in the lab environment. For instance, Fig. 9.4 shows an introductory item that requires students to mix chemicals to determine whether a solid forms or not. Students are able to demonstrate inquiry skills by mixing chemicals and evaluating the results to determine what compounds are produced. In the example below, the student response requires both interaction with the virtual

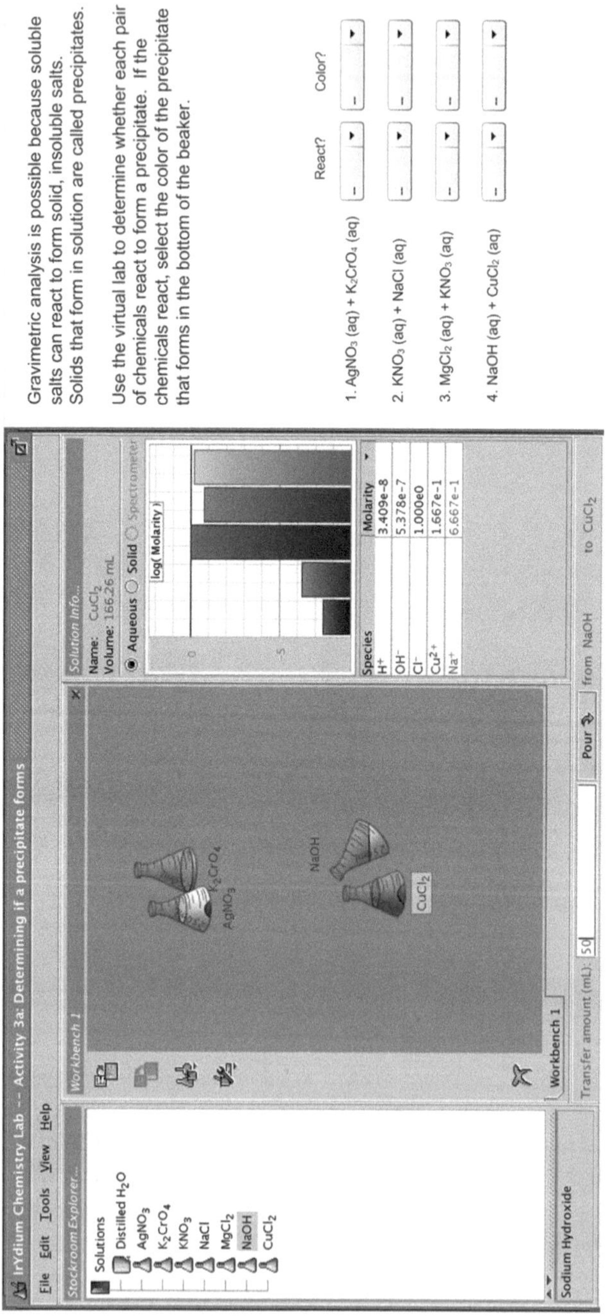

Fig. 9.4 Students carry out experiments using a virtual chemistry lab in the ChemVLab+ system

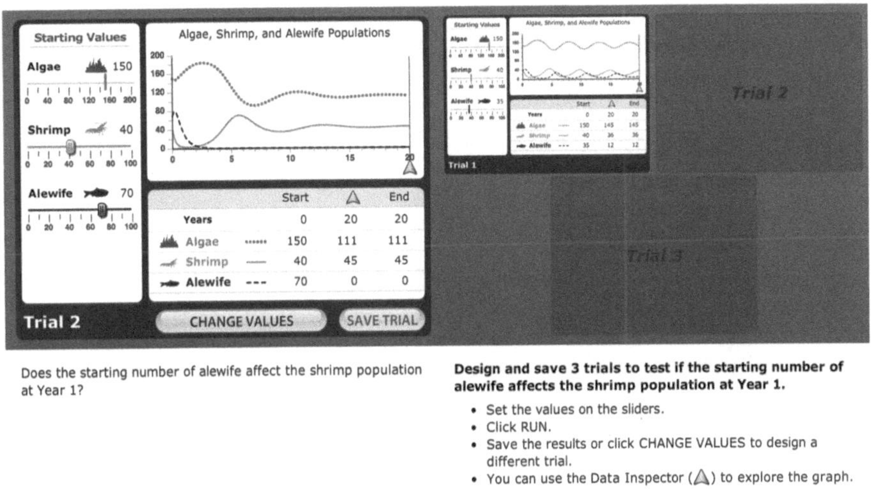

Fig. 9.5 Students design an experiment by setting parameters, running, and saving trials

chemistry lab as well as the selection from alternatives to ensure they correctly interpret the results of their experiments. The design of the virtual chemistry lab facilitates task-relevant processing by using color-coding to link the concentrations with the chemical names and proximity of the labels to the chemicals so students can identify the substances in the beakers.

Another example of an interactive assessment environment is the SimScientists computer simulation of an ecosystem for determining student proficiency on science inquiry skills. Students are able to adjust sliders to set the parameters of the model, run trials, and select which trials to save (see Fig. 9.5). This task requires both procedural knowledge about how to design experiments and strategic knowledge about how to set the particular values to address the research question. Visual cues, such as color-coding of the graphs and labels and proximity of the labels to sliders are used to support processing.

9.5 Evidence of the Utility of Dynamic and Interactive Assessments

Having a broader range of possible task types should broaden the range of science knowledge and skills that can be discriminated during assessment. To test this hypothesis, we carried out an empirical study that addressed the following research question: Are test formats that include dynamic visualizations and interactive simulations more effective for discriminating between three types of science practice skills described in the NAEP *Science Framework*: identifying principles, using principles and conducting inquiry (NAGB, 2010) than formats that are static? We summarize findings from a large-scale evaluation of computer-based science tests that differed in the level of interactivity.[1] More detailed descriptions of these investigations are available in Quellmalz et al. (2013) and DeBoer et al. (2014).

To determine whether different levels of interactivity led to richer assessment information about student proficiency, we created three parallel versions of a middle school ecosystem science assessment. Each student undertook each of the three versions of the assessment and we used analytic techniques to determine whether the three test formats differed in their capacity to discriminate between proficiency on the science practice skills. The static assessment was most similar to traditional tests in that the items consisted of still text and images. For static assessments, all students view the same information presented in images and text and the information remains unchanged as students process it and respond. The response options could be either selected or constructed responses and could include selecting from drop-down menus, yes/no, and writing words or sentences. The active assessments took advantage of the computer-based format to include animations that students could play (and replay) to make observations while viewing dynamic processes in action. In active assessments, not all information was available at the outset and students had to initiate the dynamic presentations. The information in the item changed over time (e.g., as the animation ran), however the information available was not contingent on a specific response by the student. Compared to the static response options, active items provide the additional response facility of allowing students to initiate animations and pause or replay animations. Finally, the interactive assessment included simulations that allowed students to manipulate parameters and observe the resulting output. Interactive items allow students to get feedback on their actions by observing changes in the portrayal of the science phenomena and information that is contingent on student input.

9.5.1 Test Design

All test formats were designed using the evidence-centered assessment design process described earlier in the chapter. To ensure content knowledge differences were not confounded with science practices, we created three parallel versions of the assessment using three ecosystem contexts (tundra, grasslands, and mountain lake) and maintained the structure of the tasks across the three item types. The tundra context was presented in the static format, the grasslands in the active format, and the mountain lake in the interactive format. Each assessment consisted of twenty-four tasks; six testing *identifying principles*, six *using principles*, and twelve *conducting inquiry*. Because there were more component skills related to conducting inquiry (e.g., design, interpret, explain) we developed more of these tasks and items to gather evidence of the scope of inquiry skills.

[1] From Quellmalz et al., 2013. Copyright 2014 by American Psychological Association. Adapted with permission.

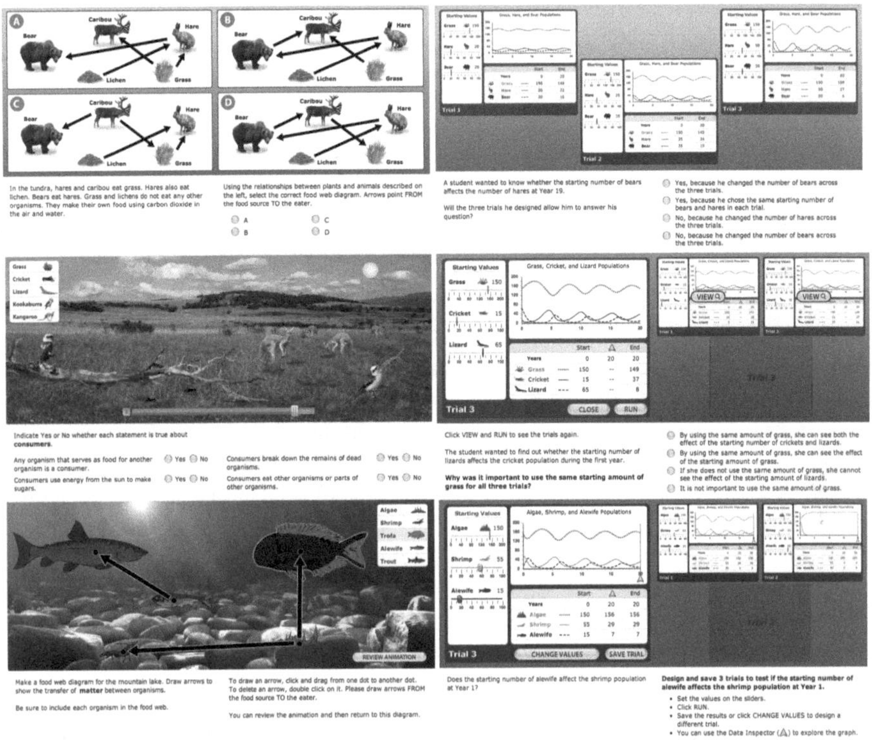

Fig. 9.6 Examples of tasks tapping into using principles (*left*) and using inquiry (*right*). The *top row* shows static items, the *middle row* shows active items, and the *bottom row* shows interactive items (From Quellmalz et al.,2013. Copyright 2014 by Amercian Psychological Association. Adapted with permission)

Figure 9.6 provides an example of parallel items in the different formats. The left column shows items tapping into the science practice of using principles by applying knowledge about predator and prey relationships to identify or create a food web. In the static format (top left), students read text descriptions about the interactions of organisms (components) in an ecosystem and were asked to select the static image of the correct food web. In the active format (middle left), students observed an animation of organisms in an ecosystem to infer organism roles (consumers, producers) and then drew a food web diagram. Students could replay the animation. In the interactive format (bottom left), students observed the animation of an ecosystem and could take advantage of additional interactivity that used highlighting on demand to cue the connection between the names and pictures of the organisms. The potential confounding effects of nesting the assessments in a modality within a specific ecosystem (tundra, grasslands) were minimized both by maintaining the same structure of tasks and items for each ecosystem and by the focus on the assessment of the science practices, not on knowledge of features of specific organisms or interactions that would differ between ecosystems.

The right column shows items tapping into using inquiry by demonstrating knowledge of designing an experiment. In the static format (top right), students viewed the outcomes of an investigation and were asked to evaluate an experimental design. In the active format (middle right), the student evaluated the design of an investigation after watching an animation of data being generated. In this active format, students did not select inputs for the simulation, they only watched the simulation run. Finally, in the interactive format (bottom right), students designed their own investigations by setting the inputs for the simulation and running and saving their own trials.

The design process involved refinement of the items based on expert reviews and think-alouds. For the expert reviews, we engaged three experts from the American Association for the Advancement of Science (AAAS). These experts independently reviewed the items and judged whether each item was aligned with one of the targeted science practices of the NAEP Science Framework: *identifying principles*, *using principles*, or *conducting inquiry* (NAGB, 2010). These experts also verified that the items were scientifically accurate, grade-level appropriate, usable, and comparable across the static, active, and interactive versions. Initially, AAAS staff reviewed the storyboards of draft items and provided detailed comments and feedback. An additional iteration of review and revision was carried out with the programmed items to ensure the final items remained aligned with targeted science practices.

The think-aloud studies allowed us to determine whether items elicited the targeted science inquiry practices. Each of ten students completed all three forms of the assessments (static, active, and interactive). As students completed the assessments, they "thought aloud" by saying everything they were thinking while screen capture software recorded students' verbalizations and actions on the screen and researchers coded whether the items elicited the targeted construct. The think-aloud studies had two goals: (1) to ensure the usability of the assessments as deployed, and (2) to provide evidence of construct validity by determining whether the questions were eliciting student thinking and reasoning about the intended science practice constructs. To ensure the items would be usable in the field test, researchers took detailed notes of usability issues that arose (e.g., navigation, difficulty running experimental trials) and modified the items to address these issues. To examine the items' construct validity, the observing researcher coded whether the item prompted student thinking related to the targeted science practice constructs. These data provided one form of evidence that the items were aligned with their intended content and inquiry targets.

9.5.2 Study Design

A total of 1566 students from the classrooms of twenty-two middle school teachers in twelve states in the United States took all three versions of the assessments on three consecutive days. The order of the assessments was counterbalanced between students. The assessments were delivered online using the SimScientists Learning Management System (Quellmalz et al., 2011). The SimScientists system provides detailed reports to teachers and allows researchers to download de-identified student data.

9.5.3 Data Analysis

Following data collection, we analyzed student responses using three different analytic techniques: (1) a generalizability study (G-study), (2) multitrait-multimethod confirmatory factor analysis, and (3) a multidimensional item response theory (IRT) model. Each of these methods takes a different approach to modeling the data with successively stronger statistical assumptions. A G-study treats items as randomly sampled from all possible items that could have been created and makes no statistical assumptions beyond minimal assumptions that certain error components are uncorrelated. The Multitrait-Multimethod Confirmatory Factor Analysis uses a theoretical model about the relationships between the science practices and assessment formats and operates on nine composite scores for item/assessment clusters (e.g., a component score for each of the practices at each assessment format). Finally, the Multidimensional Item Response Theory models item difficulty on the same scale as student ability and invokes assumptions about how the items are grouped into scales. Because Quellmalz et al. (2013) provides detailed information about these analyses, this chapter is limited to a summary of the techniques and key findings from our study.

9.5.4 G-Study

A G-study is an analytic technique that models the reliability of responses and helps to identify how much error in a data set can be attributed to different factors (e.g., students and items) or the interaction between those factors (cf. Webb & Shavelson, 2005). Multivariate G-studies allow us to investigate measures tapping multiple constructs (e.g., conducting inquiry likely also taps into using principles), accounting for sources of correlated error across constructs. In the studies reported here, G-studies, conducted using the mGENOVA computer program (Brennan, 2001) indicate the magnitude of error variance components attributable to items as well as the interaction of persons by items plus residual variance.

The first set of G-study analyses treated the nine (3 assessment format × 3 science practice) combinations as separate constructs and ignored the fact that all assessments were designed to measure the same practices. Table 9.3 shows the estimated correlations between the science practice skills separately for each of the assessment formats. Our assumption in creating the tests with differing levels of interactivity was that more interactive assessments provide better indicators of the

Table 9.3 Estimated correlations among the three science practices for the three formats of presentation

Correlations	Static	Active	Interactive
Identifying-using	.92	**.80**	.82
Identifying-conducting	.80	.80	**.72**
Using-conducting	.91	1.00	**.84**

cognitive demands associated with using principles and using inquiry. If the different practices do indeed tap into different skills, we expect lower correlations between practices for the assessment format that most clearly distinguishes between those distinct skills.

The lowest correlation for each pair of science practice constructs is shown in bold in Table 9.3. Two of the three lowest correlations among the three science practice constructs (.72 for *Identifying/Conducting* and .84 for *Using/Conducting*) are from the interactive assessment (mountain lake) format. The active (Grasslands) assessment format produced the lowest correlation of .80 for *Identifying/Using*, which was slightly lower than the .82 for the interactive format. Conversely, two of the highest correlations (.92 *Identifying/Using* and .91 *Using/Conducting*) are for the static format (tundra), suggesting the static format was not effective at distinguishing between the science practice skills. Further, the highest correlation (1.0 *Using/Conducting*) is for the active format (grasslands), suggesting that using dynamic visualizations (without interactivity) was not able to distinguish between students' abilities to apply science knowledge and their abilities to conduct inquiry.

Overall, the results from the G-study suggested that the interactive format measured the science practice constructs more distinctly and was particularly able to distinguish *Conducting Inquiry* as a clear construct, which the static and active formats measured less well.

9.5.5 Multitrait-Multimethod Confirmatory Factor Analysis

The second analytic technique was a multitrait-multimethod confirmatory factor analysis (Campbell & Fiske, 1959; Loehlin, 1998) that separates out the variance due to the traits (underlying abilities on the three science practices skill) from the variance due to the method of assessment (static, active, or interactive format). The confirmatory factor analyses correlates measurements of the three practices across the three assessment formats and produces a multitrait-multimethod matrix that depicts the likelihood for the different assessment formats to converge on the same science practices (convergent validity) and the likelihood for the different formats to distinguish between the different science practices (discriminant validity).

Consistent with our hypothesis, the format of the assessment affected how well items were able to draw out students' knowledge and skills in the three science practices. The results indicated that the assessment formats affected student scores more than the science practice construct. Overall, the factor loadings from the test format to the three constructs were slightly less for the interactive formats (.698 on average) than for the static (.732 on average) and active formats (.719 on average). These findings supported our hypothesis that the task items in the interactive modality measure the science practice constructs more distinctly than items in the other two modalities. In particular, the loading for the interactive modality (.673) was considerably lower than for the static (.702) and active (.739) modalities. These results are consistent with the finding in the G-study that the interactive modality more distinctly measures the Conducting Inquiry construct.

9.5.6 Multidimensional IRT Model

The final technique we used was a multidimensional IRT model to determine how well each assessment format could distinguish student performance on three science practice constructs. These probabilistic models simultaneously estimate the difficulty of items (based on the number of students who responded correctly) and the proficiency of the students (based on the number of items individual students responded to correctly). The estimates produce a scale that maps both students and items onto the science practice constructs. As seen in the G-study and the confirmatory factor analysis, the differences between the assessment formats were relatively small for the *identifying principles* and *using principles* constructs, but the interactive modality had a higher reliability coefficient (.82) for the *conducting inquiry* science practice. This result provided additional evidence that the interactive assessment format was better able to distinguish between using inquiry and the other two science practices than the other two assessment modalities.

9.6 Discussion

The study provided rare, large-scale evidence that interactive assessments may be more effective than static assessments at discriminating student proficiencies across different types of science practices. Previous studies comparing item formats have primarily been within the static format, using fixed images and text but comparing selected versus constructed responses, or between complex performance assessments and conventional tests. This study extended the comparison of task and item design to complex tasks involving inquiry practice constructs and the dynamic and interactive affordances of technology-based complex science assessment tasks. The three assessment formats (static, active, and interactive) were carefully constructed to keep the representations of the science ecosystem phenomena parallel. All three assessments depicted the ecosystems with similar styles of pictures of the organisms, tables, graphs, and screen layouts, so that the level of interactivity (static, active, or interactive) was the primary variable.

As the use of technology in education increases, so does the opportunity to create assessments that measure skills that are hard to assess in traditional, static formats. Our findings suggested that engaging students through interactive assessments may provide better estimates of their ability to apply complex science practices than assessments relying solely on active or static formats. In addition, the project offered initial guidelines for: (1) designing the next generation of innovative science assessments, and (2) processes to ensure the affordances of dynamic, interactive, complex assessment tasks are used effectively to represent dynamic science phenomena and elicit evidence of the targeted science knowledge and practices tested. The design methodology for developing effective assessments must begin with a deep understanding of the processing demands of the to-be-assessed domain knowledge and skills and then proceed to design task and evidence models to ensure that student performance on the tasks reflects proficiency on the targeted concepts and skills rather than being a reflection of the processing challenges imposed by traditional forms of assessment.

The research literature to date has very much focused on dynamic visualizations and simulations as learning environments, with the main focus on the *stimulus* demands of these representations. However, for assessment these types of representations also offer novel *response* formats that have the potential to yield previously elusive evidence of understanding and proficiency for a wide range of science practice skills. To fulfill this potential, research is needed into the perceptual and cognitive processing involved in using such representations as tools for eliciting learner responses as evidence of specified knowledge and skills rather than just as stimuli for instruction. There is considerable research to be done on the functions of multiple representations and interactive interfaces in learning and assessment of science systems and practices (Buckley & Quellmalz, 2013). We believe ongoing research can make a significant contribution to the field by moving the state of technology-based assessment development to more principled practice. Research addressing model-based reasoning and multimedia learning can inform the design of tests that go beyond retrieval of declarative knowledge to capture evidence of integrated system thinking, active problem solving, and transfer to new domains.

Acknowledgements This material chapter is based upon work supported by the National Science Foundation under Grant No. DRL-0814776 awarded to WestEd, Edys Quellmalz, Principal Investigator and work supported by the Institute of Education Sciences, through Grant R305A100069 to WestEd, Jodi Davenport, Principal Investigator. Any opinions, findings and conclusions or recommendations expressed in this material chapter are those of the author(s) and do not necessarily reflect the views of the National Science Foundation or the U.S. Department of Education.

References

Adams, W. K., Reid, S., Lemaster, R., McKagan, S. B., Perkins, K. K., Dubson, M., et al. (2008). A study of educational simulations part 1: Engagement and learning. *Journal of Interactive Learning Research, 19*, 397–419.

Ainsworth, S. (2008). The educational value of multiple-representations when learning complex scientific concepts. In J. K. Gilbert, M. Reiner, & M. Nakhleh (Eds.), *Visualization: Theory and practice in science education* (pp. 191–208). New York: Springer.

Bétrancourt, M. (2005). The animation and interactivity principles in multimedia learning. In R. E. Mayer (Ed.), *The Cambridge handbook of multimedia learning* (pp. 287–296). New York: Cambridge University Press.

Boucheix, J.-M. (2008). Young learners' control of technical animations. In R. K. Lowe & W. Schnotz (Eds.), *Learning with animation: Research implications for design* (pp. 208–234). New York: Cambridge University Press.

Boucheix, J.-M., & Lowe, R. K. (2010). An eye tracking comparison of external pointing cues and internal continuous cues in learning with complex animations. *Learning and Instruction, 20*, 123–135.

Bransford, J. D., Brown, A. L., & Cocking, R. R. (2000). *How people learn: Brain, mind, experience, and school*. Washington, DC: National Academy Press.

Brennan, R. (2001). mGENOVA. Retrieved from http://www.education.uiowa.edu/centers/casma/computer-programs.aspx.

Buckley, B. C., Gobert, J., Horwitz, P., & O'Dwyer, L. (2010). Looking inside the black box: Assessing model-based learning and inquiry in BioLogica. *International Journal of Learning Technologies, 5*, 166–190.

Buckley, B. C., & Quellmalz, E. S. (2013). Supporting and assessing complex biology learning with computer-based simulations and representations. In D. Treagust & C. Y. Tsui (Eds.), *Multiple representations in biological education* (pp. 247–267). New York: Springer.

Campbell, D. T., & Fiske, D. W. (1959). Convergent and discriminant validation by the multitrait-multimethod matrix. *Psychological Bulletin, 56*(2), 81–105.

Clark, R. C., & Mayer, R. E. (2011). *E-learning and the science of instruction: Proven guidelines for consumers and designers of multimedia learning*. San Francisco, CA: Pfeiffer.

Cuadros, J., Leinhardt, G., & Yaron, D. (2007). One firm spot: The role of homework as lever in acquiring conceptual and performance competence in college chemistry. *Journal of Chemical Education, 84*, 1047–1052.

Davenport, J. L., Rafferty, A., Timms, M. J., Yaron, D., & Karabinos, M. (2012a). ChemVLab+: Evaluating a virtual lab tutor for high school chemistry. In J. van Aalst, K. Thompson, M. J. Jacobson, & P. Reimann (Eds.), *Proceedings of the 10th international conference of the learning sciences* (pp. 381–385). Sydney, Australia: International Society of the Learning Sciences.

Davenport, J. L., Timms, M. J., Yaron, D., & Karabinos, M. (2012b) *ChemVlab+: Supporting science learning using a virtual chemistry lab with embedded assessments*. Paper presented at the 2012 Annual Meeting of the American Educational Research Association in Vancouver, BC.

Davenport, J.L., Powers, J., Rafferty, A., Timms, M., Karabinos, M., & Yaron, D. (2014a, April). *Classroom factors and students learning with online virtual chemistry lab activities*. Paper presented at the 2014 Annual Meeting of the American Educational Research Association in Philadelphia, PA.

Davenport, J. L., Rafferty, A., Yaron, D, Karabinos, M., & Timms, M. (2014b, April). *ChemVLab+: Simulation-based lab activities to support chemistry learning*. Paper presented at the 2014 Annual Meeting of the American Educational Research Association in Philadelphia, PA.

De Jong, T. (2006). Technological advances in inquiry learning. *Science, 312*(5773), 532–533.

De Koning, B. B., & Jarodzka, H. (2017). Attention guidance strategies for supporting learning from dynamic visualizations. In R. Lowe & R. Ploetzner (Eds.), *Learning from dynamic visualization – Innovations in research and application*. Berlin: Springer (this volume).

DeBoer, G. E., Quellmalz, E. S., Davenport, J. L., Timms, M. J., Herrmann-Abell, C. F., Buckley, B. C., et al. (2014). Comparing three online testing modalities: Using static, active, and interactive online testing modalities to assess middle school students' understanding of fundamental ideas and use of inquiry skills related to ecosystems. *Journal of Research in Science Teaching, 51*, 523–554.

Gallegos, L., Jerezano, M., & Flores, F. (1994). Preconceptions and relations used by children in the construction of food chains. *Journal of Research in Science Teaching, 31*, 259–272.

Goldman, S. R. (2003). Learning in complex domains: When and why do multiple representations help? *Learning and Instruction, 13*, 239–244.

Hegarty, M. (2004). Dynamic visualizations and learning: Getting to the difficult questions. *Learning and Instruction, 14*, 343–351.

Hegarty, M., & Just, M. A. (1993). Constructing mental models of machines from text and diagrams. *Journal of Memory and Language, 32*, 717–742.

Hmelo-Silver, C. E., Jordan, R., Liu, L., Gray, S., Demeter, M., Rugaber, S., et al. (2008). Focusing on function: Thinking below the surface of complex science systems. *Science Scope, 31*(9), 27–35.

Horwitz, P., Gobert, J. D., Buckley, B. C., & O'Dwyer, L. M. (2010). Learning genetics with dragons: From computer-based manipulatives to hypermodels. In M. J. Jacobson & P. Reimann (Eds.), *Designs for learning environments of the future: International perspectives from the learning sciences* (pp. 61–87). New York: Springer.

Ioannidou, A., Repenning, A., Webb, D., Keyser, D., Luhn, L., & Daetwyler, C. (2010). Mr. Vetro: A collective simulation for teaching health science. *International Journal of Computer-Supported Collaborative Learning, 5*, 141–166.

Krajcik, J., Marx, R., Blumenfeld, P., Soloway, E., & Fishman, B. (2000). *Inquiry-based science supported by technology: Achievement and motivation among urban middle school students.* Paper presented at the American Educational Research Association, New Orleans, LA.

Kriz, S., & Hegarty, M. (2007). Top-down and bottom-up influences on learning from animations. *International Journal of Human-Computer Studies, 65*, 911–930.

Kühl, T., Scheiter, K., Gerjets, P., & Edelmann, J. (2011). The influence of text modality on learning with static and dynamic visualizations. *Computers in Human Behavior, 27*, 29–35.

Larkin, J., & Simon, H. (1987). Why a diagram is (sometimes) worth ten thousand words. *Cognitive Science, 11*, 65–100.

Lee, H., Plass, J. L., & Homer, B. D. (2006). Optimizing cognitive load for learning from computer-based science simulations. *Journal of Educational Psychology, 98*, 902–913.

Lehrer, R., Schauble, L., Strom, D., & Pligge, M. (2001). Similarity of form and substance: Modeling material kind. In D. K. S. Carver (Ed.), *Cognition and instruction: 25 years of progress* (pp. 39–74). Mahwah, NJ: Lawrence Erlbaum Associates.

Loehlin, J. C. (1998). *Latent variable models: An introduction to factor, path, and structural analysis*. Mahwah, NJ: Lawrence Erlbaum Associates.

Lowe, R. K. (2008). Learning from animation: Where to look, when to look. In R. K. Lowe & W. Schnotz (Eds.), *Learning with animation: Research implications for design* (pp. 49–70). New York: Cambridge University Press.

Lowe, R., & Boucheix, J.-M. (2017). A composition approach to design of educational animations. In R. Lowe & R. Ploetzner (Eds.), *Learning from dynamic visualization – Innovations in research and application*. Berlin: Springer (this volume).

Lowe, R., Boucheix, J.-M., & Fillisch, B. (2017). Demonstration tasks for assessment. In R. Lowe & R. Ploetzner (Eds.), *Learning from dynamic visualization – Innovations in research and application*. Berlin: Springer (this volume).

Lowe, R., & Schnotz, W. (Eds.). (2008). *Learning with animation: Research implications for design*. New York: Cambridge University Press.

Lowe, R. K., & Schnotz, W. (2014). Animation principles in multimedia learning. In R. E. Mayer (Ed.), *The Cambridge handbook of multimedia learning* (2nd ed., pp. 279–315). New York: Cambridge University Press.

Mayer, R. E. (Ed.). (2005). *The Cambridge handbook of multimedia learning*. New York: Cambridge University Press.

Mayer, R. E. (Ed.). (2014). *The Cambridge handbook of multimedia learning* (2nd ed.). New York: Cambridge University Press.

Mayer, R. E., & Fiorella, L. (2014). Principals for reducing extraneous processing in multimedia learning: Coherence, signaling, redundancy, spatial contiguity, and temporal contiguity principles. In R. E. Mayer (Ed.), *The Cambridge handbook of multimedia learning* (2nd ed., pp. 279–315). New York: Cambridge University Press.

Mayer, R. E., & Johnson, C. (2008). Revising the redundancy principle in multimedia learning. *Journal of Educational Psychology, 100*, 380–386.

Mislevy, R. J., Almond, R. G., & Lukas, J. F. (2003). *A brief introduction to evidence-centered design* (Research Report No. RR-03-16). Princeton, NJ: Educational Testing Service.

Mullis, I. V. S., & Martin, M. O. (Eds.). (2013). *TIMSS 2015 assessment frameworks*. Chestnut Hill, MA: TIMSS & PIRLS International Study Center at Boston College.

National Assessment Governing Board (NAGB). (2010). *Science framework for the 2011 national assessment of educational progress*. Washington, DC: National Assessment Governing Board.

National Center for Education Statistics (NCES). (2012). *The nation's report card: science in action: Hands-on and interactive computer tasks from the 2009 science assessment* (NCES 2012–468). Washington, DC: Institute of Education Sciences, U.S. Department of Education.

NGSS Lead States (NGSS). (2013). *Next generation science standards: For states, by states*. Washington, DC: The National Academies Press.

Organization for Economic Cooperation and Development (OECD). (2009). *PISA 2009 assessment framework: Key competencies in reading, mathematics and science*. Paris, France: OECD.

Papaevripidou, M., Constantinou, C. P., & Zacharia, Z. C. (2007). Modeling complex marine ecosystems: An investigation of two teaching approaches with fifth graders. *Journal of Computer Assisted Learning, 23*, 145–157.

Pedone, R., Hummel, J. E., & Holyoak, K. J. (2001). The use of diagrams in analogical problem solving. *Memory and Cognition, 29*, 214–221.

Pellegrino, J. W. (2013). Proficiency in science: Assessment challenges and opportunities. *Science, 340*(6130), 320–323.

Petre, M., & Green, T. R. G. (1993). Learning to read graphics: Some evidence that "seeing" an information display is an acquired skill. *Journal of Visual Languages & Computing, 4*, 55–70.

Plass, J. L., Homer, B. D., & Hayward, E. O. (2009). Design factors for educationally effective animations and simulations. *Journal of Computing in Higher Education, 21*(1), 31–61.

Quellmalz, E. S., Davenport, J. L., Timms, M. J., DeBoer, G. E., Jordan, K. A., Huang, C., et al. (2013). Next-generation environments for assessing and promoting complex science learning. *Journal of Educational Psychology, 5*, 1100.

Quellmalz, E. S., & Pellegrino, J. W. (2009). Technology and testing. *Science, 323*(5910), 75–79.

Quellmalz, E. S., Timms, M. J., & Buckley, B. (2010). The promise of simulation-based science assessment: The Calipers project. *International Journal of Learning Technology, 5*, 243–263.

Quellmalz, E. S., Timms, M. J., Silberglitt, M. D., & Buckley, B. C. (2011). Science assessments for all: Integrating science simulations into balanced state science assessment systems. *Journal of Research in Science Teaching, 49*, 363–393.

Reiner, M., & Eilam, B. (2001). Conceptual classroom environment: A system view of learning. *International Journal of Science Education, 23*, 551–568.

Rieber, L. P., Tzeng, S., & Tribble, K. (2004). Discovery learning, representation, and explanation within a computer-based simulation. *Computers and Education, 27*, 45–58.

Schwartz, D. L., & Black, T. (1999). Inferences through imagined actions: Knowing by simulated doing. *Journal of Experimental Psychology: Learning, Memory & Cognition, 25*, 116–136.

Schwartz, D. L., & Heiser, J. (2006). Spatial representations and imagery in learning. In R. K. Sawyer (Ed.), *The Cambridge handbook of the learning sciences* (pp. 283–298). New York: Cambridge University Press.

Schweingruber, H., Keller, T., & Quinn, H. (Eds.). (2012). *A framework for K-12 Science Education: Practices, crosscutting concepts, and core ideas*. Washington, DC: National Academies Press.

Slotta, J. D., & Chi, M. T. H. (2006). Helping students understand challenging topics in science through ontology training. *Cognition and Instruction, 24*, 261–289.

Stewart, J., Cartier, J. L., & Passmore, C. M. (2005). Developing understanding through model-based inquiry. In M. S. Donovan & J. D. Bransford (Eds.), *How students learn* (pp. 515–565). Washington, DC: The National Academies Press.

Tversky, B., Heiser, J., Lozano, S., MacKenzie, R., & Morrison, J. (2008). Enriching animations. In R. K. Lowe & W. Schnotz (Eds.), *Learning with animation: Research implications for design* (pp. 263–285). New York: Cambridge University Press.

Van Merriënboer, J. J. G., & Kester, L. (2005). The four-component instructional design model: Multimedia principles in environments for complex learning. In R. E. Mayer (Ed.), *The Cambridge handbook of multimedia learning* (pp. 71–93). New York: Cambridge University Press.

Vattam, S. S., Goel, A. K., Rugaber, S., Hmelo-Silver, C. E., Jordan, R., Gray, S., et al. (2011). Understanding complex natural systems by articulating structure-behavior- function models. *Journal of Educational Technology & Society, 14*, 66–81.

Webb, N. M., & Shavelson, R. J. (2005). Generalizability theory: Overview. In B. Everitt & D. Howell (Eds.), *Encyclopedia of statistics in behavioral science* (Vol. 2, pp. 717–719). Chichester, UK: Wiley.

Zacharia, Z. C. (2007). Comparing and combining real and virtual experimentation: An effort to enhance students' conceptual understanding of electric circuits. *Journal of Computer Assisted Learning, 23*, 120–132.

Chapter 10
Design of Effective Dynamic Visualizations: A Struggle Between the Beauty and the Beast? Commentary on Parts I and II

Katharina Scheiter

10.1 Introduction

Gottfried Wilhelm Leibniz (1646–1716), a German philosopher and mathematician, known amongst other things for his discovery of calculus (simultaneously discovered by Sir Isaac Newton) was a strong advocate for *theoria cum praxi*, that is, for theory to be combined with application. The present book adopts this approach by bringing together researchers and professional designers to reflect upon their insights and practices regarding the development of dynamic visualizations that are effective – be it for learning, reasoning, communication, or problem solving.

As can be seen from the chapters in Parts I and II, bringing these two perspectives together is by no means a trivial task. Not only is visualization design a multifaceted problem, it is also constrained by very different forces in the world of research than apply in the world of application. Research is often considered as an end in itself, where the dynamic visualizations are created for the purpose of conducting empirical investigations on how people learn from them. As a consequence, the major outcome of this process are scientific insights on how to design effective visualizations, with such insights being evaluated according to whether or not they constitute sound scientific knowledge. On the other hand, professional designers create visualizations in response to a request from a client (such as a curator of a museum or a developer of multimedia educational software). Although considerations with regard to effectiveness typically play a major role in the formulation of the request and how it is fulfilled by the designer, in the end, the visualization must sell, that is, the client must be convinced that it is good value for money. Key differences in the design and research contexts for developing visualizations tend to yield

K. Scheiter (✉)
Leibniz-Institut für Wissensmedien, Tübingen, Germany

University of Tübingen, Tübingen, Germany
e-mail: k.scheiter@iwm-tuebingen.de

R. Lowe, R. Ploetzner (eds.), *Learning from Dynamic Visualization*,
DOI 10.1007/978-3-319-56204-9_10

rather different types of visualization that could be characterized as 'the beauty' and 'the beast'. Professionally designed visualizations – the beauty – typically privilege realism and aesthetic quality, representing content comprehensively with considerable visual detail and thus delivering a high fidelity illustration. On the other hand, visualizations used in research – the beast – often have a more 'homespun' appearance with the content depicted in a way that is intended to reduce complexity so that users will not become overwhelmed. Although the comparison suggested here is undoubtedly somewhat exaggerated in order to make a point, the question posed in the present chapter is whether or not the beauty and the beast of visualization design can be united, and if so, how that might be done. In the first section of this commentary I discuss the main aspects of the multifaceted design challenge before addressing the question of what and how researchers and practitioners might learn from each other.

10.2 Visualization Design: A Multifaceted Challenge

When designing visualizations, researchers and practitioners have to consider at least four dimensions: the user, the features of the content to be displayed, the context in which the visualization is to be used, and the main objective to be achieved with a visualization. Each of these dimensions (along with their interactions) implies an agenda for research and development of its own. In addition, visualization design is influenced by external factors such as resource availability (time, money, expertise) or constraints imposed by research agendas and clients. However, in the following discussion, I will not consider the latter factors further unless explicitly noted otherwise. Rather, I will focus on a largely neglected but from my point of view most important issue, namely, the question of what constitutes a high-quality visualization for its end-users (e.g., learners, visitors of a science exhibit).

10.2.1 The User

As emphasized in the Animation Processing Model (APM; Lowe & Boucheix, 2017, this volume), extracting information from a dynamic visualization and building a mental model from this information is a complex endeavor that involves both perceptual and cognitive processes. Visualization design needs to take account of the fact that the human perceptual system is severely limited with regard to the amount of (changing) information to which we can simultaneously devote attention. In addition, organizing and integrating the extracted information over time while combining it with prior knowledge involves the storage and manipulation of elements in working memory, a resource that is also limited in its capacity. Thus, from the information processing perspective that guides much research on learning with dynamic visualizations, crucial goals for visualization design are to develop visualizations that attract a user's attention towards the most relevant information elements at just the right moment and help her/him to properly represent this information in a suitable mental model. Frequently, the pursuance of these goals results in very parsimonious visualizations designs with highly schematized representations whose depicted elements bear little visuospatial resemblance to their real-world referents (Scheiter, Gerjets, Huk, Imhof, & Kammerer, 2009).

In addition to these general limitations of the human information processing system, there may also be individual differences in the cognitive resources that users have available for processing (Wiley, Sanchez, & Jaeger, 2014). Such differences may require a visualization design that is specifically tailored to address the information processing characteristics of a particular target audience.

As discussed in the APM (Lowe & Boucheix, 2017, this volume), perception and cognition are affected not only by bottom-up processes but also by top-down processes, that is, influences of a user's prior domain knowledge. There is abundant research evidence with respect to both static and dynamic visual displays suggesting that users with higher domain knowledge are more likely to focus their attention largely on relevant information while ignoring information that is less relevant for the task at hand (for a review see Gegenfurtner, Lehtinen, & Säljö, 2011). In contrast, users with lower domain knowledge are more likely to be distracted by such low relevance features if the happen to be visually salient. For instance, they could be strongly drawn towards elements just because they are moving, even if these movements are not directly relevant to the to-be-conveyed principle. Comparable effects with respect to the level of prior knowledge have been observed across a variety of different content domains (e.g., meteorology, biology, radiology), suggesting that this is a pervasive phenomenon. This implies that novice learners in particular are likely to require attentional guidance regarding the processing of dynamic visualizations. In addition, prior knowledge affects the processes of interpreting and comprehending a dynamic visualization in that learners with more prior knowledge will be better able to relate their observations to cognitive schemata acquired earlier. However, learners with less prior knowledge may have problems in recognizing objects depicted in a visualization (especially if these have no observable counterparts in the real world) as well as in inferring the kinematics and principles that govern the illustrated phenomenon. As a consequence, designing dynamic visualizations for novice learners may warrant both perceptual and cognitive support to ensure their adequate processing.

Design implications are less clear-cut when considering users' spatial ability, the influence of which has mainly been investigated regarding the question of for whom dynamic visualizations should be used (see Wagner & Schnotz, 2017, this volume). In general, spatial ability has been shown to be relevant for learning from visualizations, with high-spatial ability students showing better learning outcomes (for a meta-analysis see Höffler, 2010). It seems that having sufficient spatial ability is more crucial for static than for dynamic visualizations which could suggest that dynamic visualizations have the potential to compensate for a lack of spatial ability. However, such conclusions would be based on studies that used either static visualizations or dynamic visualizations. It is quite possible that these studies also differed on dimensions other than their spatiotemporal character (e.g., content domain, complexity of the illustrated process) and that these dimensions drove the aforementioned aptitude-treatment interaction. When controlling for differences in content features by investigating the effects of spatial ability on learning from static versus dynamic visualizations within a single study, the evidence for spatial ability serving as moderator for the format's effectiveness is much less abundant. For instance, in five studies involving learning about different types of fish motion, there was no interaction between learners' spatial ability and visualization format (Brucker, Scheiter, & Gerjets, 2014; Imhof, Scheiter, & Gerjets, 2011; Imhof, Scheiter, Edelmann, & Gerjets, 2012, 2013). The same lack of interaction was found for three studies using either static or dynamic visualizations where the to-be-learned content was the underlying physical principles of fish locomotion, (Kühl, Scheiter, & Gerjets, 2012; Kühl, Scheiter, Gerjets, & Edelmann, 2011; Kühl, Scheiter, Gerjets, & Gemballa, 2011). Although the domain in these two sets of studies was the same (biological motion), the learning objectives were very different (recognition of movement patterns versus understanding Newton's laws). Also different were the visualization designs within and across the two study series, ranging as they did from animated black-and-white line drawings to videos and 2D-animations rendered from 3D models. Despite these differences in content and design, the finding that spatial abilities are equally important for learning from static and dynamic visualizations is highly robust. Moreover, in all cases the aforementioned studies yielded a strong main effect of spatial ability indicating that for the learning tasks, spatial ability is beneficial regardless of how those tasks are presented. In contrast to these findings, Sanchez and Wiley (2017, this volume) report findings from a study suggesting that the ability to integrate visuospatial information over time and space (i.e., multiple-object dynamic spatial ability) loses its power to predict better learning outcomes in dynamic compared with static visualizations. In other words, in their study animation seems to compensate for a lack of dynamic spatial ability, at least when this type of reasoning ability is required by the learning task. Further research is needed to study the role of multiple-object dynamic spatial ability compared with spatial visualization ability for single objects, which is the type of spatial ability assessed in most of the aforementioned studies.

A third individual difference concerns skills and knowledge regarding the interpretation of notations used in visualizations. McGill (2017, this volume) refers to this capacity as *graphicacy*, which can be seen as one element of the broader idea of *representational competence* (e.g., Kozma & Russell, 1997). I will use this latter term in the discussion that follows. According to Kozma and Russell (1997), representational competence comprises a set of skills for constructing, interpreting, transforming, and coordinating external representations used in scientific discourse. As a prelude to illustrating the challenges that students face when learning from dynamic visualizations, I will start with a discussion of what constitutes representational competence in extracting information from descriptive representations such as written text (Schnotz & Bannert, 2003). Texts are made up of a sequence of words, which (at least in languages using the Latin alphabet) are in turn constructed by recombining a very restricted number of discrete symbols (26 letters in the standard English alphabet) into strings of different length. The meaning of words is determined through convention, that is, by cultural agreement, with a relationship between the word and what it stands for being an arbitrary one. Despite those word meanings having to be learnt one-by-one, the task is manageable: adult native English speakers typically understand the meaning of 17,000 words or more (Zechmeister et al., 1995) although effective communication requires far less vocabulary. Importantly, in written language there is a clearly marked distinction between the content it conveys and the way additional information *about* the text is communicated to the reader (cf. Lemarié, Lorch, Eyerolle, & Virbel, 2008). For instance, printing a word in bold can denote its importance but leaves its meaning unaffected, something that young readers learn early on.

The whole situation changes when moving to depictive representations (Schnotz & Bannert, 2003). With pictures, there is no equivalent to a standardized alphabet and no conventional sequence in which visual elements are to be processed. One might argue that the extraction of meaning should be easier in depictive representations because the constituent visual elements resemble their real-world referents. However, visualizations are often also used to represent phenomena that have no observable real-world referents (e.g., molecules; see McGill, 2017, this volume). In such cases, the distinction between representation of content and communication about content can be ambiguous that can result in a literal interpretation of observations not intended by the animation's author (Jenkinson, 2017, this volume). For instance, with an animation showing a galloping horse, all but very young children would have sufficient world knowledge to perceive a hoof turning red as a signal that denotes that we should attend to this hoof – we know that a horse's hooves do not change color by themselves. However, in the case of visualizations of the non-observable world, we lack comparable background knowledge. Colors and shapes may be chosen by designers simply to make elements more discernable, even if their referents do not have any color or their color is unknown (see McGill, 2017, this volume). It can be very difficult for a viewer to decide whether specific aspects regarding the visual appearance of objects reflect the actual real world appearance of their referents, are used as highlights to focus attention, or are simply arbitrary because the designer had to make a choice when sketching the object. Similarly, using highly schematized visualizations may lead to misconceptions if students lack experience with the real world referents. Imagine the depiction of a cell as a circular object which, if interpreted literally by a user, would lead to the misconception that cells are perfectly round.

There may be many more as yet unknown individual characteristics that will affect the way users perceive a dynamic visualization. From an application-oriented point of view, this will mean that in some cases visualization design needs to be tailored towards different user groups in order to be most effective. This is possible only to the extent to which relevant features can be identified beforehand and to which there are user groups being homogenous with respect to these features. The latter is clearly not the case in outreach contexts, where the wide range of visitors who patronize museums and science centers possess very different levels of key attributes such as prior knowledge.

10.2.2 Features of the Content

In published research on learning from dynamic visualizations, one often reads statements suggesting that dynamic visualizations were expected to be superior to static ones because of the dynamic nature of the to-be-conveyed content. However, when looking at some examples of content delivered through dynamic visualization, it is clear that changes over time can take many forms. Earthworm locomotion is a repetitive, relatively slow process involving a single object where the motion pattern (i.e., the contraction and expansion of its body segments) re-occurs again and again. The galloping of a horse is similar regarding its repetitiveness, but the movements are much faster and more varied. A piano mechanism consists of one device composed of multiple, simultaneously moving sub-elements that are linked to each other via causal chains which determine the event structure. For some types of content, it is crucial to know about the (relative) speed by which events unfold (e.g., the fast movements of the pendulum in a pendulum clock versus the slow movements of its weight, or a fish moving forwards at a speed depending on the rate of its fin strokes), whereas for others only the fact that a change occurs is important. Finally, for some types of content, the change that occurs in a scene over time is limited to just a very few elements (e.g., the flapping of a fish's tail fin), whereas for others there is a complete change to the object in question (e.g., a plant growing from a seed or oxygen flowing from lungs through the body). Some changes are characterized by their continuity (e.g., the moving of an earthworm), whereas others occur suddenly (e.g., a volcanic eruption).

The list of examples given here is by no means exhaustive, but it shows that repetitiveness, simultaneity of moving elements, speed, importance of (relative) speed and speed changes, and comprehensiveness of change are all dynamic variables that can vary from one type of content to another and that are thus are specific to the design of a particular dynamic visualization. Recently, Ploetzner and Lowe (2012) provided a systematic account of such attributes that are concerned with the spatiotemporal arrangements in an animation. Their analysis offers a most useful starting point for characterizing a visualization's content. Lowe and Boucheix (2017, this volume) provide an even finer-grained way of describing a content's spatiotemporal features. The manifoldness of changes over time that can be portrayed in a visualization is also the reason why research that targets overly general questions, such as whether dynamic visualizations are more effective than static ones, is doomed to failure. As rightly pointed out by Wagner and Schnotz (2017, this volume), an answer to this question is likely to depend on all of the aforementioned aspects and will furthermore be affected by users' spatial abilities (Sanchez & Wiley, 2017, this volume). For instance, if changes in dynamic features such as speed variations are crucial for a proper understanding of the content, a static visualization is less likely to serve this purpose (Tversky, Bauer-Morrison, & Betrancourt, 2002). On the other hand, if a process can be well represented by a limited number of key frames and if changes between those states occur in a linear fashion, then there is little reason to assume that dynamic visualizations should be superior, given that they often impose additional and potentially unnecessary processing demands. Jenkinson (2017, this volume) presents some very nice examples of how well static visualizations can be used to convey dynamic content – if they are properly designed.

Similarly, the role of task features needs to be taken into account when considering the use of 3D for the design of animations (Schwan & Papenmeier, 2017, this volume). The mere fact that the referent subject matter involves three-dimensional objects does not necessarily mean that the representing visualization has to be three-dimensional as well. Rather, the effectiveness of 3D in dynamic visualizations is likely to depend on whether those aspects that are obligatory for comprehension require an internal representation of 3D information and on whether students possess the necessary (visuospatial) skills for apprehending the external representation (Huk, Steinke, & Floto, 2010; Khooshabeh & Hegarty, 2010). Thus, in order for animations and 3D representations to be effective, the structure and the content of the visualization should correspond to the desired mental representation (Tversky et al., 2002; see also Wagner & Schnotz, 2017, this volume).

10.2.3 Context of Use and Objectives

Educational research is mostly concerned with the use of visualizations for learning. However, as can be seen from the chapters of this section, there are many more purposes for which dynamic visualizations can be used, such as problem solving in science, assessment of students' understanding, or communication and public outreach. Often, the context of use is associated with one particular objective that is to be achieved with the visualization, although there is not necessarily a one-to-one mapping between the two. Nevertheless, for the sake of simplicity, in the following discussion I will consider the context of use and objectives together.

In line with Ploetzner (2016), I will refer to dynamic visualizations designed for learning as explanatory visualizations (cf. Sharpe, Lumsden, & Wooldridge, 2008; expository animation according to Ploetzner & Lowe, 2012) which aim to help students construct a comprehensive mental model of some kind of phenomenon or process (objective). Very frequently, their design is motivated by a wish to simplify the learning task – that is, the representation of the content is limited to what is deemed essential for mental model construction, while 'irrelevant' details are left out. Explanatory visualizations are typically tailored for students who have little prior knowledge of the to-be-learned content and who require guidance in order to learn it successfully.

Dynamic visualizations intended to be used for fostering reasoning and problem solving in science are aimed at allowing for novel insights into the targeted problem. Their design can involve a high fidelity treatment, which is often achieved by using real data to model the phenomenon in question (McGill, 2017, this volume). Such scientific visualizations reflect the complexity and comprehensiveness of the modeled phenomenon and may entail ambiguities and unknown aspects which in turn pose particular challenges in terms of visualization. According to Sharpe, Lumsden, and Wooldridge (2008), they satisfy the purpose of simulation and speculation (cf. Jenkinson, 2017, this volume). These visualizations typically address expert users who wish to visually explore complex phenomena, generate hypotheses, and make inferences. These activities would be more difficult if based on massive amounts of numerically presented quantitative data alone.

Dynamic visualizations for assessment are used with the aim of providing a valid estimate of a student's scientific understanding and reasoning abilities when interacting with dynamic systems represented in animations and simulations. As discussed by Davenport and Quellmalz (2017, this volume), the challenge in designing animations for the purpose of assessment is to create visualizations that do not overwhelm students, while at the same time offering sufficient degrees of freedom regarding their use and interpretation so that they are indicative of a wide range of students' reasoning abilities (for a similar challenge regarding the design of demonstration tasks see Lowe, Boucheix, & Fillisch, 2017, this volume). When using animations for assessment, careful consideration must be given as to whether students' failure to perform on a dynamic assessment task is an expression of her/his lack of scientific reasoning abilities or is also indicative of their inability to adequately handle this specific assessment format. In the latter case, the performance measured would not provide a valid estimate of the construct that one wishes to assess. On the other hand, animation-based assessments (and demonstration tasks) offer a major advantage when compared with verbal, paper-based assessments in that they can provide much richer data regarding the way students solve the problems (e.g., number and type of wrong steps taken). This data can be used to identify gaps in students' knowledge and model different types of erroneous conceptions they have.

Finally, the objective of dynamic visualizations designed for communication and outreach is often twofold in that they are supposed to attract attention, arouse curiosity, and entertain while also fostering some degree of understanding of scientific phenomena. The visualizations that are found in many science centers and museums therefore need to keep a careful balance between being visually rich and complex versus conveying content in a sufficiently simplified manner so that visitors' understanding of the explanation in a short viewing time lies within their zone of proximal development. Sharpe et al. (2008) consider that visualizations used for communication and public outreach also serve the purpose of explanation and thus fall into the same category as educational visualizations. However, because of their twofold objectives I believe that it is better to treat them separately. Although there are examples of science centers and museums where the objectives of motivation and learning both appear to be well integrated into the exhibits, the tension between these aspects frequently results in exhibits where one goal (most often learning) is sacrificed for the sake of another. In those cases, as noted by Jenkinson (2017, this volume), "capturing … attention takes priority" (p. 111). For instance, this becomes evident when considering the large number of immersive 3D visualizations that are used for outreach purposes in light of the contribution of Schwan and Papenmeier

(2017, this volume), according to which there is little evidence for any strong learning benefits associated with 2.5 or 3D representations. To conclude, visualization designs that are suited to attract attention may not necessarily be the ones that foster understanding.

10.3 What Can Researchers and Designers Learn from Each Other?

There tend to be fundamental differences in the view of researchers and designers as to what constitutes a high-quality visualization for end-users. Researchers try to answer this issue by means of empirical studies that compare different visualization versions with respect to how well a target population achieves the desired outcome (e.g., performance in a comprehension test). The ultimate aim of this research is to generate evidence-based design guidelines on how to design an effective dynamic visualization. Professional designers, on the other hand, have primarily relied on their treasure trove of craft-based experience. This experience has accumulated over the years, is augmented by usability tests and other, more informal ways of testing its validity and has been passed on from generation to generation in a well-documented fashion. In the past, these two types of knowledge have developed by-and-large independently of each other. One way to move closer to the Leibniz goal of *theoria cum praxi,* would be for designers to learn more about the empirical evidence regarding certain design aspects and apply it in their work. However, it is equally important for researchers to incorporate craft-based knowledge in their studies and evaluate it empirically. The good news is that both of these things are beginning to happen (albeit slowly; cf. Jenkinson, 2017, this volume).

However, there are barriers that prevent such cross-fertilization, some of which have to do with the fact that researchers and designers still seldom intermingle so that communication about design knowledge remains largely within each community (see below). There can also be a certain skepticism regarding the other profession's work. On one hand, researchers are typically reluctant to accept that a design is sound unless there is empirical evidence regarding its effectiveness. On the other, one critique expressed by McGill (2017, this volume) is that in many cases educational research on learning with dynamic visualizations may have yielded invalid conclusions because of the low quality of the visualizations used in this research. As a result, research based on such poor materials may not be regarded as worthwhile by professional visualization designers. This view needs to be examined with respect to at least two aspects of visualization quality, namely, content-related accuracy and design attributes.

10.3.1 Visualization Accuracy

Do the visualizations used in research correctly represent the phenomenon in question? Although I agree that there is research where simplification of the content matter is achieved at the expense of accuracy, this flaw is not confined to research visualizations – there are also many commercially available visualizations that contain inaccuracies. Thus, the question is not so much whether it was researchers or designers who are responsible for the visualizations, but whether domain experts were involved in the process of their production to ensure that state-of-the-art domain knowledge is represented. Ideally, the ability to create high-fidelity visualizations and content expertise are united in one person as is the case for many people working on science visualizations who are trained in specialized science visualization programs and who have a science background (Jenkinson, 2017, this volume). However, when looking at who produces visualizations for commercially available products (e.g., educational software, museum displays), this is by no means always the case. Thus, both researchers and professional designers can profit from input by content matter experts in order to ensure accuracy of their visualizations.

10.3.2 Design Attributes

The view that 'low-quality' visualizations are used in research may also be a reference to the design of the visualization, which gets us back to the distinction between the beauty and the beast. I agree that many visualizations used in research are homespun in that they lack the aesthetic appeal and visual richness frequently associated with professionally designed visualizations. To some extent this has to do with the fact that these visualizations are often produced by the researchers who lack professional skills in visualization design. In most research projects, financial and temporal constraints leave no other option but to use these home-made, low cost visualizations. Doing research on professionally designed visualizations is possible only if one has access to the source materials for the visualization and is able to manipulate its characteristics for experimental purposes. This is typically not an option for commercially available products.

Using highly schematized and simplified visualizations in research can however also be justified by findings that suggest that learners struggle with more complex visualizations. This is because such visualizations are less well matched with a human's limited capabilities for processing simultaneously presented and transient information. I believe that the question of whether researcher-based visualizations or professionally designed visualizations are more effective can be answered only by running joint empirical studies, where consensus is reached about the design of different visualization versions and the dependent variables to be assessed as evaluation criteria.

In particular, in such a comparison researcher-based visualizations should not just look homespun because of a lack of money and skill required to design them in a better way. Rather, they should be characterized by the simplification that has been intentionally applied to the visualizations in order to make them more comprehensible for learners. For professionally designed visualizations involved in such a comparison, a key consideration would have to be that these materials typically not only appear visually more complex but also often include attention-guiding devices to reduce processing demands potentially resulting from this visual complexity. In professional visualization production, many attention-guiding techniques have evolved, including some borrowed from the movies industry (cf. McGill, 2017, this volume; Schwan & Papenmeier, 2017, this volume). These are smoothly built into visualizations in a way that a user might not even become aware of them. Such approaches include the use of a virtual camera to provide users with canonical viewpoints (Garsoffky, Schwan, & Huff, 2009), zooming, lighting, as well as techniques to highlight the event structure such as cuts or speed alterations (for examples see Jenkinson, 2017, this volume).

In a comparison of researcher-based and professionally designed visualizations, both motivational and perceptual/cognitive aspects should be considered. With respect to motivation, one might argue that professionally designed visualizations tend to be more engaging and that this engagement is a key prerequisite for learning, especially in situations that are less constrained than the classical experimental set-up students encounter in the laboratory (e.g., during free-choice learning situations). Thus, in situations where motivation is a primary consideration, researcher-designed visualizations may fail simply because students will not invest sufficient effort into learning. On the other hand, it may be that the fancier looking a visualization is, the more likely it will be perceived as little more than entertainment. Salomon's (1984) seminal studies regarding the amount of invested mental effort (AIME) suggest that the investment that students are willing to make in processing depends on how they perceive the medium used for information delivery. Media associated with entertainment such as television are less likely to result in effortful and thus deeper learning (cf. underwhelming effect; Lowe, 2004). The same might apply for professionally designed visualizations which often include the same visual features and storytelling elements that are also found in movies and other screen-based forms of entertainment (McClean, 2007; Jenkinson, 2017, this volume).

As to cognitive outcomes, there is the question of how effective the attention-guiding techniques built into professional visualizations are with regard to compensating for the potentially harmful effects of visual complexity. On the one hand, there is empirical research evidence that these techniques can promote learning and understanding (e.g., use of canonical viewpoints: Schwan & Papenmeier, 2017, this volume; speed alterations: Fischer, Lowe, & Schwan, 2008). In this case, extracting information from dynamic visualizations that effectively deploy these techniques should not be a problem. On the other hand, one has to keep in mind that movies most often deal with subject matter, environments, and activities that are familiar to viewers and thus represent events for which they have extensive prior knowledge (e.g., scripts that help us anticipate an agent's actions). Explanatory visualizations, on the other hand, often deal with situations where the target audience lacks suitable prior knowledge and thus may fail to correctly interpret attention-guiding techniques like film cuts (cf. Schwan & Ildirar, 2010).

10.3.3 Influencing Factors

Synergistic effects regarding the design and development of dynamic visualizations are possible if we simultaneously consider factors that are the main foci of not only researchers but also designers. Because research tends to focus on how humans will process information conveyed through a dynamic visualization, the way individual characteristics of the users will moderate the effectiveness of a particular visualization design receives much attention in studies of learning with dynamic visualizations. In contrast, it is comparatively rare that task features and their influence on visualization effectiveness are explicitly addressed in visualization research (for exceptions see Sanchez & Wiley, 2017, this volume; Wagner & Schnotz, 2017, this volume). As discussed previously, one reason for this neglect may be that until recently research lacked accounts that would help to systematically describe task or domain features (Ploetzner & Lowe, 2012). However, professional designers are very focused on the best way of displaying information and thus give particular consideration to task features. Unfortunately, as a consequence too little consideration can sometimes be given to the fact that the visualization has to be handled by human beings and that our processing capacities are limited. Indeed, there is evidence that in some cases users' processing capabilities may actually be more predictive of a visualization's effectiveness than its design alone (Khooshabeh & Hegarty, 2010; Sanchez & Wiley, 2017, this volume). Moreover, the example offered by Lowe and Boucheix (2017, this volume) suggests that devising a design that is grounded on assumptions regarding a user's information processing capabilities and the mental model that s/he has to construct (i.e., presenting multiple simultaneously occurring events sequentially) compared with one that takes the nature of the phenomenon as a guiding principle (i.e., presenting simultaneous events as they occur) might actually be more effective for learning. As a consequence, researchers and designers should both strive for a more balanced, comprehensive approach to visualization design that takes into account a wider range of influencing factors – namely, the user, the features of the content to be displayed, the context in which the dynamic visualization is to be used, and the main objective to be achieved with an visualization. An important implication is that the research agenda should be widened so that is not limited to just explanatory visualizations designed for learning.

10.3.4 Evaluation of Visualization Effectiveness

It seems that two differing sets of criteria for ensuring a visualization's effectiveness are currently used by researchers on one hand and by professional designers on the other. In research, a key concern is with the final outcome of the visualization – objective measures of how well it has fostered student learning. However, with professional design, much of the focus on the process of producing the visualization (e.g., whether its design is aligned with the available craft-based knowledge), and on whether different stakeholders (i.e., designers, clients, and users) are satisfied with the outcome.

The process of developing visualizations seems to be much more systematic and explicit in the world of professional design than is commonly the case in the world of research (cf. the descriptions of multi-step animation production in McGill, 2017, this volume; Jenkinson, 2017, this volume). Although the professional design approach cannot guarantee high-quality visualizations as an outcome, it does provide useful guidance as to necessary steps in visualization design that should not be overlooked. In research, on the other hand, the procedures of developing a visualization are usually not very systematic and are rarely if ever documented in detail (Jenkinson, 2017, this volume). Research papers tend not to include information on why the visualization was designed in a particular way (apart from rationalizing the experimental manipulation). This is due, at least in part, to the fact that space constraints in publishing outlets do not allow for elaborate descriptions of the process of designing a visualization. However, providing a thorough analysis of the task to be performed or the knowledge to be acquired from a visualization and describing the consequences thereof for the design would be yet another quality indicator beyond a visualization's effect on learning outcomes (see Lowe & Boucheix, 2017, this volume). Such information would help reveal far more about the design of effective visualizations than is possible on the basis of current research papers (where the guiding principles for designing the visualization under investigation are often left unstated).

A further potential advantage of the multiple stages in professional design, where prototypes and beta versions are discussed with the clients, is that this activity constitutes a type of formative evaluation. Formative evaluations have the potential to provide powerful insights into how a visualization might be improved. However, at present most of these evaluations tend to rely on qualitative data such as interviews with a small number of users. A possible alternative approach for collecting more far-reaching insights could be based on the work of Lowe, Boucheix, and Fillisch (2017, this volume) who discuss demonstration tasks as a way of assessing learning with dynamic visualizations in a summative sense (i.e., regarding the final product). I believe that these tasks could also serve very well as the basis for highly revealing formative evaluations. Compared with other more traditional forms of evaluation, demonstration tasks provide few constraints on a user's responses, thereby allowing for the expression of a wide range of behaviors that reflect the user's mental model acquired from the visualization. In addition, performance in these tasks is likely to have a far more direct link to the visualization input compared with performance in conventional verbal tasks where users first have to translate their depiction-based internal representations into a linear, descriptive output (cf. the linearization problem; Levelt, 1981). Demonstration tasks appear to be very well suited to assessing possible misconceptions of users that have been evoked by an inadequate visualization design. Whereas summative assessments often focus on whether students ultimately learned what they were supposed to learn from a visualization (i.e., the learning objective), whether or not students developed misconceptions from watching a dynamic visualization is seldom assessed (at least in psychological research). Using demonstration tasks for formative evaluation appears well suited to helping avoid the danger that a visualization's final version will actually evoke misconceptions. This possibility for misconceptions may otherwise go undetected if summative evaluations alone are the predominant focus.

As well as researchers benefiting from the approaches of professional designers, there are also potential advantages in the other direction. Professional visualization design could benefit from empirical studies that test the effectiveness of particular visualization designs with respect to the desired outcome, both in the laboratory and in the context for which it was developed. This would complement the solid basis of intuition and craft-based knowledge currently used for developing visualizations with what is typically missing in the professional design process described by McGill (2017, this volume) – a feedback loop involving end-users such as those learning with an explanatory visualization in a digital textbook or those trying to understand an animated scientific explanation in a museum.

10.3.5 Devising an Animation-Based Instructional Strategy

Empirical research on learning from explanatory dynamic visualizations reveals that the associated learning outcomes are often far from optimal, even when these visualizations have been designed against the backdrop of state-of-the-art knowledge. There are likely to be multiple reasons for such deficiencies. In most laboratory studies, the learning required during an experimental task may have little personal relevance for students. As a consequence, participants may not invest the necessary effort into processing the visualization. Further, in the name of experimental control, learning with a visualization is often expected to result from its one-time, system-controlled presentation. This is despite learning about complex processes being likely to require multiple viewings of a visualization. Finally, mental model construction for many dynamic phenomena involves comprehension of multiple aspects, including an adequate mental representation of a system's visuo-spatial arrangement of its main components, the way these components behave over time (i.e., their kinematics), and the more abstract principles that are reflected in these motions (Hegarty, 1992). A single dynamic visualization is unlikely to be capable of acting as a multi-purpose educational tool that can adequately address all of these aspects at the same time. Thus, whereas from a purely research-oriented perspective experimental control and focus on a single instructional component like dynamic visualization may be necessary, from the point of view of educational practice we need to take a broader approach that addresses how learning from dynamic visualizations can be embedded into a comprehensive instructional strategy (cf. Kombartzky, Ploetzner, Schlag, & Metz, 2010; Ploetzner & Lowe, 2017, this volume). This has resonances with the workflow applied in professional animation design where designing a visualization requires consideration of far more elements than just the illustration itself (e.g., integration of sound and voice to augment the narrative; Jenkinson, 2017, this volume).

Instructional design theories such as Elaboration Theory (Reigeluth & Stein, 1983) could provide some guidance as to how to devise such a strategy. According to Elaboration Theory, instruction should start with a comprehensive and concrete overview regarding the to-be-conveyed content at a relatively coarse level (i.e., the epitome) so that students can first acquire a schema to which details can be added later. In the case of an animation-based instructional strategy, the epitome could be a static picture of the components of a mechanical system that serves as pre-training concerning its visuo-spatial structure as well as a first observation of the whole animated process (cf. Kombartzky et al., 2010). Reigeluth and Stein (1983) propose then to teach each aspect of the content in more detail in separate units. This assumes that the content is first broken down into these individual pieces, an approach which when applied to learning with dynamic visualizations has parallels with the preliminary decomposition involved in the composition approach devised by Lowe and Boucheix (2017, this volume). According to the composition approach, complex event structures that consist of multiple, simultaneously occurring sub-events should be broken down in an animation by focusing on each sub-event separately and in succession. Thus, the sub-events can be considered as separate units of a learning sequence. As shown by Lowe and Boucheix (2017, this volume), even on its own, this sequential approach can produce notable improvements in learning from complex animation. However, it has the potential for even further improvements by including summarizers and synthesizers that according to Elaboration Theory should be included after each unit, to help students understand how the aspects addressed in each unit (e.g., each micro event) contribute to the process as a whole. When learning with dynamic visualizations, this would mean not only illustrating micro events in succession, but also showing and explaining after each micro event what has been covered so far and how the micro event is linked to the overall macro structure. This approach is consistent with the Animation Processing Model (Lowe & Boucheix, 2017, this volume). Finally, Elaboration Theory suggests using embedded strategy activators, which require students to engage in a more elaborate processing of the content. For learning with dynamic visualizations, this might imply interspersing a dynamic visualization with self-assessment tasks (McGill, 2017, this volume) or self-explanation prompts (e.g., van der Meij & de Jong, 2011; Renkl & Scheiter, in press). When implementing such a procedure, an instructional sequence would contain not just dynamic representations but rather would alternate static and dynamic segments, depending on what was being explained at a given moment. A comprehensive instructional strategy would not of course consist of only visual representations, but could also contain, for instance, text to explain the principles that govern the dynamics of a complex system (Hegarty, 1992).

From the perspective of experimental research, evaluating the effectiveness of a comprehensive animation-based instructional strategy of the type suggested here offers some challenges. If the goal is to evaluate the strategy as a whole rather than just its individual components, determining what a fair control condition would consist of may be less than straightforward. Thus, empirical approaches for addressing research questions such as 'are static representations more effective than animated ones?' or 'does color coding aid learning from animations?' are in many ways far clearer because appropriate research designs are well established and unproblematic. However, this type of research can have serious limitations when it comes to finding the most effective instructional strategy because it excludes factors without which learning will likely remain suboptimal. In this regard, Ploetzner and Lowe (2017, this volume) present an interesting discussion in which they contrast different types of an animation-based instructional strategy that either contains or does not contain verbal explanations and that additionally consists of different ways of presenting the visualization (as one instance or as multiple segments presented either sequentially or simultaneously).

10.4 How Can Researchers and Designers Learn from Each Other?

It is clear that researchers and designers have a lot to learn from each other, but how can this best be achieved? Communication, training, and collaborative work teams are perhaps most important ways of ensuring that the present community boundaries are broken down. With respect to communication, it is essential that research-based and craft-based knowledge is synthesized and documented in a way that is easily accessible for a broader audience. The present book as well as Ploetzner (2016) provide very good examples for communicating knowledge on visualization design. Similarly, conferences that attract an interdisciplinary audience interested in science visualization such as the bi-annual Gordon Research Conference: Visualization in Science and Education (www.grc.org) are important for bringing together researchers, instructors, science communicators, and designers to learn from each other. This conference has not only stimulated discussions amongst representatives from very different communities, but has also instigated collaborative projects whose outcomes are also addressed in this book (e.g., the ant animation; McGill, 2017, this volume). In addition, training programs that cover visualization design from multiple perspectives can help to foster a new generation of designers and researchers who have a similar familiarity with theory and practice. The International Cognitive Visualization Program (ICV, icvprogram.net) is one example where research on visualizations meets practice. There is also an emerging field of training programs that address the intersection of visualization design and the sciences (e.g., biomedical visualization programs). These might benefit from the inclusion of courses in empirical research, a development that is beginning to occur in some of the more innovative programs. Finally, interdisciplinary work teams that design and research dynamic visualizations collaboratively potentially provide the most promising opportunities for challenging each other's assumptions and learning from each other.

References

Brucker, B., Scheiter, K., & Gerjets, P. (2014). Learning with dynamic and static visualizations: Realistic details only benefit learners with high visuospatial abilities. *Computers in Human Behavior, 36*, 330–339.

Davenport, J. L., & Quellmalz, E. S. (2017). Assessing science inquiry and reasoning using dynamic visualizations and interactive simulations. In R. Lowe & R. Ploetzner (Eds.), *Learning from dynamic visualization – Innovations in research and application*. Berlin: Springer (this volume).

Fischer, S., Lowe, R. K., & Schwan, S. (2008). Effects of presentation speed of a dynamic visualization on the understanding of a mechanical system. *Applied Cognitive Psychology, 22*, 1126–1141.

Garsoffky, B., Schwan, S., & Huff, M. (2009). Canonical views of dynamic scenes. *Journal of Experimental Psychology: Human Perception and Performance, 35*, 17–27.

Gegenfurtner, A., Lehtinen, E., & Säljö, R. (2011). Expertise differences in the comprehension of visualizations: A meta-analysis of eye-tracking research in professional domains. *Educational Psychology Review, 23*, 523–552.

Hegarty, M. (1992). Mental animation: Inferring motion from static displays of mechanical systems. *Journal of Experimental Psychology: Learning, Memory, and Cognition, 18*, 1084–1102.

Höffler, T. (2010). Spatial ability: Its influence on learning with visualizations – A meta-analytic review. *Educational Psychology Review, 22*, 245–269.

Huk, T., Steinke, M., & Floto, C. (2010). The educational value of visual cues and 3D-representational format in a computer animation under restricted and realistic conditions. *Instructional Science, 38*, 455–469.

Imhof, B., Scheiter, K., Edelmann, J., & Gerjets, P. (2012). How temporal and spatial aspects of presenting visualizations affect learning about locomotion patterns. *Learning and Instruction, 22*, 193–205.

Imhof, B., Scheiter, K., Edelmann, J., & Gerjets, P. (2013). Learning about locomotion patterns: Effective use of multiple pictures and motion-indicating arrows. *Computers & Education, 65*, 45–55.

Imhof, B., Scheiter, K., & Gerjets, P. (2011). Learning about locomotion patterns from visualizations: The effects of presentation format and degree of realism. *Computers & Education, 57*, 1961–1970.

Jenkinson, J. (2017). The role of craft-based knowledge in the design of dynamic visualizations. In R. Lowe & R. Ploetzner (Eds.), *Learning from dynamic visualization – Innovations in research and application*. Berlin: Springer (this volume).

Khooshabeh, P., & Hegarty, M. (2010). Inferring cross-section: When internal visualizations are more important than properties of external visualizations. *Human-Computer Interaction, 25*, 119–147.

Kombartzky, U., Ploetzner, R., Schlag, S., & Metz, B. (2010). Developing and evaluating a strategy for learning from animations. *Learning and Instruction, 20*, 424–433.

Kozma, R. B., & Russell, J. (1997). Multimedia and understanding: Expert and novice responses to different representations of chemical phenomena. *Journal of Research in Science Teaching, 34*, 949–968.
Kühl, T., Scheiter, K., & Gerjets, P. (2012). Enhancing learning from dynamic and static visualizations by means of cueing. *Journal of Educational Multimedia and Hypermedia, 21*, 5–22.
Kühl, T., Scheiter, K., Gerjets, P., & Edelmann, J. (2011a). The influence of text modality on learning with static and dynamic visualizations. *Computers in Human Behavior, 27*, 29–35.
Kühl, T., Scheiter, K., Gerjets, P., & Gemballa, S. (2011b). Can differences in learning strategies explain the benefits of learning from static and dynamic visualizations? *Computers & Education, 56*, 176–187.
Lemarié, J., Lorch, R. F., Eyrolle, H., & Virbel, J. (2008). SARA: A text-based and reader-based theory of text signaling. *Educational Psychologist, 43*, 1–23.
Levelt, W. J. M. (1981). The speaker's linearization problem. *Philosophical Transactions of the Royal Society, Series B, 295*, 305–315.
Lowe, R. (2004). Interrogation of a dynamic visualization during learning. *Learning and Instruction, 14*, 257–274.
Lowe, R., & Boucheix, J.-M. (2017). A composition approach to design of educational animations. In R. Lowe & R. Ploetzner (Eds.), *Learning from dynamic visualization – Innovations in research and application*. Berlin: Springer (this volume).
Lowe, R., Boucheix, J.-M., & Fillisch, B. (2017). Demonstration tasks for assessment. In R. Lowe & R. Ploetzner (Eds.), *Learning from dynamic visualization – Innovations in research and application*. Berlin: Springer (this volume).
McClean, S. T. (2007). *Digital storytelling*. Cambridge, MA: MIT Press.
McGill, G. (2017). Designing instructional science visualizations in the trenches: Where research meets production reality. In R. Lowe & R. Ploetzner (Eds.), *Learning from dynamic visualization – Innovations in research and application*. Berlin: Springer (this volume).
Ploetzner, R. (2016). *Understanding explanatory animation – An introduction to cognitive design and user support*. Freiburg: eXplanatory!media.
Ploetzner, R., & Lowe, R. (2012). A systematic characterization of expository animations. *Computers in Human Behavior, 28*, 781–794.
Ploetzner, R., & Lowe, R. (2017). Looking across instead of back and forth – How the simultaneous presentation of multiple animation episodes facilitates learning. In R. Lowe & R. Ploetzner (Eds.), *Learning from dynamic visualization – Innovations in research and application*. Berlin: Springer (this volume).
Reigeluth, C. M., & Stein, F. S. (1983). The elaboration theory of instruction. In C. M. Reigeluth (Ed.), *Instructional-design theories and models: An overview of their current status* (pp. 335–381). Hillsdale, NJ: Erlbaum.
Renkl, A., & Scheiter, K. (in press). Studying visual displays: How to instructionally support learning. *Educational Psychology Review*.
Salomon, G. (1984). Television is "easy" and print is "tough": The differential investment of mental effort in learning as a function of perceptions and attributions. *Journal of Educational Psychology, 76*, 647–658.
Sanchez, C. A., & Wiley, J. (2017). Dynamic visuospatial ability and learning from dynamic visualizations. In R. Lowe & R. Ploetzner (Eds.), *Learning from dynamic visualization – Innovations in research and application*. Berlin: Springer (this volume).
Schnotz, W., & Bannert, M. (2003). Construction and interference in learning from multiple representation. *Learning and Instruction, 13*, 141–156.
Scheiter, K., Gerjets, P., Huk, T., Imhof, B., & Kammerer, Y. (2009). The effects of realism in learning with dynamic visualizations. *Learning and Instruction, 19*, 481–494.
Schwan, S., & Ildirar, S. (2010). Watching film for the first time: How adult viewers interpret perceptual discontinuities in film. *Psychological Science, 21*, 970–976.

Schwan, S., & Papenmeier, F. (2017). Learning from animations: From 2d to 3d? In R. Lowe & R. Ploetzner (Eds.), *Learning from dynamic visualization – Innovations in research and application*. Berlin: Springer (this volume).
Sharpe, J., Lumsden, C., & Woolridge, N. (2008). *In silico: 3D animation and simulation of cell biology with Maya and MEL*. New York: Morgan Kaufmann.
Tversky, B., Bauer-Morrison, J., & Betrancourt, M. (2002). Animation: Can it facilitate? *International Journal of Human-Computer Studies, 57*, 247–262.
van der Meij, J., & de Jong, T. (2011). The effects of directive self-explanation prompts to support active processing of multiple representations in a simulation-based learning environment. *Journal of Computer Assisted Learning, 27*, 411–423.
Wagner, I., & Schnotz, W. (2017). Learning from static and dynamic visualizations: What kind of questions should we ask? In R. Lowe & R. Ploetzner (Eds.), *Learning from dynamic visualization – Innovations in research and application*. Berlin: Springer (this volume).
Wiley, J., Sanchez, C. A., & Jaeger, A. J. (2014). The individual differences in working memory capacity principle for multimedia learning. In R. Mayer (Ed.), *The Cambridge handbook of multimedia learning* (2nd ed., pp. 598–619). New York: Cambridge University Press.
Zechmeister, E. B., Chronis, A. M., Cull, W. L., D'Anna, C. A., & Healy, N. A. (1995). Growth of a functionally important lexicon. *Journal of Reading Behavior, 27*, 201–212.

Part III
Innovations in Scaffolding

Dynamic visualizations are widely considered to be an inherently effective way of helping students to learn about subject matter that involves temporal change. Further, there are compelling theoretical arguments as to why visualizations that present dynamic information explicitly should be superior to their static counterparts for which such information must be inferred. However, recent research has shown that dynamic visualizations too often fail to deliver the anticipated educational benefits. Although inadequacies in the basic design of the representation can be used to account for some of these failures (see Part I of this volume), there are clearly also other factors at work.

For example, if a dynamic visualization presents many and varied changes simultaneously or in rapid succession, limitations on our human capacity to process information can prevent learners from dealing with all the information that is presented. A typical learner response in this situation is to select a subset of the available information and neglect the rest. Unfortunately, learners who lack background knowledge about the depicted content domain are ill-equipped to make selections amongst the available information that are optimal for a specific learning task. This can result in them missing aspects that are crucial to developing a proper understanding of the referent subject matter.

An established way to address this problem is to augment the basic visualization with ancillary features intended to help ensure that learners do not miss key information. The role of these *scaffolding ancillaries* is to support the learner in dealing with the demands of the visualization – they function as supplementary 'add-ons' and so are not inherent aspects of the visualization itself. Scaffolding can be provided in a range of ways – from providing forms of support that are intended to guide the learners' attention to providing information that helps the learners' to develop an appropriate interpretation of the visual information being presented.

This part includes two chapters on enhancing learning from animation by means of scaffolds. De Koning and Jarodzka (2017, this volume) provide an overview of three approaches to attention guidance and summarize the findings of empirical research on the effectiveness of these approaches with respect to the perceptual and cognitive processing of animations. The overview covers the traditional approach of

cueing – in which verbal or visual signals indicate which entities or relationships between entities in the display should receive particular attention – as well as innovative approaches that rely on ***eye movements*** and ***gestures***. Such approaches entail efforts to support the processing of existing animations designs which contrasts with the approach suggested by Lowe and Boucheix (2017, this volume) of fundamentally re-thinking how animations are designed in the first place. In an empirical study, Berney and Bétrancourt (2017, this volume) investigate whether the learning of animated three-dimensional anatomical structures can be improved by providing the learners with specific scaffolds, namely, ***referential ancillaries*** such as axes. Because the learning of three-dimensional structures can perceptually and cognitively be demanding (see Schwan and Papenmeier, 2017, this volume), the referential ancillaries aim to support the learners in orienting the displayed structures in space so that the required mental operations can be performed more effectively.

References

Berney, S., & Bétrancourt, M. (2017). Learning three-dimensional anatomical structures with animation: Effects of orientation references and learners' spatial ability. In R. Lowe & R. Ploetzner (Eds.), *Learning from dynamic visualization – Innovations in research and application*. Berlin: Springer Publishers (this volume).

De Koning, B. B., & Jarodzka, H. (2017). Attention guidance strategies for supporting learning from dynamic visualizations. In R. Lowe & R. Ploetzner (Eds.), *Learning from dynamic visualization – Innovations in research and application*. Berlin: Springer (this volume).

Lowe, R., & Boucheix, J.-M. (2017). A composition approach to design of educational animations. In R. Lowe & R. Ploetzner (Eds.), *Learning from dynamic visualization – Innovations in research and application*. Berlin: Springer (this volume).

Schwan, S., & Papenmeier, F. (2017). Learning from animations: From 2d to 3d? In R. Lowe & R. Ploetzner (Eds.), *Learning from dynamic visualization – Innovations in research and application*. Berlin: Springer (this volume).

Chapter 11
Attention Guidance Strategies for Supporting Learning from Dynamic Visualizations

Björn B. de Koning and Halszka Jarodzka

11.1 Introduction

A defining characteristic of dynamic visualizations (e.g., animation, video) is that they present multiple changes simultaneously and/or in rapid succession within a single display (Lowe, 2003). This obviously offers the benefit of explicitly and realistically depicting natural (e.g., lightning formation), biological (e.g., human circulatory system), and technical (e.g., toilet cistern) systems (Ploetzner & Lowe, 2012). However, the down side is that for dynamic visualizations of even moderate complexity, learners are faced with the challenging task of extracting the information that is most relevant to their construction of a high-quality mental representation of the referent subject-matter (Lowe & Boucheix, 2008a, 2017, this volume). That is, learners are required to identify all relevant elements, the changes in these elements across time and space (i.e., behavior), and the relations that are involved. Novices in a domain, who lack the background knowledge to guide their processing top-down, have to rely primarily on the perceptual characteristics of the display (i.e., bottom-up) for attention allocation during the fulfillment of these tasks (in contrast to experts, who attend to relevant areas in dynamic visualizations: e.g., Balslev et al., 2012; Jarodzka, Scheiter, Gerjets, & Van Gog, 2010; Lowe, 2003). The situation for learners who are domain novices can be especially difficult if dynamic visualizations contain large numbers of elements that differ substantially with respect to their perceptual features, such as color, size, and placement (i.e., visuospatial contrast), and their behavior, such as when element movements contrast greatly with surrounding movements (i.e., dynamic contrast). Under these circumstances, such

B.B. de Koning (✉)
Erasmus University Rotterdam, Rotterdam, The Netherlands
e-mail: b.b.dekoning@fsw.eur.nl

H. Jarodzka
Open University of The Netherlands, Heerlen, The Netherlands
e-mail: halszka.jarodzka@ou.nl

R. Lowe, R. Ploetzner (eds.), *Learning from Dynamic Visualization*,
DOI 10.1007/978-3-319-56204-9_11

learners must deal with many and varied aspects that simultaneously compete for their limited attentional resources (Schnotz & Lowe, 2008).

Given that in instructional dynamic visualizations, the thematic relevance of depicted elements is often not well aligned with their perceptual salience, effective information extraction is likely to be less than optimal (Lowe, 2003; Lowe & Boucheix, 2017, this volume). Indeed, novices tend to pay much attention to perceptually salient elements of a dynamic visualization, even though these elements are not necessarily most relevant for constructing a coherent mental representation (Lowe, 1999, 2003). Given that dynamic visualizations present learners with a continuous flux of information, attending to salient but less relevant information implies that relevant information is attended to only partially or is missed completely (De Koning, Tabbers, Rikers, & Paas, 2009; Lowe, 2004). Moreover, by processing low relevance information the chance is increased that ineffective working memory load is created, which likely compromises subsequent processing (Sweller, Ayres, & Kalyuga, 2011). Consequently, learners will lack the necessary "building blocks" required to build an accurate and coherent mental model of the depicted content (Lowe & Boucheix, 2008a, 2017, this volume).

One way to avoid problems with information extraction is to provide scaffolds that guide learners' attention to the most relevant information in the dynamic display. The attention-guidance principle (Bétrancourt, 2005; Van Gog, 2014) holds that directing learners' attention to specific parts of the learning material may support learning. Further, according to the Animation Processing Model (APM; Lowe & Boucheix, 2008a, 2017, this volume) and the Cognitive Theory of Multimedia Learning (CTML; Mayer, 2009), attention guidance can be particularly useful in the first stages of animation processing involving the selection of relevant elements and/or the spatio-temporal relations between them. The potential benefit of attention guidance is not limited to making thematic relevant information more salient. De Koning et al.'s (2009) attention-guidance framework suggests that attention guidance may additionally serve to support the organization and integration of relevant information, both of which are also necessary to develop an accurate understanding of the content depicted in dynamic visualizations.

In the past decade, attention guidance has become an ubiquitous strategy among researchers to support learners in extracting the relevant information from dynamic visualizations (De Koning et al., 2009). Within this work, many of the techniques used for guiding attention in dynamic visualizations have essentially been 'borrowed' from those long used in static educational visualizations (Tversky, Heiser, Mackenzie, Lozano, & Morrison, 2008; for an overview, see Van Gog, 2014). For example, in static graphics there is a long tradition of using a bright color for indicating key information so as to draw attention to relevant aspects of the graphic (e.g., Jamet, 2014). Applied to dynamic visualizations, bright colors that establish a contrast with less relevant elements could be used for indicating relevant information (De Koning et al., 2009). However, it is important to note that what works with static visualizations may not necessarily be effective for dynamic visualizations (Tversky et al., 2008). Rather than just directing the learners' attention 'where to look and what to look at' to deal with (traditional) visuospatial challenges, attention guidance in dynamic visualizations is also concerned with the temporal challenge of 'when to look' (Schnotz & Lowe, 2008).

Therefore, researchers have started to develop more comprehensive and sophisticated attention guidance approaches, such as gradually progressing color cues on a path that is aligned with the unfolding causal chains of events (Boucheix & Lowe, 2010). These research-based approaches go far beyond the traditionally used attention guidance techniques in which instructional designers relied solely on their intuition and design experience to make decisions about how to help learners deal with dynamic visualizations. In the new approaches, researchers have designed principled ways of directing attention that are based both on the actual behaviors viewers exhibit during their processing and on associated theoretical explanations of such behaviors (see Lowe & Boucheix, 2017, this volume). By using eye tracking, for example, it is possible to investigate how domain experts allocate their attention to the information in a dynamic visualization and how this differs from what novices do. Such information can in turn be used to inform the design of attention guidance. So, a further approach includes "looking through the expert's eye" by highlighting the focus points of the expert while studying. Another development that can be observed in attention guidance strategies is to find ways to exploit the potential of a natural form of human communication, namely gesturing. An approach that is receiving increasing attention in this respect is to add human-like gestures to a dynamic display to guide learners' attention (De Koning & Tabbers, 2011).

This chapter summarizes the different approaches to attention guidance that have hitherto been explored in research on learning from dynamic visualizations. In doing so, we categorize them into three types of attention guidance based on the currently available literature: (1) cueing, (2) eye movement modeling examples, and (3) gesturing. For each type of attention guidance, we provide research-based examples, consider its strengths and weaknesses, and discuss its effectiveness in terms of perceptual and/or cognitive processing. Like the *referential ancillaries* investigated by Berney and Bétrancourt (2017, this volume), these types of support can be considered as a form of visual scaffolds.

11.2 Three Types of Attention Guidance

11.2.1 Cueing

Cueing, sometimes referred to as signaling, is the use of non-content information that is intended to direct attention to task-relevant elements or relations in a visual display (De Koning, Tabbers, Rikers, & Paas, 2007; Mautone & Mayer, 2001). Cueing can be done verbally (e.g., varying the speaker's intonation in a narration accompanying an animation to draw attention to key terms; Mautone & Mayer, 2001, Experiment 3) or graphically (e.g., arrows added to the dynamic visualization to direct attention to task-relevant information; Kriz & Hegarty, 2007). Irrespective of how cues are designed, most cues aim to fulfill the same information function (cf. Lemarié, Lorch, Eyerolle, & Virbel, 2008), namely, directing attention to specific parts of a dynamic visualization. So, in fact, as long as cues are able to guide attention to task-relevant information, it should not matter which specific cue is used.

The most widely used cueing approach, borrowed from the field of static graphics (Tversky et al., 2008), is graphical and concerns the use of visual cues that target specific elements in the dynamic display (De Koning et al., 2009). This is hence the type of cueing that we will focus on in this section, and we constrain our consideration to the processing of dynamic visualizations. Even though the attention that visual cues have received in research on learning from dynamic visualizations is more recent compared to the long tradition of using visual cues to direct learner attention to key information in static visualizations, a similar multitude of visual cues exists. Conventional visual cues can, for example, consist of arrows (Boucheix & Lowe, 2010; Lin & Atkinson, 2011; Lin, Atkinson, Savenye, & Nelson, 2016), colored circles (Jarodzka, van Gog, Dorr, Scheiter, & Gerjets, 2013), or colored lines (Boucheix & Lowe, 2010). Such visual cues are referred to as extrinsic cues because they are separate from the target entities and are only added to the dynamic display to draw attention to relevant elements. Other visual cues (e.g., color, texture) are embedded within the graphic entities of the referent content depicted in dynamic visualizations (i.e., intrinsic cues; Lowe & Boucheix, 2011). For intrinsic cueing, a range of techniques is available for increasing the salience of specific displayed elements relative to surrounding information. For example, giving a relevant element in the dynamic display a bright color helps making it stand out against the rest of the display. An alternative approach to visual cueing, which Lowe and Boucheix (2011) refer to as anti-cueing, is to reduce the salience of less relevant information while leaving the relevant information unchanged. As a result, the relevant information has higher salience than the information in the rest of the display. This is the case, for example, when color contrasts are used to shade everything except for the element of interest, which gives the impression as if there was a spotlight on it (De Koning et al., 2007, 2010a). In that sense, designers, who nowadays often also include teachers, not only make decisions on what aspects of a dynamic visualization should receive attention, but also have a variety of visual cues from which they can choose for offering attention guidance.

Empirical Findings on the Effectiveness of Cueing The potential of visual cueing to support learner processing of dynamic visualizations has received much attention from researchers in recent years (for an overview, see De Koning et al., 2009). Although results regarding the educational effectiveness of visual cues have been mixed, several interesting findings have surfaced. First, eye-tracking studies have provided useful insights into the perceptual processing of visual cues. Across several studies, it has been shown that conventional visual cues in the form of spotlight-cueing (De Koning et al., 2010a) and arrows (Boucheix, Lowe, Putri, & Groff, 2013; Kriz & Hegarty, 2007) can effectively guide learners' attention to specific, relevant information in an unnarrated animation. That is, given equal study time, learners inspected the relevant information in an animation longer and more often when this information was visually cued than when it was uncued. This is in line with the notion that visual cues may serve to facilitate selecting relevant information, which is one of the essential processes underlying learning from dynamic visualizations (Lowe & Boucheix, 2008a; Mayer, 2009). A caveat here is that the effect of the attention guidance may be short-lived. Studies by Boucheix et al. (2013) and De Koning et al. (2010a) indicate that learners initially focus their attention on the visually cued information, but shortly after the onset of the cue learners may redirect their attention to surrounding information. Presumably, this occurs because learners are either distracted by ongoing perceptually conspicuous elements or because they try to relate the visually cued information to adjacent elements and/or higher-order event units of which the depicted system is composed (Boucheix et al., 2013). Directing learners' attention productively for a longer period of time may thus require a more temporally distributed form of visual cueing, such as provided by dynamic cues as discussed below (e.g., Boucheix & Lowe, 2010; Boucheix et al., 2013).

Despite guiding the learners' attention, in contrast to visual cues in static graphics, in dynamic visualizations conventional (i.e., static) visual cues do not necessarily improve cognitive processing leading to enhanced learning outcomes (e.g., De Koning et al., 2010a; Lin & Atkinson, 2011; Lin et al., 2016; for an overview, see De Koning et al., 2009). De Koning et al. (2007), for example, showed that using a spotlight-cue to make the heart valves in a non-narrated animation of the cardiovascular system more salient, resulted in better comprehension and transfer performance than studying the animation without visual cues. Similarly, adding conventional arrow cues (i.e., one arrow was presented at the moment attention needed to be directed to specific information) to an animation about the cardiovascular system accompanied by a narration resulted in better comprehension of the depicted content than students who studied the animation without these cues. Lin and Atkinson (2011), using arrows to direct attention to relevant information in a narrated animation on the formation of rocks, demonstrated that arrow-cueing decreased learning times compared to a no-cueing condition. Because learning outcomes did not differ between cued and uncued conditions, they suggested that studying an animation with conventional visual cues is more efficient than studying an animation without such cues. In a similar vein, Amadieu, Mariné, and Laimay (2011) report that, in trying to understand a non-narrated animation about long-term potentiation in synaptic transmission, participants experienced less mental effort, as measured by perceived difficulty ratings, when attention was directed to task-relevant information by zooming in, compared to studying an uncued version of the animation. But this difference was found only after several exposures of the animation. Those in the visually cued condition also showed better comprehension and problem solving performance. This suggests that under appropriate conditions, providing visual cueing can be instructionally more effective.

Although the studies mentioned so far may give the impression that conventional visual cues can be effective for enhancing cognitive processing (here: learning), there are many studies that indicate otherwise (e.g., Boucheix & Lowe, 2010; Boucheix et al., 2013; De Koning et al., 2010a, 2011; Lin & Atkinson, 2011; Lowe

& Boucheix, 2011; also see De Koning et al., 2009 for an overview). De Koning et al. (2009) showed that in 7 out of 8 studies comparing conventional visual cues to no cues, visual cueing was ineffective for learning. Consistent with this, studies by Kriz and Hegarty (2007) and De Koning et al. (2010a), for instance, both showed that participants spent more time looking at relevant information when attention was drawn to it in a non-narrated animation by means of conventional visual cues (i.e., arrows and spotlight-cueing respectively) compared to studying the same information in an uncued animation. So, the visual cues undoubtedly helped learners to look at the animation's most relevant elements. In spite of that, learners failed to show improved understanding and transfer performance. This suggests that merely getting someone to look at something does not guarantee that the person actually engages in relevant cognitive processing activities and learns something as a result. Similar results have been reported by Boucheix et al. (2013) and Boucheix and Lowe (2010) using arrow-cueing whereby multiple arrows could be simultaneously presented, compared to no cueing, in a non-narrated animation on the workings of a piano mechanism. Moreover, several studies have shown that, given equal learning outcomes in cued and uncued conditions, conventional visual cues also do not have any cognitive benefits in terms of experiencing lower mental effort (e.g., De Koning et al., 2007, 2010a; Lin et al., 2016; for an overview see De Koning et al., 2009). Together, these findings suggest that more appropriate direction of attention due to the effect of conventional visual cues does not automatically lead to enhanced cognitive processing of dynamic visualizations. So, although conventional visual cues are capable of guiding attention, they may not be sufficient to enhance understanding of a dynamic visualization (cf. De Koning et al., 2009).

The findings described above have led several researchers to explore ways to improve the effectiveness of conventional visual cueing by complementing its *perceptual* guidance with other techniques that encourage deeper *cognitive* processing. An approach that has become increasingly popular in this respect is asking learners to generate self-explanations (e.g., De Koning et al., 2010b, 2011; Lin & Atkinson, 2011; Lin et al., 2016). Self-explaining has been shown to be very effective at promoting the comprehension of text and static graphics (Fonseca & Chi, 2010). De Koning et al. (2011) demonstrated that learning from animations containing conventional visual cues (i.e., spotlight-cueing) can be improved by prompting students to provide self-explanations. Participants studied a non-narrated animation of the cardiovascular system either with or without visual cueing and did or did not provide self-explanations. Results showed that the attention guidance provided by the visual cues helped learners to generate more detailed and more meaningful self-explanations about the components and events depicted in the animation. On a test of inference making and transfer performance, the cueing with self-explanation group outperformed participants who self-explained without visual cues or those who did not self-explain. However, in a follow-up study, De Koning et al. (2010b) were unable to replicate this finding. Similarly, using a narrated animation on the formation of rocks containing arrow-cueing, Lin et al. (2016) did not show improved learning outcomes of self-explaining prompted by the computer, irrespective of whether the self-explanations were provided during or after the animation. It seems that even engendering cognitive activity in learners may not guarantee that they will make more effective use of the guidance provided by conventional visual cues for learning from dynamic visualizations.

New Directions in Research on Cueing The research discussed thus far indicates that conventional visual cues do not consistently improve learning from dynamic visualizations, even when additional measures like self-explanation are taken. Nevertheless, these findings are not completely inexplicable if one takes into account the fact that conventional visual cues primarily address "where" to look and "what" to look at. Such cues provide little or no guidance as to "when" to look at this information in the dynamic display, unless they are displayed only for specific phases during which the learner should pay attention to the cued content (Schnotz & Lowe, 2008; Boucheix & Lowe, 2010). According to Lowe and Boucheix (Boucheix & Lowe, 2010; Boucheix et al., 2013; Lowe & Boucheix, 2011), conventional visual cues are limited in the sense that they are static pointing aids providing attention guidance to only the individual elements in the display. Further, these cues are insufficiently powerful to compete with the dynamic contrasts present in the display (Schnotz & Lowe, 2008), that is, the very strong attention-directing effects of the movements and changes of the animation itself (Lowe & Boucheix, 2011). In addition, conventional visual cues do not continuously provide information about how the different events spread sequentially through the depicted system. They therefore do not provide precise, explicit information about unfolding causal chains of events (Boucheix & Lowe, 2010). It follows that (re)designing visual cues that are spatially and temporally aligned with the presentation of key elements in a dynamic visualization and also convey relational information among these elements may better match the perceptual and cognitive processing challenges induced by dynamic visualizations.

In order to address the shortcomings of conventional static visual cues, Boucheix and Lowe (Boucheix & Lowe, 2010; Boucheix et al., 2013; Lowe & Boucheix, 2011) devised an alternative to such external point cueing which they refer to as internal continuous cueing (Boucheix & Lowe, 2010). A particular instantiation of this dynamic form of visual cueing are spreading color cues (or progressive path cues; Boucheix et al., 2013). In this approach, highly contrasting colored ribbons are progressively overlaid on the dynamic visualization's entire chain of relevant events in synchrony with the propagation of the depicted system's events (Boucheix & Lowe, 2010). If multiple chains of events take place simultaneously, different colors can be used to distinguish these different events chains. In contrast to conventional cues, dynamic visual cueing not only draws attention to relevant individual elements, but also signals the events to which these elements belong and how different events are related. Local coordinated cues (Boucheix et al., 2013) are a further type of dynamic cueing that are a refined implementation of the principles embodied in progressive path cues. Rather than being applied to the entire chain of events, they provide more tightly focused guidance directed at specific locations that present relevant interactions and/or functional relations. The close spatial and temporal fit of these dynamic forms of visual cueing to the changing information that is the defining characteristic of a dynamic visualization is designed to help learners to extract the relevant information from this demanding form of representation.

Empirical studies show that both forms of dynamic visual cueing do indeed help learners to direct their attention to relevant elements and events, and that they do so for a longer period of time than with conventional visual cueing (i.e., arrows; Boucheix & Lowe, 2010; Boucheix et al., 2013). This suggests that dynamic visual cueing may more effectively outcompete the dynamic challenges posed by dynamic visualizations. Moreover, in contrast to conventional visual cues or no cueing, both forms of dynamic cueing have been found to enhance comprehension of the depicted system. From the six (out of 13) studies reviewed by De Koning et al. (2009) that showed a positive effect of cueing on cognitive processing, three studies involved dynamic visual cues and in two studies dynamic relations and events were cued, suggesting that dynamic cueing of elements and relations is an effective cueing approach. For example, Boucheix et al. (2013), using a non-narrated dynamic visualization of the piano mechanism, showed that participants who studied the animation with local coordinated cues or with progressive path cues had a better comprehension of the kinematic and functional information than those who studied the uncued or the conventionally cued animation.

In short, both conventional and dynamic visual cues can be used effectively for directing attention to relevant information in a dynamic visualization. The continuous guidance and possibility to counteract dynamic contrast in dynamic visualizations offered by dynamic visual cues makes this type of cueing preferable for directing attention in dynamic visualizations involving complex perceptual interplay between events. With respect to cognitive processing, the failure of conventional visual cues to provide consistent beneficial results on learning, also speaks in favor of incorporating this latter type of visual cues. Indeed, dynamic visual cues do seem to be related to improved learning outcomes. So, tailoring (dynamic) visual cues to the demands of dynamic visualizations seems a promising way to effectively guide attention and improve learning from dynamic visualizations. However, the efficacy of even dynamic visual cues can be limited when animations present complex content that is unfamiliar to the target learners. Under these circumstances, more fundamental matters such as the basic assumptions underpinning the design of the animated presentation itself may need to be addressed (see Lowe & Boucheix, 2017, this volume).

11.2.2 Eye Movement Modeling Examples

The previous section has shown how visual cueing can influence learners' attention allocation and comprehension of dynamic visualizations. It also demonstrated that many traditional cueing techniques were based on the designers' experience and intuition on which elements to cue in a visualization. It is easy to imagine that a lack of domain knowledge or experience in teaching a certain topic could lead to inappropriate cueing that would prejudice students' understanding. Consequently, instructional designers should (and often do) consult domain and/or instructional experts from the target domain to design cues within an instructional visualization. Some visualizations, however, introduce additional demands on viewers because they depict perceptually complex subject matter. Examples include portrayals of animal locomotion (such as fish swimming styles) or the operation of sophisticated devices (such as the working of a traditional piano mechanism). In such cases, the observer's perceptual processing plays a crucial role in dealing effectively with the presented information. These processes of attention allocation involve fast, unconscious decisions and thus are not readily accessible and difficult to put into words (Ericsson & Simon, 1993). Under these circumstances, an alternative way to access information on perceptual processing, and hence what information should receive attention, is to use eye tracking to directly measure the attentional focus of a domain-expert, while executing a task (Holmqvist et al., 2011).

Grant and Spivey (2003) implemented this approach by presenting participants with a graphic showing an insight problem (Dunker's radiation problem) while their eye movements were recorded. They then used the eye movement data of those participants who successfully solved the problem to design a cue (i.e., pulsing colored lines). In a follow-up experiment, participants received the same insight problem either with a cue based on the eye movements of successful problem-solvers, or of unsuccessful problem-solvers, or an uncued visualization. Participants who studied the visualization with the 'successful' cue solved the problem significantly more often than the other two groups. Litchfield and Ball (2011) followed up on this study and, instead of creating a cue, used eye movements directly. Their study showed that indeed, those eye movements that indicated a successful viewing strategy, helped participants to solve the insight problem. In a different study, Litchfield, Ball, Donovan, Manning, and Crawford (2010) have shown that viewing another person's eye movements on a visualization not only enhances insight problem solving, but also the performance on complex cognitive tasks, such as visual search of nodules in medical images. These studies show that displaying a successful person's eye movements (or cues based on them) can foster task performance, as indicated by reaction times and problem solving success, when dealing with static visualizations. Drawing upon these findings, other researchers have begun to explore whether such an approach (i.e., using an expert's eye movements to direct attention) might also be beneficial for *learning* from *dynamic* visual displays. Their approach is based on example-based learning as outlined below.

Learning from Examples in Dynamic Visualizations Learning by imitating another person is a very fundamental and safe way of learning how to do something. It (a) shows how to approach a problem or a task without having to search for one's own approach that might be inefficient or suboptimal, and (b) immediately shows the consequences of the modelled approach. This form of learning is inherent to humans as evidenced by the fact that 3 week old babies are already able to imitate the mimic and gesturing of grown-up models (Meltzoff & Moore, 1977). However, imitation itself does not guarantee that the imitated behavior will be displayed later once the model is no longer present. Bandura has shown in a series of experiments how children learn by observing a model during task performance (Bandura, 1977). In his famous 'bobo doll' studies, he showed children videos of an adult model interacting with a so-called 'bobo doll' (e.g., Bandura, Ross, & Ross, 1961). These light-weighted dolls can be pushed down but, due to their low center of mass, get up again 'by themselves'. Participating children watched adult models interacting with these dolls either in an aggressive or a non-aggressive way (another group of children watched no model at all). These children then entered a room where such a bobo doll was present. Because they imitated the previously observed behavior, it was concluded that learning had taken place. This approach has since been taken up by educational researchers under the title of 'example-based learning' or 'modeling' (for an overview see Van Gog & Rummel, 2010). The general principle involved is that learning by studying examples of a worked-out, successful task performance is more efficient than learning by problem-solving alone (Kirschner, Sweller, & Clark, 2006).

In education, we are often interested in helping learners to develop skills, such as those involving cognitive processes, that are not directly observable. A standard approach to developing these skills is for a model to verbalize her or his internal states (cf. cognitive apprenticeship: Collins, Brown, & Newman, 1989; process-oriented worked-examples: Van Gog, Paas, & Van Merriënboer, 2004). Until recently, only skills that are primarily cognitive, such as reading, writing, and calculating have been modeled. However, for tasks with a highly visual component, such as the processing of dynamic visualizations, perceptual skills would also need to be modeled. Van Gog, Jarodzka, Scheiter, Gerjets, & Paas (2009a) therefore proposed that process-oriented worked-examples be enriched by recordings of an expert model's eye movements as a means of also externalizing successful *perceptual* processes. When applied to visualizations, this approach involves overlaying on the presented visualization the eye movements of an expert model who is explaining while he or she completes the task. These so-called Eye Movement Modeling Examples (EMME; Van Gog et al., 2009a) thus refer to instructional videos that include (1) audio of an expert model's explanation of how to interpret a dynamic visualization and (2) a graphic representation of the moving attentional focus (i.e., the eye movements) recorded from that model during the explanation.[1] This form of attention guidance can be considered to be a specific type of dynamic visual cueing as discussed in the previous section. In fact, the way EMME is displayed is similar to that used in other types of cueing. It involves adding artificial visual information about the model's eye movements in the form of a dot or circle (i.e., extrinsic cues), or by subordinating existing information and displaying the model's eye movements as a spotlight (i.e., anti-cues). The crucial difference between this approach and visual cueing that is based solely on an instructional designer's experience and intuition is that the decision about which elements of a visualization to cue is based on actual and successful processing behavior. The EMME approach is consistent with

[1] It should be noted that not all researchers see the verbal explanation as an inherent part of EMME. The initial idea by Van Gog et al. (2009a) expected both aspects to be necessary, assuming that people cannot "read" another person's eye movements easily.

theoretical perspectives regarding cognitive processing of media information (cf. Cognitive Theory of Multimedia Learning: Mayer, 2009). In a nutshell, CTML assumes that the learner has to (a) *select* the relevant information from the visualization (and other information sources such as written or spoken text) and then (b) *integrate* all information found (preferably together with prior knowledge) into one coherent mental model. In line with these assumptions, eye movement-based guidance should (a) capture the learner's attention and guide it to the relevant aspect of the visualization at the right time to help the learner in selecting the relevant information from the dynamic visualization. Next, (b) this guidance should maintain the learner's attention while processing the audio information to help the learner integrate the information from both sources. Both steps are necessary for successful information processing and thus, for learning to take place. Another advantage of EMME is that it should also improve comprehension of the model's verbal explanation. This assertion is supported by the work of Richardson and Dale (2005), who showed that when a listener simultaneously looks at the spot a speaker is looking at, the listener understands the speaker's speech far better than if their gazes are less tightly coupled. A better understanding of the model's explanation should in turn have a positive influence on learning.

Empirical Findings on the Effectiveness of EMME Although there has as yet been relatively little research into the effectiveness of EMME on learners' processing of dynamic visualizations, several interesting findings have already emerged. First, EMME have been shown to be effective in guiding attention. This effect, however, is bound to design particularities of how the eye movement records are presented. In one approach, eye movement information is presented via additional elements superimposed onto the visualization (i.e., extrinsic cues). However, adding yet more information to visualizations that are already perceptually complex may overwhelm inexperienced learners. Another option is to present eye movements using a spotlight-like design (i.e., anti-cues), which subdues existing information of these visualizations (Dorr, Jarodzka, & Barth, 2010). Research not directly related to learning has already shown that displaying eye movement data on videos by reducing the local spectral energy on the parts that are not automatically looked at (i.e., a form of spotlight) guides the attention of new viewers inspections of the presented information (Dorr, Vig, Gegenfurtner, Martinetz, & Barth, 2008; Nyström & Holmqvist, 2008). Extending this work to a learning context, Jarodzka and colleagues (Jarodzka et al., 2012; Jarodzka et al., 2013) investigated two aspects of attention guidance. They investigated whether EMME guide learners' visual attention during learning by tracking the learners' eye movements while they studied EMME videos. Additionally, they investigated whether learners' visual processes were trained to the extent that they were able to transfer them to novel examples. This was done by having these learners watch new, but comparable, videos without any guidance while their eye movements were recorded. In both studies, (Jarodzka et al., 2012, 2013) participants studied four modeling videos either with (a) a verbal explanation only, (b) with a verbal explanation and the model's eye movements displayed as a dot, or (c) with a verbal explanation and the model's eye movements displayed as a spotlight. Results showed that displaying the model's eye movements as a spotlight was best for guiding the learner's visual attention to the right places at the right time. Furthermore, the spotlight display also trained learners' perceptual skills insofar that their visual search for relevant elements in new videos was facilitated most. Thus, EMME have the potential to guide learners' attention to the right spots at the right time. In addition, learners can adapt such perceptual skills so as to successfully apply them to new examples.

Second, with respect to the effects of EMME on learning, findings have been mixed. Although some studies show that EMME can foster learning (Jarodzka et al., 2012, 2013), there is also evidence that EMME is sometimes not helpful for learning (Van Gog et al., 2009a). This suggests that, similar to cueing, increased attention for relevant information in a dynamic display does not necessarily coincide with improved cognitive processing and hence learning. A close examination of the EMME studies reveals some aspects that may have contributed to the divergent findings on learning. One crucial aspect appears to be the nature of the task in which EMME are used. Van Gog and colleagues (2009a), for example, investigated EMME using a so-called 'frog leap' game, which is a logic game resembling the tower-of-Hanoi game (a popular task used in psychology to investigate problem solving). It presented three brown frogs facing three green frogs with each frog sitting on one stone plus one empty stone in the middle. The aim of the game is that both frog groups switch sides, given that (a) only one frog can sit on each stone and (b) a frog can only jump over one other frog to sit on the empty stone. This game was presented as an animated flash-game. Participants studied a modeling example consisting of a video recording of a successful model performing the game, either with or without an audio explanation of why s/he chose each step and with or without the model's eye movements superimposed onto the video recording. A fifth group received no modeling examples and learned by problem-solving alone. Results showed that observing a model led to better learning outcomes and lower reported mental effort than mere problem-solving. For the modeling examples, learning from an example with verbal explanations, but without additional eye movements displayed on the screen yielded the best learning results, while learning from an EMME example with verbalizations and eye movement display yielded the worst results. These findings suggest that studying EMME with additional attention guidance offered by eye movements superimposed on the display was prejudicial to learning. Solving of the frog-leap problem required an understanding of the reason behind each step. This was provided very well by the verbal explanations in their own right which is a likely explanation for the findings. These explanations did not pose any perceptual challenges to the learner. Further, perceptual guidance (as provided by the eye movements) might not actually have been relevant to this particular type of task.

In contrast, for tasks requiring learners to deal with perceptually complex subject-matter, such as classifying locomotion patterns of reef fish and diagnosing epileptic seizures of infants, the effects of EMME on learning performance are more promising (Jarodzka et al., 2012, 2013). These tasks used by Jarodzka et al. (2012, 2013) involved visualizations that are (a) information dense so that it is difficult to select the relevant out of the irrelevant information and (b) dynamic so that it is important to attend to the relevant information at the right time. To execute these two tasks it is important to (1) visually search and detect relevant elements (locomotion classification: specifying body parts that were used to produce propulsion; epileptic seizure: specifying body parts that might be affected by the disease), (2) interpret the motion of the detected elements, and (3) categorize these observations according to the appropriate locomotion class (Lindsey, 1978) or diagnosis (International League Against Epilepsy, 2010). It is important to note that the first two steps – which form the basis for executing the third step – require a perceptual input, and are thus referred to as 'perceptual skills' (e.g., Chi, 2006). These are not trivial skills because they cause novices considerable problems (Balslev et al., 2012; Jarodzka et al., 2010). In the studies by Jarodzka et al. (2012, 2013), learning to classify the fish and diagnosing epileptic seizures was facilitated by EMME in that learners were better able to interpret the relevant elements in the dynamic display. Thus, learning to execute tasks that have a significant perceptual component was more likely to benefit from EMME than learning to execute a task in which the challenges are mainly cognitive (i.e., problem-solving).

Another aspect that may play a role in the effectiveness of EMME on learning is the way in which the guidance offered by EMME is designed. In the study by Van Gog et al. (2009a), for example, the model's eye movements were displayed as yellow solid dots 'jumping' across the screen. This form of eye movement display increased the visual complexity of the original visualization without adding relevant information and probably explains the worse performance of the modeling groups. In contrast, Jarodzka et al. (2012, 2013) found that participants benefited from studying EMME when they classified or diagnosed new videos presenting examples comparable to those they previously studied with EMME. However, interpretation of the relevant elements was improved for new videos only if the eye movement display still allowed for a holistic impression of the entire visualization (i.e., when the spotlight did not blur out too much information). This suggests that the learning effectiveness of EMME is influenced by how the eye movements are displayed in EMME.

New Directions in EMME Research Positive findings regarding the potential utility of EMME have recently been obtained in various domains in which perceptually complex visualizations play a role, such as X-rays and histological slides (Vitak, Ingram, Duchowski, Ellis, & Gramopadhye, 2012), text-picture integration when learning from science text (Pluchino, Tornatora, & Mason, 2012), or learning geometry (Van Marlen, Van Wermeskerken, Jarodzka, & Van Gog, 2014). Accompanying these developments with respect to static visualizations, the use of EMME in dynamic visualizations is also increasing with a number of novel directions being pursued. First, researchers have started to investigate the usefulness of EMME in different types of learning environments. In an interactive environment, for instance, Wilson et al. (2011) have shown that using expert's eye movements as a basis for training the gazes of less experienced surgeons on a simulator improved learning in comparison to a group with movement training or discovery learning.

Further, Bednarik, Gowases, and Tukiainen (2013) have used a specific form of EMME in a computer-based collaborative problem-solving scenario in which they displayed the eye movements of both cooperating partners to each other (in contrast to displaying only their mouse movements or nothing at all). Their study indicates that the EMME-group outperformed the other two groups in problem-solving in that they made fewer mistakes and reported a better user experience. Second, recent research suggests that EMME is not only helpful as a processing aid during a task, but may also effectively support learners' perceptual and cognitive processing when used as pre-training. In their study, Skuballa and Renkl (2014) presented students with an EMME pre-training before learning from (either static or dynamic) visualizations of a solar power plant. They found that such pre-training led to more homogenous perceptual processes during inspection of the actual visualizations and resulted in better learning outcomes compared to no pre-training.

To summarize, EMME have potential to guide students in learning from complex, dynamic visualizations and thus foster learning from them. However, EMME are not a universal remedy. On the contrary, they are helpful only if at least the following two conditions are fulfilled: First, the visualization that has to be learned is perceptually complex to such an extent that a novice learner would benefit from perceptual guidance. Second, the display of the model's eye movements should not pose additional challenges onto the learner, but instead still allow for a holistic impression of the entire visualization. If EMME complies with these aspects, it can be a helpful strategy for effectively processing dynamic displays, even in interactive environments or as pre-training.

11.2.3 Gesturing

Another way to draw attention to task-relevant elements in a dynamic visualization is to use pointing gestures (typically via the index finger, hand or arm) that can direct learners' attention to a specific location within a dynamic display (De Koning & Tabbers, 2011). The specific pointing gestures discussed here are limited to human-based processing aids that are purposefully aimed at directing learners' attention. Our discussion therefore excludes gestures that are spontaneously produced as part of human communication and learning which often occurs during the completion of educational tasks such as when reasoning about diagrams (Hegarty, Mayer, Kriz, & Keehner, 2005) or spatial problem solving (Chu & Kita, 2011). The pointing gestures we consider have some similarity to EMME in the sense that they are based on natural and purposeful human behavior. In this sense, it seems a small step to designing pointing gestures as attention directing aids on the basis of what experts point at with their gestures. Nevertheless, empirical literature shows that thus far researchers seem to only use pointing gestures for information that they themselves identify as important for learners to look at. Pointing gestures may be provided by inserting an image of a pointing gesture, for instance in the form of a pointing hand (De Koning & Tabbers, 2013), into a dynamic visualization, or by a human or virtual agent (often referred to as a pedagogical agent; Mayer, 2014) positioned next to a dynamic visualization. In this chapter, the emphasis will be on on-screen additions of pointing gesture to dynamic visualizations in order to facilitate comparisons amongst gesturing, visual cueing, and EMME.

Empirical Findings on the Effectiveness of Pointing Gestures The idea that pointing gestures can be helpful for processing dynamic visualizations is being increasingly acknowledged (De Koning & Tabbers, 2011). In dynamic visualizations, considerable research attention has been devoted to the relative processing benefits of pointing gestures produced by on-screen animated pedagogical agents and pointing gestures added to the dynamic visualization itself. There has been substantial research into the effects of on-screen pedagogical agents on cognitive performance. This research generally demonstrates that adding an on-screen agent displaying human-like pointing gestures, movements, facial expressions, and eye contact facilitates learning, a finding which has been referred to as the "embodiment principle" (Mayer, 2014). According to Mayer (2014), 11 out of 11 empirical comparisons showed that performance on transfer tasks was higher for learners who learned with a high-embodied agent than for those who learned with a low-embodied agent ($d = 0.36$). Although most of the studies involved in this comparison did not include dynamic visual materials, several studies suggest that the same findings apply to dynamic visualizations (e.g., Lusk & Atkinson, 2007; Mayer & DaPra, 2012; Moreno, Reisslein, & Ozogul, 2010). In a study by Lusk and Atkinson (2007), for example, learners obtained higher transfer scores when studying animated worked-out examples of mathematical word problem solving with a fully embodied on-screen agent (i.e., involving pointing gesture, gaze, movement) than with a minimally embodied agent (i.e., not involving pointing gesture, gaze, movement) or no on-screen agent. Despite these promising findings regarding the effectiveness of incorporating pointing gestures in dynamic visual displays, several aspects remain unclear.

First, there has been little empirical work that has specifically compared pointing gestures of an animated pedagogical agent with other types of attention guidance, such as conventional visual cues (e.g., arrows) as an appropriate control, so their relative effectiveness is unclear. Moreno et al. (2010), for example, asked learners to study a computer-based simulation on electrical circuits including either an on-screen pedagogical agent that pointed to relevant elements in the dynamically presented visual information using gesture, conventional arrow-cues signaling relevant information, or no attention-directing method. Results showed that students learning with pointing gestures produced by an on-screen agent (together with eye gaze and facial expressions) outperformed the other groups on transfer performance. Similarly, Johnson, Ozogul, and Reisslein (2015), using animated visualizations on Ohm's law, provided evidence that learners developed a better understanding and had higher electrical circuit problem solving performance when an on-screen agent used gesture to point to relevant elements than when elements were identified via arrow-cueing. Their results further suggest that the attention guidance provided by on-screen agents is especially beneficial for learners with low, rather than high, prior knowledge. These findings suggest that on-screen agents using pointing gestures to direct (novice) learners' attention to relevant information can facilitate learning from dynamic visualizations (for similar findings, see Johnson, Ozogul, Moreno, & Reisslein, 2013).

The findings discussed thus far, however, still do not justify the conclusion that pointing gestures themselves are responsible for enhanced understanding. On-screen pedagogical agents usually use a combination of different human-like behaviors (facial expressions, eye gaze, etc.), not just pointing gesture alone. In line with social agency theory (Mayer, 2009), it has been proposed that these artificial cues mimic the social cues given by real humans and so allow learners to characterize on-screen pedagogical agents as having a social dimension. Hence, learners are more likely to take a social stance in which they are more willing to invest effort to learn in order to make sense of the material. Without experimental studies that try to distinguish between these different human-like attributes of on-screen agents, it is not currently possible to isolate the precise contributions of pointing gesture. Recent research on learning from a gesture-enhanced dynamic display, however, provides preliminary evidence that pointing gesture may serve to promote learning independently of social and motivational influences. In a study by De Koning and Tabbers (2013), learners studied an animation on the formation of lightning, with attention guidance to relevant information provided either by conventional arrow cueing or by a real picture of a pointing finger (with hand and arm being shown). The arrow and pointing finger that provided dynamic guidance by tracing the movements in the animation were presented on-screen with identical size and position. Results showed that studying the animation with the on-screen human hand yielded higher retention and transfer performance than studying the animation with arrow-cueing. So, even when pointing gestures are used in the absence of any other human-related behaviors (e.g., facial expressions) that might engender social influences on cognitive processing, they seem to be more beneficial for learning than conventional arrow-cues. Although a decisive explanation for this finding has yet to be developed, it has been proposed (De Koning & Tabbers, 2011, 2013) that pointing gestures recruit another (i.e., psychomotor) modality through which the presented information can be processed. According to this suggestion, pointing gestures could enable learners to establish a link between the presented information and their own body movements, thereby helping them to understand the system's dynamics based on the movements they can make with their own body rather than just serving as an extra retrieval cue (De Koning & Tabbers, 2013). This would be consistent with more general ideas on the embodied or situated nature of cognition which posit that mental representations are grounded in perceptual and motor experiences (for an overview, see Barsalou, 2008). It may, however, very well be possible that a simpler explanation underlies this finding. Van Gog, Marcus, Ayres, and Paas (2009b) have speculated that in addition to auditory and visual working memory processing channels, the psychomotor system may have its own working memory processing channel dedicated specifically to processing psychomotor information. Pointing gestures added to a dynamic visualization may therefore be a way to further expand working memory capacity available for learning.

New Directions in Research on Pointing Gestures The research conducted thus far provides little definitive indication as to the plausibility of these or other perceptual and/or cognitive processing mechanisms underlying the observed effects of pointing gestures. As with conventional visual cueing, it seems reasonable to suggest that pointing gestures primarily function to direct learners' attention to specific parts of the dynamic display, so that the relevant information is selected for further processing (Mayer, 2014). However, specific studies have not yet been conducted into whether pointing gestures, either produced by on-screen pedagogical agents or incorporated into the dynamic visualization itself, effectively guide attention. Whereas some studies with an on-screen pedagogical agent using pointing gestures have involved eye-tracking methodology, these studies only investigated learners' general distribution of visual attention over the dynamic display in relation to the on-screen agent presented along with it (e.g., Louwerse, Graesser, McNamara, & Lu, 2008). Evidence from other fields of research suggest that human (−like) pointing gestures can effectively guide visual attention across various perceptual (learning) tasks (e.g., Kuhn & Martinez, 2011). For example, Gregory and Hodgson (2012) showed that a pointing hand in an anti-saccade task affected reaction times, whereas a pointing arrow did not. Findings with regard to dynamic visualizations provide indirect evidence only for similar perceptual processing effects due to pointing gestures. For example, lower perceived difficulty ratings regarding the learning task were obtained for pointing gestures (and arrow-cueing) by Johnson et al. (2015), which may suggest that extraneous cognitive load associated with distinguishing relevant from less relevant information was decreased (also see Moreno et al., 2010). This may have helped learners to attend to the relevant information. It may also be suggested that pointing gesture, as well as other types of visual cueing, can establish a referential connection between the dynamic display portraying the content and a narration, which in the majority of studies with on-screen agents accompanies the visual material. However, no direct evidence for this is available yet (e.g., using eye-tracking investigating the number of integrative saccades) and, despite higher learning outcomes, difficulty ratings do not lend support to interpretations of increased generative processing (e.g., Johnson et al., 2015).

In sum, empirical evidence is accumulating that pointing gestures are capable of improving learners' understanding of dynamic visual content and may even be more effective than conventional visual cues in facilitating interpretation of conceptual material (e.g., De Koning & Tabbers, 2013). Several aspects, particularly regarding the effects of pointing gestures on perceptual processing and separating these effects from social influences of on-screen agents, require further investigation.

11.3 Conclusions and Outlook

11.3.1 Theoretical Implications

This chapter discussed three attention guidance strategies that are currently receiving increasing attention as processing supports for learning from dynamic visualizations. The studies reviewed here showed that offering attention guidance to learners can contribute to improved perceptual and cognitive processing of dynamic visualizations. In particular, the empirical evidence regarding learners' perceptual processing consistently shows that inserting visual cues and EMME can be helpful for focusing attention on specific information in the dynamic display. This is in line with, and provides support for, the assumptions of relevant theoretical models such as the Animation Processing Model and the Cognitive Theory of Multimedia Learning that the provision of attention guidance can help learners to attend to task-relevant information from dynamic visualizations (Lowe & Boucheix, 2008a; Mayer, 2009). It is important to realize that attention direction of itself does not automatically result in corresponding cognitive benefits with respect to learning outcomes, such as better memory for and a deeper understanding of the materials. This needs to be taken into account when considering theoretical suggestions that attention guidance enables learners to engage in meaningful cognitive processing of the presented information (e.g., De Koning et al., 2009; Mayer, 2009). In many of the studies discussed above, attention guidance inserted in the dynamic display did not result in improved learning outcomes or else attention guidance alone could not account for the observed cognitive benefits. On this latter point, improved learning from the dynamic visualization could also be the result of the visual guidance serving functions other than just directing attention (such as the organization and integration of information; De Koning et al., 2009; Lowe & Boucheix, 2008b). In several studies showing enhanced learning performance from the addition of visual cues or pointing gestures to the dynamic display, verbal explanations also accompanied the dynamic visualization. It is likely that mention of specific information in a narration can function to direct learner attention to that information in the dynamic display and facilitates making connections between corresponding pieces of information (De Koning et al., 2009). The research reviewed on EMME further showed that cognitive benefits are most likely to show up when learners can still perceive all aspects of the dynamic visualization while their attentional focus is being directed at one of its specific constituent elements. This suggests that even though attention guidance to one part of the display that should receive attention at a certain moment in time is important, there should also be possibilities for learners to place this information in the context of the surrounding information. Integrating information this way likely helps them to identify what function a specific component has in service of key events and/or the entire causal chain of events. In this respect, progressive path cues and local coordinated cues (Boucheix et al., 2013) provide good examples of how to draw attention to important events and combine these into gradually more complex dynamic chains of events in order to build a coherent mental representation of the depicted content.

11.3.2 Educational Implications

The research findings reviewed in the previous sections show that attention guidance is a potentially useful approach for those who design dynamic visualizations. However, careful consideration on a case-by-case basis is required in order for attention guidance to fulfill its potential for supporting perceptual and/or cognitive processing. For example, depending on the particularities involved, guidance of attention may need to be continuous, aligned with the causal chain of events, and (in some cases) reflective of the specific viewing behavior of successful and experienced performers. We have seen that there are many and varied ways in which attention guidance can be designed (e.g., colored dots, progressing color ribbons, arrows, pointing hands, and luminance changes to the dynamic display) and that there are marked differences across the range of possible dynamic visualizations (e.g., content domain, narration or not). As a consequence, it is difficult to identify definitive, generalizable guidelines for instructional designers regarding (i) which elements to guide attention to, (ii) at what time to provide such guidance, and (iii) which form of attention guidance is most useful for which dynamic visualization. Decisions about the suitability of a particular form of attention guidance likely depend on the educational goal that instructional designers have in mind and the resources, topic knowledge and time they have at their disposal. For example, in deciding which combination of elements and/or processes to identify as relevant for understanding a dynamic system and hence to (dynamically) guide attention to, EMME can only be used if instructors' or domain experts' eye movements can be recorded and displayed overlaid on the dynamic visualization. This requires both access to domain and instructional expertise and the availability (and technical knowledge) of eye tracking equipment. Although this may be possible for certain specialist fields such as X-ray interpretation or advanced medical diagnosis, it is less likely for mainstream school-based education. In the latter case, visual cues or pointing gesture (which is part of natural communication) may be more time and cost effective. Nevertheless, it should be noted that low-cost and easy-to-use eye trackers already are starting to make their way into our daily working environment. Many companies have recently offered eye trackers for below $500 (e.g., Tobii, EyeTribe, Gazepoint) and simple eye trackers are being built into objects of everyday use, such as cellphones (e.g., to activate the screen upon looking at it: Samsung), cars (e.g., to warn drivers from falling asleep: VW), or laptops (e.g., to navigate: Lenovo). Notably, in deciding which form of attention guidance to use, it is important to realize that instructional designers have a wider array of possibilities that they can choose from if the general goal is to support perceptual processing than if one aims to support learners' cognitive processing of dynamic visualizations. That is, even though dynamic forms of attention guidance (either based on designer intuitions or expert viewing behavior) are more specific and better able to continuously guide attention, non-dynamic forms of attention guidance (i.e., static point cues like arrows) also appear to be able to direct learner attention to the key information in the display. Benefits obtained regarding cognitive processing seem to be related to dynamic forms of attention guidance, while the research on the effectiveness of non-dynamic forms of attention guidance is less conclusive in this respect.

11.3.3 Future Research

Several key issues with respect to the design and use of attention guidance strategies remain to be addressed in future research. First, drawing upon the notion that expert viewing behavior can provide valuable insights about which information to focus on and at what time, taking account of the eye movements of instructors or domain experts in the process of designing visual cues or pointing gesture is warranted. It should be noted however that because novices and experts have qualitatively different approaches to problem solving, it is not always appropriate to model expert performance *per se*. Instead, the model has to have substantial expertise in teaching the subject of interest, too. Furthermore, many of the studies discussed vary in the specific attention guidance strategy used, but very few attempts have been made to directly compare these different approaches. For example, it is uncertain whether progressive color ribbons or colored circles displaying an experts' eye movements may best be used for providing dynamic attention guidance. Importantly, this likely also depends on aspects like the type of content, learning task, and target audience. This makes it difficult to decide on how to best design the attention guidance offered to learners, even if it is known what and when to focus learner attention on. Indeed, it is highly unlikely that any single 'best approach' exists that is generalizable across all situations. Studies directly investigating the relative effectiveness of different types of attention guidance under different circumstances could shed light on this issue. Another aspect worth exploring in more detail is the extent to which the visual complexity of dynamic visualizations influences the effectiveness of the provided attention guidance. It has been suggested that attention guidance is most effective for visually complex displays that place high demands on learners' perceptual processing (De Koning et al., 2009). However, it has yet to be investigated whether and how attention guidance needs to be adapted in response to increasing or decreasing visual complexity of the dynamic display and varying levels of learner expertise. Making these aspects the focus of investigation in future research endeavors allows for a more comprehensive account of the effectiveness of attention guidance strategies to support learning from dynamic visualizations.

References

Amadieu, F., Mariné, C., & Laimay, C. (2011). The attention-guiding effect and cognitive load in the comprehension of animations. *Computers in Human Behavior, 27*, 36–40.

Balslev, T., Jarodzka, H., Holmqvist, K., De Grave, W., Muijtjens, A., Eika, B., et al. (2012). Visual expertise in paediatric neurology. *European Journal of Paediatric Neurology, 16*, 161–166.

Bandura, A. (1977). *Social learning theory*. Englewood Cliffs, NJ: Prentice-Hall.

Bandura, A., Ross, D., & Ross, S. A. (1961). Transmission of aggression through imitation of aggressive models. *The Journal of Abnormal and Social Psychology, 63*, 575.

Barsalou, L. W. (2008). Grounded cognition. *Annual Review of Psychology, 59*, 617–645.

Bednarik, R., Gowases, T., & Tukiainen, M. (2013). Gaze-augmented interaction improves problem-solving: New evidence from verbal protocols. *Behaviour & Information Technology, 32*, 836–844.

Berney, S., & Bétrancourt, M. (2017). Learning three-dimensional anatomical structures with animation: Effects of orientation references and learners' spatial ability. In R. Lowe & R. Ploetzner (Eds.), *Learning from dynamic visualization – Innovations in research and application*. Berlin: Springer (this volume).

Bétrancourt, M. (2005). The animation and interactivity principles in multimedia learning. In R. E. Mayer (Ed.), *The Cambridge handbook of multimedia learning* (pp. 287–296). New York: Cambridge University Press.

Boucheix, J.-M., & Lowe, R. K. (2010). An eye tracking comparison of external pointing cues and internal continuous cues in learning with complex animations. *Learning and Instruction, 20*, 123–135.

Boucheix, J.-M., Lowe, R. K., Putri, D. K., & Groff, J. (2013). Cueing animations: Dynamic signaling aids information extraction and comprehension. *Learning and Instruction, 25*, 71–84.

Chi, M. T. H. (2006). Two approaches to the study of experts' characteristics. In K. A. Ericsson, N. Charness, R. R. Hoffman, & P. J. Feltovich (Eds.), *The Cambridge handbook of expertise and expert performance* (pp. 21–30). Cambridge: University Press.

Chu, M., & Kita, S. (2011). The nature of gestures' beneficial role in spatial problem solving. *Journal of Experimental Psychology: General, 140*, 102–115.

Collins, A., Brown, J. S., & Newman, S. E. (1989). Cognitive apprenticeship: Teaching the crafts of reading, writing, and mathematics. In L. B. Resnick (Ed.), *Knowing, learning, and instruction* (pp. 453–494). Hillsdale, NJ: Erlbaum.

De Koning, B. B., & Tabbers, H. K. (2011). Facilitating understanding of movements in dynamic visualizations: An embodied perspective. *Educational Psychology Review, 23*, 501–521.

De Koning, B. B., & Tabbers, H. K. (2013). Gestures in instructional animations: A helping hand to understanding non-human movements? *Applied Cognitive Psychology, 27*, 683–689.

De Koning, B. B., Tabbers, H. K., Rikers, R. M. J. P., & Paas, F. (2007). Attention cueing as a means to enhance learning from an animation. *Applied Cognitive Psychology, 21*, 731–746.

De Koning, B. B., Tabbers, H. K., Rikers, R. M. J. P., & Paas, F. (2009). Towards a framework for attention cueing in instructional animations: Guidelines for research and design. *Educational Psychology Review, 21*, 113–140.

De Koning, B. B., Tabbers, H. K., Rikers, R. M. J. P., & Paas, F. (2010a). Attention guidance in learning from a complex animation: Seeing is understanding? *Learning and Instruction, 20*, 111–122.

De Koning, B. B., Tabbers, H. K., Rikers, R. M. J. P., & Paas, F. (2010b). Learning by generating vs. receiving instructional explanations: Two approaches to enhance attention cueing in animations. *Computers & Education, 55*, 681–691.

De Koning, B. B., Tabbers, H. K., Rikers, R. M. J. P., & Paas, F. (2011). Improved effectiveness of cueing by self-explanations when learning from a complex instructional animation. *Applied Cognitive Psychology, 25*, 183–194.

Dorr, M., Jarodzka, H., & Barth, E. (2010). Space-variant spatio-temporal filtering of video for gaze visualization and perceptual learning. In C. Morimoto & H. Instance (Eds.), *Proceedings of Eye tracking research & applications ETRA '10* (pp. 307–314). New York: ACM Press.
Dorr, M., Vig, E., Gegenfurtner, K. R., Martinetz, T., & Barth, E. (2008). *Eye movement modelling and gaze guidance*. Fourth International Workshop on Human-Computer Conversation.
Ericsson, K. A., & Simon, H. A. (1993). *Protocol analysis: Verbal reports as data*. Cambridge, MA: MIT Press.
Fonseca, B. A., & Chi, M. T. H. (2010). Instruction based on self-explanation. In R. E. Mayer & P. Alexander (Eds.), *The handbook of research on learning and instruction* (pp. 296–321). New York: Routledge Press.
Grant, E. R., & Spivey, M. J. (2003). Eye movements and problem solving: Guiding attention guides thought. *Psychological Science, 14*, 462–466.
Gregory, N. L., & Hodgson, T. L. (2012). Giving subjects the eye and showing them the finger: Socio-biological cues and saccade generation in the anti-saccade task. *Perception, 41*, 131–147.
Hegarty, M., Mayer, S., Kriz, S., & Keehner, M. (2005). The role of gestures in mental animation. *Spatial Cognition and Computation, 5*, 333–356.
Holmqvist, K., Nyström, N., Andersson, R., Dewhurst, R., Jarodzka, H., & Van de Weijer, J. (2011). *Eye tracking: A comprehensive guide to methods and measures*. Oxford, UK: Oxford University Press.
International League Against Epilepsy. (2010). Revised terminology and concepts for organization of the epilepsies: Report of the commission on classification and terminology. *Epilepsia, 51*, 676–685.
Jamet, E. (2014). An eye-tracking study of cueing effects in multimedia learning. *Computers in Human Behavior, 32*, 47–53.
Jarodzka, H., Balslev, T., Holmqvist, K., Scheiter, K., Nyström, M., Gerjets, P., et al. (2012). Conveying clinical reasoning based on visual observation via eye-movement modelling examples. *Instructional Science, 40*, 813–827.
Jarodzka, H., Scheiter, K., Gerjets, P., & Van Gog, T. (2010). In the eyes of the beholder: How experts and novices interpret dynamic stimuli. *Learning and Instruction, 20*, 146–154.
Jarodzka, H., van Gog, T., Dorr, M., Scheiter, K., & Gerjets, P. (2013). Learning to see: Guiding students' attention via a model's eye movements fosters learning. *Learning and Instruction, 25*, 62–70.
Johnson, A. M., Ozogul, G., Moreno, R., & Reisslein, M. (2013). Pedagogical agent signaling of multiple visual engineering representations: The case of the young female agent. *Journal of Engineering Education, 102*, 319–337.
Johnson, A. M., Ozogul, G., & Reisslein, M. (2015). Supporting multimedia learning with visual signalling and animated pedagogical agent: Moderating effects of prior knowledge. *Journal of Computer Assisted Learning, 31*, 97–115.
Kirschner, P. A., Sweller, J., & Clark, R. E. (2006). Why minimal guidance during instruction does not work: An analysis of the failure of constructivist, discovery, problem-based, experiential, and inquiry-based teaching. *Educational Psychologist, 41*, 75–86.
Kriz, S., & Hegarty, M. (2007). Top-down and bottom-up influences on learning from animations. *International Journal of Human Computer Studies, 65*, 911–930.
Kuhn, G., & Martinez, L. M. (2011). Misdirection – Past, present, and the future. *Frontiers in Human Neuroscience, 5*, 172.
Lemarié, J., Lorch, R. F., Eyrolle, H., & Virbel, J. (2008). SARA: A text-based and reader-based theory of text signaling. *Educational Psychologist, 43*, 27–48.
Lin, L., & Atkinson, R. K. (2011). Using animations and visual cueing to support learning of scientific concepts and processes. *Computers & Education, 56*, 650–658.
Lin, L., Atkinson, R. K., Savenye, W. C., & Nelson, B. C. (2016). The effects of visual cues and self-explanation prompts: Empirical evidence in a multimedia environment. *Interactive Learning Environments, 24*, 799–813.

Lindsey, C. (1978). Form, function, and locomotory habits in fish. *Fish Physiology, 7*, 1–88.

Litchfield, D., & Ball, L. J. (2011). Using another's gaze as an explicit aid to insight problem solving. *The Quarterly Journal of Experimental Psychology, 64*, 649–656.

Litchfield, D., Ball, L. J., Donovan, T., Manning, D. J., & Crawford, T. (2010). Viewing another person's eye movements improves identification of pulmonary nodules in chest x-ray inspection. *Journal of Experimental Psychology, 16*, 251–262.

Louwerse, M. M., Graesser, A. C., McNamara, D. S., & Lu, S. (2008). Embodied conversational agents as conversational partners. *Applied Cognitive Psychology, 23*, 1244–1255.

Lowe, R. K. (1999). Extracting information from an animation during complex visual learning. *European Journal of Psychology of Education, 14*, 225–244.

Lowe, R. K. (2003). Animation and learning: Selective processing of information in dynamic graphics. *Learning and Instruction, 13*, 157–176.

Lowe, R. K. (2004). Interrogation of a dynamic visualization during learning. *Learning and Instruction, 14*, 257–274.

Lowe, R., & Boucheix, J.-M. (2008a). Learning from animated diagrams: How are mental models built? In G. Stapleton, J. Howse, & J. Lee (Eds.), *Diagrams 2008 – Lecture notes in computer science* (Vol. 5223, pp. 266–281). Heidelberg: Springer.

Lowe, R., & Boucheix, J.-M. (2008b). Supporting relational processing in complex animated diagrams. In G. Stapleton, J. Howse, & J. Lee (Eds.), *Diagrams 2008 – Lecture notes in computer science* (Vol. 5223, pp. 381–394). Heidelberg: Springer.

Lowe, R. K., & Boucheix, J.-M. (2011). Cueing complex animations: Does direction of attention foster learning processes? *Learning and Instruction, 21*, 650–663.

Lowe, R., & Boucheix, J.-M. (2017). A composition approach to design of educational animations. In R. Lowe & R. Ploetzner (Eds.), *Learning from dynamic visualization – Innovations in research and application*. Berlin: Springer (this volume).

Lusk, M. M., & Atkinson, R. K. (2007). Varying a pedagogical agent's degree of embodiment under two visual search conditions. *Applied Cognitive Psychology, 21*, 747–764.

Mautone, P. D., & Mayer, R. E. (2001). Signaling as a cognitive guide in multimedia learning. *Journal of Educational Psychology, 93*, 377–389.

Mayer, R. E. (2009). *Multimedia learning* (2nd ed.). New York: Cambridge University Press.

Mayer, R. E. (Ed.). (2014). *The Cambridge handbook of multimedia learning*. New York: Cambridge University Press.

Mayer, R. E., & DaPra, S. (2012). An embodiment effect in computer-based learning with animated pedagogical agents. *Journal of Experimental Psychology: Applied, 18*, 239–252.

Meltzoff, A. N., & Moore, M. K. (1977). Imitation of facial and manual gestures by human neonates. *Science, 198*(4312), 75–78.

Moreno, R., Reisslein, M., & Ozogul, G. (2010). Using virtual peers to guide visual attention during learning: A test of the persona hypothesis. *Journal of Media Psychology: Theories, Methods, and Applications, 22*, 52–60.

Nyström, M., & Holmqvist, K. (2008). Semantic override of low-level features in image viewing – Both initially and overall. *Journal of Eye Movement Research, 2*(2), 2:1–2:11.

Ploetzner, R., & Lowe, R. K. (2012). A systematic characterization of expository animations. *Computers in Human Behavior, 28*, 781–794.

Pluchino, P., Tornatora, M. C., & Mason, L. (2012). *Improving text and picture integration during reading through eye movement modeling*. Presentation during the 15th Biennial EARLI Conference, Munich, Germany.

Richardson, D. C., & Dale, R. (2005). Looking to understand: The coupling between speakers' and listeners' eye movements and its relationship to discourse comprehension. *Cognitive Science, 29*, 1045–1060.

Schnotz, W., & Lowe, R. K (2008). A unified view of learning from animated and static graphics. In R. K. Lowe & W. Schnotz (Eds.), Learning with animation: Research implications for design (pp. 304–356). New York: Cambridge University Press.

Skuballa, I. T., & Renkl, A. (2014). A non-verbal pre-training based on eye movements to foster comprehension of static and dynamic learning environments. In P. Bello, M. Guarini, M. McShane, & B. Scassellati (Eds.), *Proceedings of the 36th annual conference of the cognitive science society* (pp. 1443–1448). Austin, TX: Cognitive Science Society.

Sweller, J., Ayres, P., & Kalyuga, S. (2011). *Cognitive load theory*. New York: Springer.

Tversky, B., Heiser, J., Mackenzie, R., Lozano, S., & Morrison, J. (2008). Enriching animations. In R. K. Lowe & W. Schnotz (Eds.), *Learning with animation: Research implications for design* (pp. 263–285). New York: Cambridge University Press.

Van Gog, T. (2014). The signaling (or cueing) principle in multimedia learning. In R. E. Mayer (Ed.), *The Cambridge handbook of multimedia learning* (2nd ed., pp. 263–278). New York: Cambridge University Press.

Van Gog, T., Jarodzka, H., Scheiter, K., Gerjets, P., & Paas, F. (2009a). Attention guidance during example study via the model's eye movements. *Computers in Human Behavior, 25*, 785–791.

Van Gog, T., Paas, F., Marcus, N., Ayres, P., & Sweller, J. (2009b). The mirror-neuron system and observational learning: Implications for the effectiveness of dynamic visualizations. *Educational Psychology Review, 21*, 21–30.

Van Gog, T., Paas, F., & Van Merriënboer, J. J. G. (2004). Process-oriented worked examples: Improving transfer performance through enhanced understanding. *Instructional Science, 32*, 83–98.

Van Gog, T., & Rummel, N. (2010). Example-based learning: Integrating cognitive and social-cognitive research perspectives. *Educational Psychology Review, 22*, 155–174.

Van Marlen, T., Van Wermeskerken, M., Jarodzka, H., & Van Gog, T. (2014). *Can eye movement modeling examples foster geometry learning?* Poster presented at the EARLI Sig 6/7 conference, Rotterdam, The Netherlands.

Vitak, S. A., Ingram, J. E., Duchowski, A. T., Ellis, S., & Gramopadhye, A. K. (2012). Gaze-augmented think-aloud as an aid to learning. In *Proceedings of the 2012 ACM annual conference on human factors in computing systems* (pp. 2991–3000). New York: ACM Press.

Wilson, M. R., Vine, S. J., Bright, E., Masters, R. S., Defriend, D., & McGrath, J. S. (2011). Gaze training enhances laparoscopic technical skill acquisition and multi-tasking performance: A randomized, controlled study. *Surgical Endoscopy, 25*, 3731–3739.

Chapter 12
Learning Three-Dimensional Anatomical Structures with Animation: Effect of Orientation References and Learners' Spatial Ability

Sandra Berney and Mireille Bétrancourt

12.1 Introduction

Many scientific, technological, and engineering domains require the understanding and handling of representations of three-dimensional models. Such representations are necessary to capture the complexity of spatial or structural subject matter that has important information distributed across all three dimensions. Examples include molecules in chemistry, gears in mechanics, or joints in anatomy. Nowadays, computer programs offer the possibility to represent three-dimensional (3D) information by means of interactive and/or animated 2D visualizations of 3D models. While Schwan and Papenmeier (2017, this volume) provide a global overview of major conceptual issues concerning learning from 3D animations, this chapter reports an experimental study in which animations of 3D models were used to support learning functional anatomy.

Functional anatomy is a challenging instructional domain that involves spatial reasoning and depends on information that is distributed within the anatomical 3D body space. It requires the future anatomist "to encode, maintain, and infer information about spatial structures and processes" (Hegarty, 2010, p. 269). Traditional 2D visualizations, whether static or dynamic, such as pictures in textbooks, anatomical charts, or animations of schematic systems often do not easily portray three-dimensional content (cf. Jenkinson, 2017, this volume; Schwan & Papenmeier, 2017, this volume). This is due to the fact that they model the content by means of flat planes. This limitation can be overcome with computer-based animations of 3D anatomical models that display the content by means of volumes with a full virtual visualization of objects rotating in the three space dimensions (called 3D animations in this chapter). A 3D animation provides learners with additional spatial cues

S. Berney (✉) • M. Bétrancourt
FPSE – TECFA, University of Geneva, Geneva, Switzerland
e-mail: sandra.berney@unige.ch; mireille.betrancourt@unige.ch

R. Lowe, R. Ploetzner (eds.), *Learning from Dynamic Visualization*,
DOI 10.1007/978-3-319-56204-9_12

(Huk, 2006; Stieff, 2007), such as supplementary depth[1] information, which is believed to support learners' perception of the complex configuration of the 3D object (Hoyek et al., 2009). The rotation of the 3D model conveys different viewpoints, providing multiple perspectives similar to those observed in real in-situ conditions (Lowe, 1999; Lowe & Schnotz, 2004; Schnotz & Lowe, 2008). Using a 3D rotating model may help learners to reconstruct the third dimension of the object, as it is directly visible in the model. Even though 3D animations may provide an adequate solution to support the construction of a complete mental representation of a 3D anatomical structure (Guillot, Champely, Batier, Thiriet, & Collet, 2007; Hoyek et al., 2009), they may be more difficult to process depending on individuals' spatial ability. Learners may therefore require support from scaffolds in order to process the presented 3D structures effectively and efficiently.

12.2 Spatial Ability and Complex Information

The role of individuals' spatial ability when processing complex information is well documented in the literature. Spatial ability generally refers to an individual's capacity to mentally represent and manipulate visually perceived information as well as to understand the spatial relationships between the perceived elements (Carroll, 1993; Linn & Peterson, 1985). When performing spatial tasks, students with low spatial ability usually make more errors than students with high spatial ability (Hegarty, 2004; Höffler, 2010; Yang, Andre, & Greenbowe, 2003). The role of spatial ability and its relation to learning with animations are discussed in light of two assumptions (Mayer & Sims, 1994). Firstly, the *compensating hypothesis* (Hays, 1996; Mayer, 2002; Mayer & Sims, 1994) claims that low spatial ability learners, when learning from animation, may be better supported because animation provides an explicit external representation of the content to be learned. As opposed to the incomplete representation provided by static graphics. Animation can therefore compensate for low spatial ability learners' incapacity to perform the mental manipulations necessary to construct the required internal representation (cf. Sanchez & Wiley, 2017, this volume). In his study, Hays (1996) investigated the relationship between spatial ability (high versus low spatial ability) and the use of visual presentations (no graphics, animated graphics, static graphics) with respect to learning concepts involving time and motion. When studying with animation, low spatial ability students exhibited significantly more improvement than students who received no animation. Learners with high spatial ability from all groups made similar scores gains, with the exception of greater gains for learners receiving animation. Because an animation allows low spatial ability learners to build a more adequate mental model of the content to be learned than usual, it acts as a "cognitive prosthetic" (Hegarty & Kriz, 2008). Secondly, the *ability-as-enhancer hypothesis*

[1] Cockburn and McKenzie (2004) provide a comprehensive overview of depth cues that help learners understand 3D objects from monocular displays.

(Hegarty, 2005; Hegarty & Sims, 1994; Höffler & Leutner, 2011; Huk, 2006) claims that learners with high spatial ability are better equipped to process animations because they have sufficient cognitive capabilities to process the complex animated content but also to build a satisfactory mental model of the to be learned content. Individuals with high spatial ability should therefore have better learning outcomes from studying animated learning materials than low spatial ability individuals. This hypothesis has been supported not only with respect to learning from 2D but also with respect to learning from 3D visualizations. In his study, Huk (2006) investigated the influence of spatial ability (high versus low spatial ability) and 3D animated models of biology cells (presence versus absence of 3D models). Results indicated a significant interaction between spatial ability and presence/absence of 3D models. Only high spatial ability students benefited from using 3D animated models. This was explained in terms of the cognitive overload that low spatial ability students experienced due to the additional demands of the 3D models, thus hindering their building of a satisfactory mental model. Although the compensating and ability-as-enhancer hypotheses may appear to be contradictory, they actually reflect distinct learning situations. The overall implication is that learners with high or low spatial ability differ in their processing of dynamic visualizations. This reflects the intricate relation between the complexity of the content, students' prior knowledge, and the task to be performed (Hegarty, Canham, & Fabrikant, 2010).

12.2.1 Object-Based and Egocentric Mental Transformations

Functional anatomy requires learners to mentally manipulate, transform and/or reorganize anatomical components in order to imagine the 3D spatial relationships between them as well as their relationships with the 3D body space (Hegarty, Keehner, Cohen, Montello, & Lippa, 2007; Marks, 2000; Thiriet, 1982). These mental operations that are based on internal representations of 3D structures enable correct positioning and/or (re)locating of the anatomical structure within the body space. In undertaking such operations, learners may apply either (i) an object-based mental transformation on the structure, or (ii) an egocentric perspective mental transformation (Kosslyn, Brunn, Cave, & Wallach, 1984; Kozhevnikov & Hegarty, 2001; Zacks, Mires, Tversky, & Hazeltine, 2000; Zacks & Tversky, 2005). The object-based transformation strategy entails imagining the object in different orientations and involves a mental rotation process (Carroll, 1993; Höffler, 2010; Miyake, Friedman, Rettinger, Shah, & Hegarty, 2001; Shepard & Metzler, 1971). The egocentric perspective transformation strategy entails performing mental rotations that change the imagined viewing position in order to internally visualize the object from another perspective (Hegarty & Waller, 2004; Kozhevnikov & Hegarty, 2001; Michelon & Zacks, 2006; Zacks & Tversky, 2005). These two mental reasoning strategies entail different changes, in the form of continuous alignments, in the spatial relationship between the object and the learner/observer. Indeed, in memory,

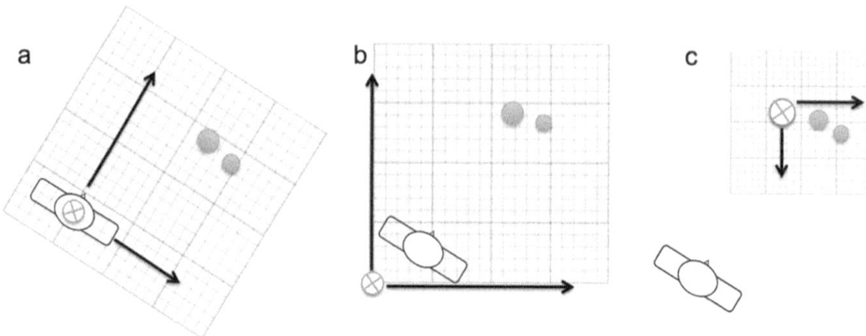

Fig. 12.1 Three frames of reference: (**a**) egocentric, (**b**) allocentric, and (**c**) intrinsic (From Waller (2013). Adapted with permission)

mental representations of objects are organized in multiple frames of reference (FORs) (Hinton & Parsons, 1988; Klatzky, 1998; Levinson, 1996; Mou & McNamara, 2002; Shelton & McNamara, 2001). These FORs are necessary for specifying the location and the orientation of an object (Klatzky, 1998; Levinson, 1996; Mou & McNamara, 2002; Pani, 1993; Parsons, 1995; Zacks & Michelon, 2005; Zacks et al., 2000). Two broad classes of FORs enable learners to encode the multiple spatial relationships between an object and its different parts, as well as between several objects. Firstly, the *egocentric FOR* (see Fig. 12.1a) is embodied and based on the viewer's point of view. It locates objects in relation to the viewer's perspective and is organized according to the viewer's body orientation. An object can quickly and accurately be located according to one's left-right, front-back, or up-down axes. Secondly, the *allocentric or environmental FOR* (see Fig. 12.1b) is based on the visual environment and uses a coordinate system that is independent of the viewer and the object location. An alternative and distinctive allocentric FOR was highlighted by Mou and McNamara (2002), namely the *intrinsic* FOR (see Fig. 12.1c). This FOR bases the reference points and axes on the object or array of objects, which are inside the environment but external to the viewer. Therefore, it defines the spatial relationships between an object and its parts independently of its location.

Generally, the FORs organize the spatial relations according to their salience (Tamborello, Sun, & Wang, 2012; Wang & Spelke, 2000). Therefore, "some relations within a FOR (…) have corresponding mental representations that are more readily accessible to the mind than others" (Tamborello et al., 2012, p. 9). Learners might maintain different FORs depending on the display, and according to the task's requirements, they are more likely to use one of them (Kozhevnikov, Royan, Blazhenkova, & Gorbunov, 2008; Ziemek, 2010). The mental transformations that can be operated on the representations make it possible to capture the spatial changes related to an object or its configuration. For both strategies, the mental transformations imply an update of the FOR. Object-based transformations involve an update of the *intrinsic* FOR whereas egocentric perspective transformations involve the update of the *egocentric* FOR, relative to the other two FORs (Kessler & Thomson, 2010; Zacks & Michelon, 2005; Zacks et al., 2000; Zacks & Tversky, 2005).

Research has shown that object-based and egocentric strategies involve two distinct abilities, namely mental rotation and spatial perspective-taking respectively (Burgess, 2006; Hegarty & Waller, 2004; Kozhevnikov & Hegarty, 2001; Wraga, Creem, & Proffitt, 2000; Zacks & Michelon, 2005). Although these abilities are somewhat related (because they both imply imagining rotations), psychometric research indicates that they are distinct spatial abilities, despite being highly correlated (Hegarty & Waller, 2004). This distinction accounts for the dissociation observed in the cognitive literature between tasks involving mental rotations and tasks involving egocentric perspective transformations (e.g., Wraga et al., 2000). On the one hand, studies investigating the ability to perform mental rotations have shown that this ability is related to successful anatomy learning (Cohen & Hegarty, 2007; Garg, Norman, & Sperotable, 2001; Guillot et al., 2007; Luursema & Verwey, 2011; Stull, Hegarty, & Mayer, 2009). On the other hand, the ability to perform egocentric perspective transformations with respect to the field of anatomy has largely been neglected by the research community. Although this ability has been discussed as originating from mental rotations (Guillot et al., 2007), both spatial abilities can play a role in learning anatomy. In addition, it is essential that future health/sport practitioners are able to adopt different imagined perspectives. For instance, in order to weigh up possible treatment options for patients suffering from disabling and diffused pain in the shoulder area, a practitioner should be able to integrate information about the scapula and the painful movements from different perspectives.

Literature on spatial cognition identifies different factors affecting both the strategies involved in such cognition and the associated spatial abilities. The decision to use an object-based transformation strategy is influenced by, among other things, the stimulus presentation format (Kosslyn, 1980; Shepard & Cooper, 1984, cited by Shepard, 1984), the type of stimulus, its shape and complexity (Bialystok, 1989; Ionta, Fourkas, Fiorio, & Aglioti, 2007; Parsons, 1994; Petit, Pegna, Mayer, & Hauert, 2003), its dimensionality – 2D versus 3D (Shepard & Metzler, 1988), and the planes on which the mental rotation is performed (Carpenter & Proffitt, 2001; Hinton & Parsons, 1988; Pani, 1993; Pani, William, & Shippey, 1995). The decision to use an egocentric perspective transformation strategy is influenced *inter alia* by the observer's movement (Amorim, Isableu, & Jarraya, 2006; Simons & Wang, 1998) and the number of stimuli involved (Wraga et al., 2000). Therefore, tasks, instructions, and stimuli are all likely to influence both object-based and egocentric strategies (Zacks et al., 2000; Zacks & Tversky, 2005). Stimuli depicting bodies or parts of body, real or avatar, are a special case. Indeed, these body-based stimuli are the only perceptual stimuli to prompt both transformation strategies in everyday life (Kessler & Thomson, 2010; Michelon & Zacks, 2006; Zacks & Tversky, 2005). Indeed, "people experience object-based transformations of bodies when they observe others' motion, and experience perspective transformations of their own body as they move around the world" (Zacks & Tversky, 2005, p. 275). In their study, Zacks and Tversky (2005, Experiment 1) investigated the effect of instructions on egocentric perspective and object-based transformations by asking participants to perform two judgment tasks regarding bodies. They manipulated the tasks' instructions, (a) to fit the judgment tasks (i.e, consistent), (b) to not fit the judgment tasks (i.e. to be inconsistent or contradictory), and (c) to be neutral to the tasks. When inconsistent with the required task's transformations, performances were slowed and altered toward the instructed transformation. However, under neutral or consistent instruction conditions, the body-based stimuli allowed both object-based and egocentric perspective transformations, depending on the judgment tasks that were performed.

Applying these findings to the learning of functional anatomy, students would be expected to be able to perform both object-based and egocentric perspective transformation strategies. However, they would be more efficient (and would exhibit superior performance) if they adopted a strategy consistent with the task.

12.2.2 Orientation Indicators: "Cognitive Handles" for Encoding Spatial Relationships

Orientation indicators, such as traditional axes, provide learners with the opportunity to determine the main axes of a structure, which in turn allows them to determine its orientation relative to the observer (Corballis, 1988). Orientation indicators can thus be regarded as visual scaffolds that aim to support learners' processing of the presented information (for other kinds of visual scaffolds see De Koning & Jarodzka, 2017, this volume). Conventional indicators that are widely used in anatomy refer to standardized orientation references. The three basic reference planes used – sagittal, frontal, and transverse – are related to three perpendicular axes – X, Y, and Z. The X-axis goes from left to right, the Y-axis goes from top to bottom, and the Z-axis goes from front to back (Kapandji, 2009). Most pictures in anatomical textbooks or charts do not include orientation references.

A study by Stull, Hegarty, and Mayer (2009) demonstrated that when the manipulation of 3D anatomical content was required, the presence of visible standard orientation references benefited learning. The presence of internal orientation references (see Fig. 12.2, left) helped learners to acquire better spatial representations, and improved the subsequent performance of learners with low mental rotation ability by raising their performance to the level of those with high mental rotation ability. The presence of visible orientation references provided distinguishable landmarks and diminished the challenge of rotation around non-canonical axes. The authors concluded that providing orientation references supports learning and contributes to the construction of coherent mental representations of anatomical structures that combine information from different perspectives. Ziemek (2010) examined the contribution of traditional orientation references (internal, external or none; see Fig. 12.2, right) with respect to mechanical and anatomical 3D stimuli when viewing multiple static visualizations. The findings revealed that providing internal aid

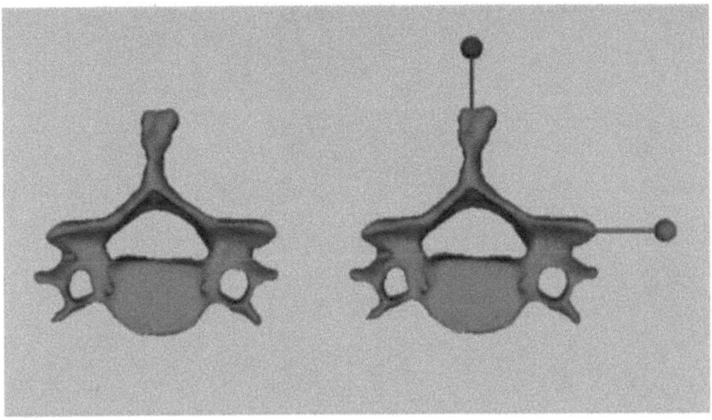

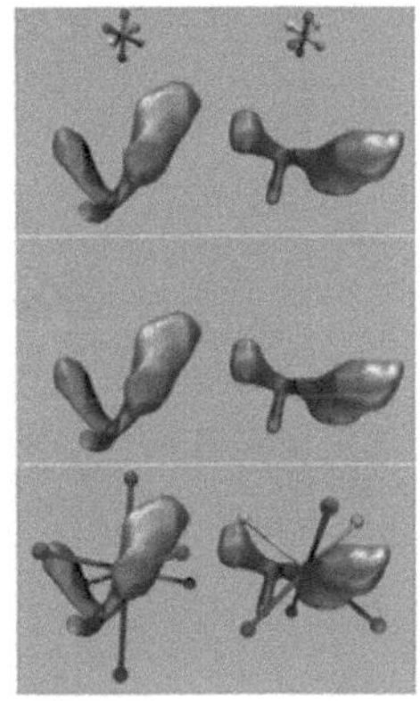

Fig. 12.2 Examples of the orientation references and stimuli used in the studies of Stull et al. (2009, *left*) and Ziemek (2010, *right*) (*Left figure* from Stull et al. (2009). Copyright 2009 by the American Psychological Association. Adapted with permission. *Right figure* from Ziemek (2010). Copyright 2010 by T.R. Ziemek. Adapted with permission)

increased the learners' ability to perceive the orientation of a 3D object. However, their performances in making orientation judgments of 3D objects were influenced by the interaction between spatial ability[2] and the availability of orientation references. Participants receiving external orientation references, compared to those receiving internal orientation references, improved their performance slightly. However, the benefit was greater for learners with high spatial ability than for learners with low spatial ability. These two studies demonstrated that the presence of orientation references, close to or even embedded in the visuals, may help learners to disambiguate the anatomical structure orientation. It is easier for learners to apprehend the structure orientation in space because it does not need to be inferred (Stull et al., 2009; Ziemek, 2010).

The two studies described above demonstrate how learners can benefit from internal orientation references when learning and manipulating an anatomical structure (Stull et al., 2009) or when performing orientation judgments (Ziemek, 2010). However, traditional internal axes can also be difficult to interpret, particularly for novice learners (Thiriet, 1982). Novice learners could interpret these internal axes as supplementary visual elements that are not part of the anatomical structure, which can disturb visual processing. An alternative to the traditional internal orientation references might be the use of external indicators, such as a human-like avatar (see Fig. 12.3). Acting as a permanent spatial anatomical reference, a human-like avatar rotates simultaneously with the 3D model. Its body position always matches the current structure position, thus providing learners with a reference that indicates the orientation of the anatomical 3D structure. An avatar offers learners the

[2]Participants' spatial ability was an aggregated measure of standardized scores for two spatial tests, namely the Paper Folding Test (Elkstrom et al., 1976) and the Mental Rotation Test (Vandenberg & Kuse, 1978).

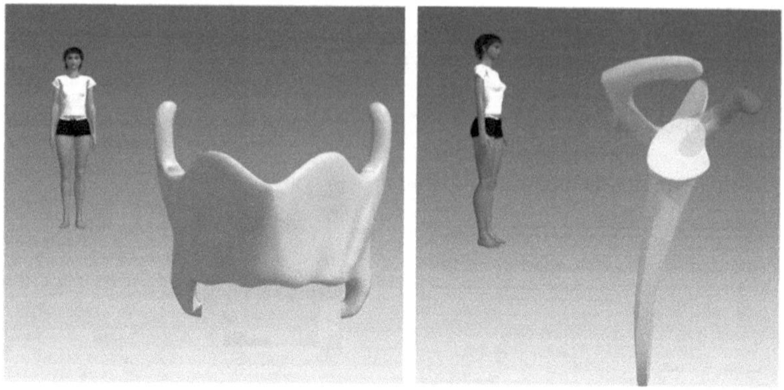

Fig. 12.3 Screenshots of two 3D animations developed by ICAP (Innovation, Conception et Accompagnement pour la Pédagogie) at the University of Lyon I for functional anatomy instruction. On the *top left corner* in both pictures is the human avatar acting as a permanent spatial orientation reference. The picture on the *left* shows a frontal anterior view of the larynx and the picture on the *right* shows a lateral view of the scapula

opportunity to link the views with their own body rather than with the conventional anatomical planes. Providing an avatar as an orientation reference should lead learners to encode the spatial relationships according to the egocentric FOR. This would enable them to process the structure's spatial information in relation to their own viewpoints. Therefore, subsequent mental transformations may be undertaken with an egocentric perspective transformation strategy.

Conversely, providing internal orientation references (as in Fig. 12.2 left) should lead learners to encode spatial information according to the intrinsic FOR of the studied structure. As a consequence, object-based transformations would be fostered, that is to say, a mental rotation strategy. To date, research has not investigated mental transformation strategies in the context of functional anatomy. Therefore, the research contribution we report below has a twofold aim. Firstly, to determine whether the presence of specific orientation indicators (external versus internal) enables learners to encode the anatomical structures in different but specific FORs. Secondly, to investigate whether these newly built mental representations support the subsequent mental transformations (object-based vs. egocentric), which in turn can influence task performance depending on the nature of the tasks and on the learners' spatial ability.

12.3 Experiment

In an experimental study, we investigated the effects of providing *orientation references* (internal axes, external avatar) or no references (control) on performance of subsequent spatial judgment tasks when studying anatomical structures through 3D animations. Two types of spatial judgment tasks were used in this study, namely judgments of structure rotation (Task 1) and judgments of structure position (Task 2). Both tasks required mental transformation strategies to be applied to newly-acquired 3D mental representations.

Evidence in the literature indicates that body-based stimuli are a special case because they could equally promote either of the two transformations strategies (e.g. Michelon & Zacks, 2006; Zacks & Tversky, 2005). Providing orientation references (internal axes, external avatar, none) when learning 3D structures should enable participants to encode mental representations in specific but different FOR. Subsequently, when spatial encoding matches the mental transformations required by the task (object-based vs. egocentric perspective transformations), the tasks performances should be more accurate and performed faster. We predicted that providing orientation references, either internal axes or an external avatar, favors the building of more complete mental representations of the 3D structures. This, in turn, fosters the subsequent mental transformations required by the tasks. The presence of orientation references when studying 3D structures should lead to more accurate and faster spatial judgments of (a) the structure rotation (Task 1), and (b) the structure position (Task 2), compared to the learning condition without orientation references (*Hypothesis 1)*. The structure rotation task (Task 1) was designed to prompt object-based mental rotation transformations. We expected the internal axes condition to foster the encoding of mental representations in the intrinsic FOR, and to subsequently favor the object-based mental transformation strategy. Learners in the internal axes condition should perform better in Task 1, compared to learners in the external avatar and control conditions (*Hypothesis 1a)*. Conversely, the structure position task (Task 2) was designed to prompt egocentric perspective transformations. We expected the external avatar condition to favor the encoding of mental representations in the egocentric FOR, and to subsequently favor the egocentric perspective transformation mental strategy. Learners in the external avatar condition should perform better in Task 2, as compared to learners in the internal axes and control conditions (*Hypothesis 1b)*.

In the literature regarding the interaction between spatial ability and learning with animations, the compensating hypothesis (Hays, 1996; Mayer, 2002; Mayer & Sims, 1994) has been well documented. We hypothesized that the learning condition will play a compensating role due to the interaction between the task performances and the associated spatial ability measure. Providing orientation references, either internal axes or external avatar, when studying 3D structures should compensate for learners' low spatial abilities, and therefore improve their performance to the level of their high spatial ability counterparts (*Hypothesis 2*). Regarding the structure rotation task (Task 1), providing internal axes should compensate for low mental rotation ability, so that performance on Task 1 does not depend on mental rotation ability scores (MRT) in this condition. In contrast, in the other two conditions (external avatar and control), the performance on Task 1 should vary with the MRT ability scores. Learners with low spatial abilities would benefit from learning with internal axes, as this condition would provide an explicit representation of the orientation indicators, which is congruent with Task 1 and fits the object-based mental transformations (*Hypothesis 2a*). Conversely, in the structure position task

(Task 2), providing an external avatar should compensate for low perspective-taking spatial ability, so that the performance on Task 2 does not depend on perspective-taking spatial ability scores (PTSO) in this condition. In contrast, in the other two conditions (internal axes and control), the performance on Task 2 should vary with the PTSO ability scores. Learners with low spatial abilities would benefit from learning with an external avatar, as this condition would provide an explicit representation of the orientation indicators, which is congruent with Task 2 and fits the egocentric perspective transformations (*Hypothesis 2b*).

12.3.1 Method

Participants and Design One hundred and forty-eight students aged between 18 and 22 years old (65 women, M = *18.82,* SD = 0.81) and enrolled in a first year kinesiology degree at the University of Lyon 1 (France) voluntarily participated in the study. Participants were asked to study the anatomy of two structures, the scapula and the larynx, in the three-dimensional body space with two 3D animations. All participants were naive with respect to the purpose of the experiment. They were randomly assigned to one of the three learning conditions: internal axes (n = 49, 25 women), external avatar (n = 57, 19 women), or no orientation references/control (n = 42, 21 women).

Material The instructional material consisted of two animations of (a) a 3D model of the scapula (84 s, see upper row of Fig. 12.4) and (b) a 3D model of the thyroid cartilage (66 s), which is part of the larynx cartilage structure (see lower row of Fig. 12.4). Four anatomical orientation views – posterior, lateral, anterior, and superior – were presented successively and the configuration of each 3D structure was emphasized by color-cueing particular anatomical features (as shown in Fig. 12.4). Three versions of the 3D animations were designed according to the type of orientation references. In the *internal axes condition,* standard reference axes were embedded in the structure (see left column in Fig. 12.4). In the *external avatar condition,* a small human-like avatar acted as a permanent spatial anatomical reference (see middle column in Fig. 12.4). Finally, in the *control condition*, no orientation references were provided (see right column in Fig. 12.4). The order in which the structures were presented was counterbalanced between participants and conditions.

Two spatial judgment tasks were designed to investigate the judgments of relative (a) structure rotation (Task 1) and (b) structure position (Task 2) within the three-dimensional body space for the scapula and the larynx. The judgment of structure relative rotation task (Task 1), presented in Fig. 12.5 (left), was a 42-item multiple-choice test on the recognition of the relative rotations of the structure. Because the focus was on an anatomical structure (object), the task was designed to prompt object-based mental rotation transformations. Each item presented a structure in a specific position (model), a rotation range statement, and five alternative images. The task instruction encouraged participants to apply an object-based

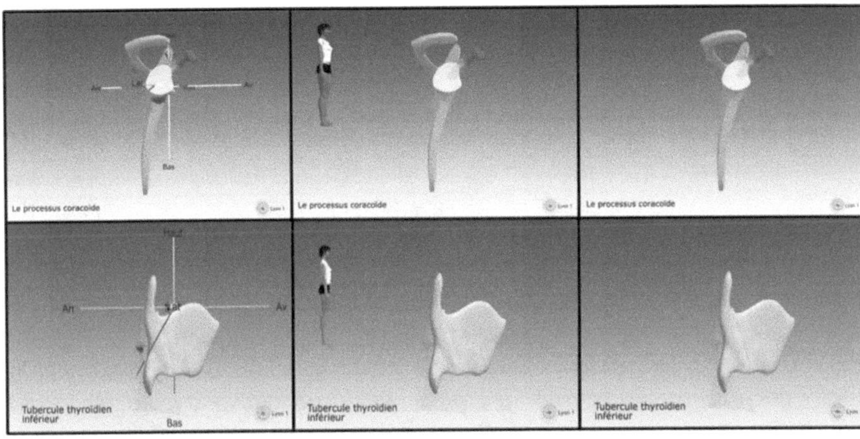

Fig. 12.4 Snapshots of the three conditions of the instructional material of the scapula structure (*upper row*) and the larynx cartilage structure (*lower row*). The *left column* represents the *internal axes* condition, the *middle column* represents the *external avatar* condition, and the *right column* represents the *control* condition. The highlighted regions in each picture represent the color cues used as the emphasis on the configuration of the 3D structures

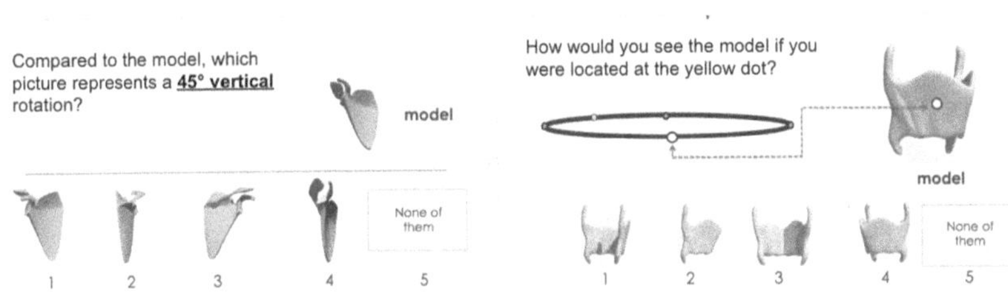

Fig. 12.5 Snapshots of the two spatial judgment tasks. On the *left* is the judgment of structure relative rotation task (*Task 1*). The rotation range statement (*bold* and *underlined* in this example) was manipulated across the items. On the *right* is the judgment of structure relative position task (*Task 2*). Participants were encouraged to adopt the perspective represented by the *yellow dot* (here a 135° horizontal shift) in order to make the position judgments

transformation of a specific and given rotation range on the model in order to be able to choose the correct answer out of five options. The dependent measures were accuracy and mean response time. The accuracy score was the number of correct answers (maximum possible score was 42). The response time measure was the average time taken to respond to the items answered correctly. The task had no time limit.

The judgment of structure relative position task (Task 2) was a 36-item multiple-choice test based on recognition of the relative positions of the structure within the 3D body space (see Fig. 12.5, right). The task was designed to engage egocentric perspective transformations of the anatomical structure by the requirement to imagine a specific view of the structure from another defined perspective. Participants had to judge, from five possible answer choices, the structure position that corresponded to the viewer's new mental perspective. The perspectives to be adopted were represented by a vertical (red) or horizontal (blue) ring. The viewer's new perspective was indicated by a yellow dot on the ring, and the model's current position by a big white dot. The dependent measures were again accuracy and response time. The accuracy score was the number of correct answers (maximum possible score was 36). The response time measure was the average time for items answered correctly. As before, there was no time limit.

Spatial ability was assessed with the Mental Rotation Test (MRT, Vandenberg & Kuse, 1978) and the Perspective-Taking Spatial Orientation Test (PTSO, Hegarty & Waller, 2004; Hegarty, Kozhevnikov & Waller, 2004; Kozhevnikov & Hegarty, 2001). The MRT was used as a measure of the spatial relation factor assessing the ability to perform object-based transformations, particularly rotations. The MRT used in this experiment is a redrawn version (Peters et al., 1995), which was computerized for the experiment. It contains two series of 12 problem sets lasting 3 min, for a total of 24 problem sets. Each problem set consists of a model and four alternatives (two correct and two incorrect alternatives). Participants must select only the two correct figures. One point is scored only if both choices are correct, no credit is given for a single correct answer, and the possible score ranges from 0 to 24.

The PTSO was used as a measure of the spatial orientation factor in order to assess the ability to adopt another imagined perspective from an egocentric perspective. A French translation of this 12-item test of 5-min duration was administered according to the specified instructions. Participants were asked to indicate the direction of an object from an imagined location, when facing another object. On the top half of each item, seven objects were drawn, and on the bottom half, a circle was drawn. At its center, an arrow started at the station point (the imagined location), and pointed to the imagined heading (object facing). Participants were asked to draw a second arrow from the center of the station point pointing in the direction of the named object. Scoring entailed the absolute deviation in degrees between the participant's response and the correct direction of the target. The participant's score was comprised of the responses' average.

Procedure The experiment took place in a computer laboratory with 14–15 students participating at a time. Each participant was seated in front of a computer and undertook the tasks individually. After signing a consent form, the computer-based experiment began. The entire experiment was presented using ParadigmExperiment© software (Perception Research System Inc.) and was system-paced. Participants had no control over the course of events. They studied the instructional material twice and performed the two spatial tasks for one structure, and then repeated the same procedure for the other structure. The order in which the two structures (scapula or larynx) were presented was counterbalanced across participants and conditions. At the end of the experiment, spatial ability was tested with a computer-based version of the MRT (Vandenberg & Kuse, 1978) and a paper-and-pencil version of the PTSO (Hegarty & Waller, 2004; Kozhevnikov & Hegarty, 2001). In total, the experimental session lasted approximately 90 min.

Table 12.1 Mean accuracies, mean response times, and standard deviations for correct answers to Tasks 1 and 2 as well as spatial ability measures according to conditions

	Internal axes ($n = 48$)		External avatar ($n = 54$)		Control ($n = 42$)	
	M	SD	M	SD	M	SD
Task 1 Structure relative rotation						
Scapula Accuracy (max = 42)	20.84	6.81	19.31	7.38	20.71	8.35
Scapula RT (ms)	12,567	6465	11,848	4313	13,840	7584
Larynx Accuracy (max = 42)	33.19	8.31	31.38	9.64	31.31	10.45
Larynx RT (ms)	9790	3449	8421	3442	10,296	3421
Task 2 Structure relative position						
Scapula Accuracy (max = 36)	16.77	5.98	16.02	6.06	17.50	6.31
Scapula RT (ms)	11,262	5412	10,094	5369	11,551	4539
Larynx Accuracy (max = 36)	24.60	6.95	22.24	8.83	24.71	7.06
Larynx RT (ms)	9187	4232	7199	3433	8887	3200
Spatial measures						
MRT	7.60	3.18	6.83	3.98	6.52	3.98
PTSO	34.17	27.65	33.46	32.86	36.75	31.04

12.4 Results

12.4.1 Spatial Ability

Descriptive statistics for the spatial measures are presented in Table 12.1. Note that as the PTSO values measure an angular deviation from the correct angle, a score near zero represents a high propensity to adopt another imagined perspective (high ability PTSO participants). Conversely, a high score indicates difficulty in adopting another imagined perspective (low ability PTSO participants). The spatial measures, the MRT and the PTSO scores, did not differ between the learning conditions ($F(1123) < 1.46$, *n.s.*), or between genders ($F(1123) < 1.78$, *n.s.*). The correlation between the MRT and the PTSO spatial ability measures was not significant (Pearson's r (137) = −0.13, *n.s.*), evidencing that these measures assess two different and separate factors.

12.4.2 Effect of the Orientation References on Accuracy

Table 12.1 presents the mean accuracy scores, response times, and standard deviations for the two tasks. A fitted factorial repeated-measures design was used with performance accuracy as the dependent variable, orientation reference (internal axes, external avatar, none) as the between-subjects variable, and task (Task 1, Task 2) and structures (scapula, larynx) as the within-subjects variables. The MRT and PTSO scores were entered as covariates. The results showed no main effect of the conditions on accuracy after controlling for the effect of spatial ability ($F(2128) = 0.84$, *n.s.*). There was a significant difference between the tasks ($F(1128) = 24.69$, $p = .001$, $\eta_p^2 = 0.16$) and the structures ($F(1128) = 26.25$, $p = 0.001$, $\eta_p^2 = 0.17$). Performances for the structure rotation task (Task 1) were better ($M = 26.270$, $SD = 0.564$) than for the structure position task (Task 2) ($M = 20.54$, $SD = 0.46$). Performances for the larynx were better ($M = 28.05$, $SD = 0.59$) than for the scapula ($M = 18.76$, $SD = 0.45$). The covariates, the MRT ($F(1128) = 55.68$, $p = 0.001$, $\eta_p^2 = 0.303$) and the PTSO ($F(1128) = 9.24$, $p = 0.003$, $\eta_p^2 = 0.07$) were significantly related to the anatomical accuracy performance.

12.4.3 Effect of the Orientation References on Response Time

The analyses of the response time on correct answers were performed with the same fitted factorial repeated-measures design. The results showed a significant effect of the conditions after controlling for the effect of spatial ability ($F(2128) = 4.01$, $p = 0.020$, $\eta_p^2 = 0.06$). Planned contrasts revealed that providing internal axes ($t(128) = -2.04$, $p = 0.043$, Cohen $d = 0.431$) or an external avatar ($t(128) = -2.74$, $p = 0.007$, $d = 0.563$) as orientation references decreased the response time needed to perform the anatomical tasks, compared to learning without any orientation references. Learning anatomical structures with orientation references, either internal axes ($M = 9748.12$, $SD = 484.68$) or an external avatar ($M = 9370.47$, $SD = 432.85$), helped learners to provide their answers more quickly in comparison to their control counterparts ($M = 11152.14$, $SD = 484.10$). Results revealed no difference between the structures ($F(1128) = 1.02$, *n.s.*) and the tasks ($F(1128) = 1.39$, *n.s.*). The covariates, the MRT ($F(1128) = 0.61$, *n.s.*) and the PTSO ($F(1128) = 3.38$, *n.s.*), were not related to response times.

12.4.4 Interactions Between Orientation References, Spatial Ability and Spatial Judgment Tasks

Correlations between the spatial ability measures (MRT and PTSO) and the performance on both tasks according to the learning conditions are presented in Table 12.2.

Table 12.2 Pearson correlations between the accuracy scores for the two tasks and the spatial ability measures

	Structure relative rotation (Task 1)				Structure relative position (Task 2)			
	Scapula		Larynx		Scapula		Larynx	
Internal axes (n = 48)								
MRT	0.37	*	0.41	**	0.47	**	0.57	**
PTSO	−0.20		−0.12		−0.29	*	−0.21	
External avatar (n = 54)								
MRT	0.33	*	0.49	**	0.50	**	0.44	**
PTSO	−0.21		−0.30	*	−0.11		−0.08	
Control (n = 42)								
MRT	0.51	**	0.46	**	0.56	**	0.55	**
PTSO	−0.49	**	−0.16		−0.37	*	−0.18	

Note: *$p < 0.05$ level; **$p < 0.001$ level. PTSO values measure an angular deviation from the correct angle. The minus symbol should thus not be considered

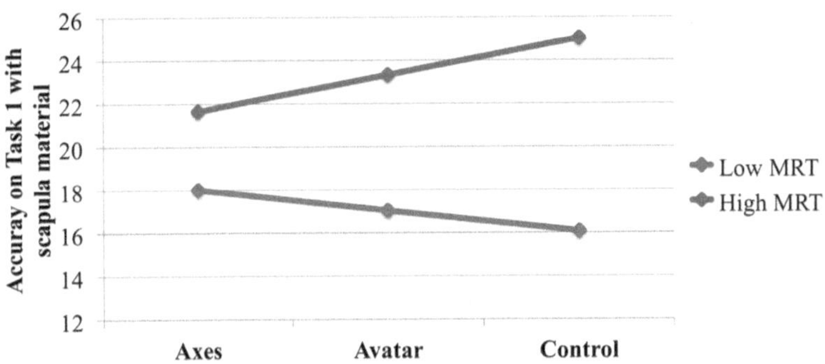

Fig. 12.6 Interaction plot between learning conditions and the accuracies in solving Task 1 for learners with high and low MRT scores

Regarding the results of Task 1 prompting object-based mental rotation transformations, the correlation analyses revealed significant and positive correlations between the MRT scores and the accuracy of both spatial tasks (Task 1 – structure rotation, Task 2 – structure position) and for both structures (scapula and larynx). Two out of the twelve correlations showed a moderate positive relationship ($r <$ 0.40), and the other ten a strong positive relationship ($r >$ 0.40). Globally, learners with high MRT ability scores outperformed learners with low MRT ability scores on both tasks.

The MRT score was examined as a moderator of the relation between the learning conditions and the performances on the structure rotation task (Task 1). The analyses were performed individually on each structure (scapula, larynx) with the interaction term between conditions and the MRT. The interaction term explained a

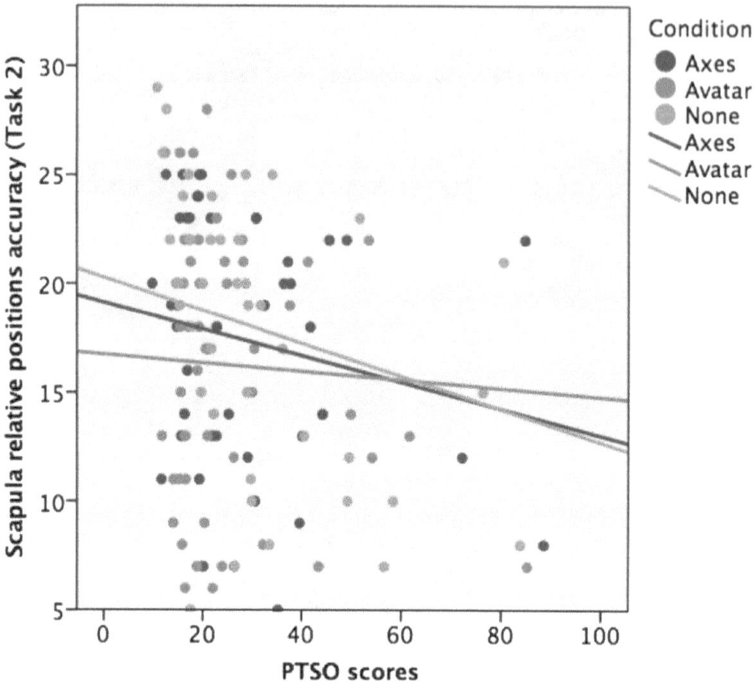

Fig. 12.7 Correlation plot of the scores for Task 2 as a function of the perspective-taking ability scores (PTSO), depending on the learning conditions. The *dots* represent the participants' individual score

significant increase in variance of judgments of the scapula's rotations task (R^2 = 0.20, $F(3134) = 12.16$, $p = 0.001$). The MRT was a significant moderator of the relationship between the learning conditions and the task performance ($t(138) = 2.25$, $p = 0.025$, $d = 0.375$). The interaction plot in Fig. 12.6 shows the direction of the moderation by distinguishing learners with high and low MRT ability scores. For learners with high MRT scores, the accuracy in solving Task 1 varied across conditions (b = 2.16, $SE_b = 1.06$, $p = 0.044$), while for learners with low MRT ability, the accuracy in the task was not affected (b = −1.22, $SE_b = 1.06$, *n.s.*). Overall, learners with high MRT scores showed better performances compared to learners with low MRT scores. Notably, learners with high MRT scores in the control condition outperformed other participants with high MRT scores. Moreover, the MRT was a significant positive predictor of accuracy performance with the scapula structure for all participants ($t(138) = 5.33$, $p = 0.001$, $d = 0.888$).

Regarding the larynx accuracy performances, the model was globally significant ($F(3134) = 9.98$, $p = 0.001$) but not the moderation effect of the MRT ($t(138) = 0.97$, *n.s.*) or the conditions ($t(138) = -481$, *n.s.*). However, the MRT was a significant and positive predictor of accuracy performance for the larynx ($t(138) = 5.37$, $p = 0.001$, $d = 0.895$).

Regarding the performance on Task 2, which prompted egocentric perspective transformations, the correlation analysis for the scapula structure (see Table 12.2) revealed significant and positive correlations between the PTSO scores and the accuracy of Task 2 for the internal axes ($p = 0.045$) and control ($p = 0.015$) conditions, but not for the external avatar condition. As shown in Fig. 12.7, the slopes for the internal axes ($R^2 = 0.09$) and the control conditions ($R^2 = 0.14$) were positive, whereas it was null for the external avatar condition ($R^2 = 0.01$). In other words, the PTSO ability did not interact with the learning conditions in the same way. No significant correlations were found for the larynx structure. The PTSO scores, examined as a moderator of the relation between the learning conditions and the performances on Task 2, revealed no significant moderator effect of PTSO ability on either structures (scapula ($F(3137) = 2.44$, *n.s.*; larynx ($F(3137) = 1.63$, *n.s.*).

12.5 Discussion

This chapter addressed two main questions. Firstly, does providing orientation references during the learning phase influence the accuracy of spatial judgments of a structure rotation or position? Secondly, is there an interaction between orientation references and learners' spatial ability when performing spatial transformation strategies?

12.5.1 *Does the Presence of Orientation References Enhance the Building of Mental Representations of 3D Structures?*

The first question addressed whether providing orientation references when learning 3D structures influences the performance on subsequent spatial judgments of a structure rotation (Task 1) and position (Task 2). Providing orientation references, either external or internal, did not enhance the accuracy of spatial judgments of rotations or positions. However, the response times indicated that provision of orientation references, either internal or external, led learners to perform accurate spatial judgments more quickly. Given this positive effect, it is important to consider the mechanisms by which learning with orientation references could yield benefits.

In the initial learning phase, we expected learners to gain the anatomical knowledge of the structures through the color-cueing elements and to gain the three-dimensional spatial relationship knowledge through the scrolling rotation of the 3D structures. Learners provided with internal or external orientation references could rely on them in order to determine the main axes of the structures, whereas learners from the control condition may have had to infer the structures' main axes. The presence of orientation indicators in the display may have provided support to define the three-dimensional location and orientation of the structures in a coordinate system. The internal orientation references may lead to anchoring the coordinate system directly on the anatomical structures, that is to say, in an intrinsic FOR. Regarding the external orientation reference, the anatomical structures could have been defined in an egocentric FOR, when learners use the avatar as a "proxy" for themselves, and thus base the spatial relationships on their own viewpoints. Learners in the control condition may have anchored the structures in an allocentric FOR, using the computer screen display as a coordinate system. However, it is not beyond the bounds of possibility that learners encoded the structures in a FOR other than the expected, more obvious one (Kessler & Thomson, 2010; Michelon & Zacks, 2006; Zacks & Tversky, 2005).

During performance of the tasks, transformations were required that called upon the newly acquired mental representations of the structures. Learners thus needed to update the FOR corresponding to the mental representations in order to perform the required judgments. The judgment of the structure rotation task (Task 1) called for object-based mental rotation transformations, in which the structure/object is updated relative to the intrinsic FOR. The judgment of the structure position task (Task 2) called for egocentric perspective mental transformations, in which the observer's perspective is updated. Thus the object-based transformations of Task 1 are more likely to have matched the spatial encoding of learners in the internal axes condition, namely the intrinsic FOR. This task may be less suitable for learners in the external avatar and control conditions, and the egocentric and allocentric FORs respectively. Similarly, the egocentric perspective transformations of Task 2 are more likely to have matched the spatial encoding of the learners in the external avatar condition based in an egocentric FOR, and be less suitable for learners in the two other conditions.

The similar accuracy performances of the three conditions for the spatial judgment tasks suggest that all learners were able to build comparable, adequate, and effective mental representations of the 3D structures during the initial learning phase. These mental representations included anatomical knowledge and spatial relationships between the multiple viewpoints. Two possible hypotheses could be suggested to account for the difference in response time across conditions. One assumption is a difference when encoding the structures in the FOR. This view is consistent with the hypothesis of a switch cost in response time when participants handle a conflict of FORs, as proposed by Tamborello et al. (2012). In this case, participants have to switch from the encoded FOR to a less salient – but task-relevant – FOR, inducing a switch cost. The second assumption is related to the mental transformation strategies. The possibility that learners, when solving the tasks, may have used a mental transformation different from the expected one, cannot be excluded. Altogether, the findings suggested that the visible presence of orientation indicators, either internal axes or external avatar, during the learning/encoding phase might have played a "cognitive handle" role (Stull et al., 2009), which in return decreased the response time needed to perform the spatial judgments.

Additionally, there was a prominent and obvious difference between the larynx and scapula structure results with participants systematically performing better with the larynx content. This result may be explained by the rather simple and symmetrical shape of the larynx cartilage, a sort of 3 cm-high V-shape. In contrast, the shape of the scapula is more complex. This triangular bone has two kinds of non-symmetrical rods on the top, which are not always visible depending on the viewpoints. Unsurprisingly, it is easier to memorize a simple versus a complex form or shape (Marr & Nishihara, 1978).

12.5.2 Does Learning with Visible Orientation Indicators Interact with Learners' Spatial Ability?

The second question examined the interaction of learning with orientation references and spatial ability when solving judgment tasks involving two distinct mental transformations. Analysis of the relation between the MRT scores and the performance on the tasks revealed that the MRT was significantly related to the judgment performance for both tasks and for all participants. Regardless of the learning conditions, participants with lower MRT ability, in comparison with those with higher MRT ability, performed more poorly on both spatial judgment tasks (Tasks 1 and 2). The results highlighted the influential role of the mental rotation ability (MRT) in learning anatomy with 3D animated models, and are in line with existing literature on anatomy. This is consistent with evidence of the interplay between the mental rotation ability and successful learning with traditional static methods (Rochford, 1985) as well as with 3D computer models (e.g., Garg et al., 2001; Garg, Norman, Eva, Spero, & Sharan 2002; Garg, Norman, Spero, & Taylor, 1999; Guillot et al., 2007; Hoyek et al., 2009; Huk, 2006; Keehner & Khooshabeh, 2002; Luursema & Verwey, 2011; Nguyen, Nelson, & Wilson, 2012; Stull et al., 2009, 2010).

Evidence for the moderator role of MRT ability was found in judgments of the scapula's rotation (Task 1). Our findings indicate that MRT ability moderated the rotation judgment performance, with performances enhanced for learners with high MRT scores only. Learners with high MRT ability scores showed better performance when compared to learners with low MRT ability scores. These results contradict the compensating hypothesis (Hays, 1996; Mayer & Sims, 1994; Mayer, 2002). However, they are in accordance with the ability-as-enhancer hypothesis (Hegarty & Sims, 1994; Hegarty, 2005; Höffler, 2010; Huk, 2006), which states that learners with high spatial ability are cognitively better equipped to process dynamic visualizations, leading to better performances than learners with low spatial ability. Another noteworthy result is that learners with high MRT ability scores performed better when they learned anatomy *without* any orientation references (control condition), compared to learning with orientation references. This is consistent with Khooshabeh and Hegarty's (2010) study. Indeed, individuals with high mental rotation ability are less dependent on external visual information during the learning process because they perform and/or rotate shapes as a whole. In the present study, this is reflected by a significant decrease in performance for participants with high MRT ability scores in the axes and avatar conditions (see Fig. 12.6), suggesting that both types of orientation references degraded rather than enhanced those learners' performances. In other words, our findings suggest that the performances of rotation judgments depend less on the type of orientation references provided in the 3D animations than on the ability to maintain and transform the mental representations of the 3D structures.

The analysis of the relation between the PTSO scores and the performance on the position judgments task (Task 2) revealed distinct patterns regarding the structures. Whereas for the larynx the judgments of positions are not related to the participants' perspective-taking ability, the results for the internal axes and control conditions showed a significant but moderate correlation between the performances for the scapula and the PTSO ability scores. When studying the scapula with internal axes or without any orientation references, high PTSO ability leads to more accurate performances on the judgments of the scapula relative position. In other words, and in these two specific conditions, participants who can better imagine transforming their actual perspective and adopt a new one are more accurate when judging of the scapula's relative positions.

These findings also highlight the beneficial effect of the avatar as an orientation reference for learners with low PTSO ability scores. They are in line with the compensating hypothesis (Hays, 1996; Mayer & Sims, 1994; Mayer, 2002) and more particularly, the compensating effect of specific types of animated display (Höffler, 2010). Thus, learning anatomical structures with the avatar provides support to learners with low PTSO ability, suggesting the possibility of embodiment processing (Amorim et al., 2006; Kessler & Wang, 2012). In that case, learners may have internalized the orientation of the anatomical structure by using either their own body's coordinates or the body of the human-like avatar as locational information. In return, this embodied locational information was then subsequently beneficial for learners with low PTSO ability, enabling them to adopt another imagined perspective more easily. The compensatory effect of a particular and specific animated display was not available, either for learners in the internal axes condition or in the control condition. However, the discrepant performances of learners with high PTSO ability scores might also be explained in terms of an expertise reversal effect (Khacharem, Zoudji, Kalyuga, & Ripoll, 2013), that is, when experts are given support that helps novices, the experts' processing is actually hindered. In this way, learners with high PTSO ability scores in the avatar condition could be hindered by the supplementary information provided by the avatar as an orientation reference. In other words, studying the scapula with the avatar reduced the differences in the judgments of positions between participants with high and low PTSO ability scores.

Overall, the findings highlight the intricate interplay between the learners' spatial ability, the 3D animations' features, and the cognitive processes of encoding and mental transformations. The findings of this study concerning mental rotation ability are in line with the ability-as-enhancer hypothesis (Hegarty & Sims, 1994; Hegarty, 2005; Höffler, 2010; Huk, 2006). Learners with high MRT ability scores performed better than learners with low MRT ability scores, irrespective of the structures and tasks. On the other hand, the findings regarding the perspective-taking ability are consistent with the compensating hypothesis (Hays, 1996; Mayer & Sims, 1994; Mayer, 2002) because the presence of an external orientation reference during learning helped learners with low PTSO ability scores to reach performances similar to or close to those of learners with high PTSO ability scores.

12.6 Conclusion

The study reported here provides no direct evidence of the impact of the presence of orientation references on the capacity to perform accurate spatial judgment tasks. However, the presence of orientation indicators in the animation of 3D structures during the learning phase, in the form of internal axes or an external human-like avatar, allowed participants to perform subsequent spatial judgment tasks faster. Our findings are consistent with the claim that studying 3D objects from 3D animations when orientation indicators are visible does not change the mental model of the structure itself, but rather influences the mental transformation strategies (object-based vs. egocentric) subsequently performed. Additionally, the intricate interplay between spatial ability and animation processing (compensating vs. enhancing role) should not be viewed as contradictory but as dependent on learning conditions. Accordingly, the findings demonstrate that design factors interact not only with learners' spatial ability but also with subsequent tasks to be performed.

It is important to acknowledge that these findings are specific to spatial judgments of 3D anatomical structures. The population was restricted to kinesiology students. Further studies should include medical education students from broader disciplines as well as other 3D structures or objects. In addition, future work using eye tracking methodology should also explore learners' visual exploration behavior in order to get insights on the specific spatial transformation strategies learners apply and how they vary across design conditions, tasks and spatial ability levels.

Acknowledgements We thank Nady Hoyek, Patrice Thiriet and Christophe Batier for their inestimable support in this research and Olivier Rastello for his precious assistance in modifying the instructional materials for the needs of this study. We would also thank the editors of this book for their helpful comments that helped us to further improve the text.

References

Amorim, M.-A., Isableu, B., & Jarraya, M. (2006). Embodied spatial transformations: "Body analogy" for the mental rotation of objects. *Journal of Experimental Psychology: General, 135*, 327–347.

Bialystok, E. (1989). Children's mental rotations of abstract displays. *Journal of Experimental Child Psychology, 47*, 47–71.

Burgess, N. (2006). Spatial memory: How egocentric and allocentric combine. *Trends in Cognitive Sciences, 10*, 551–557.

Carpenter, M., & Proffitt, D. R. (2001). Comparing viewer and array mental rotations in different planes. *Memory & Cognition, 29*, 441–448.

Carroll, J. B. (1993). *Human cognitive abilities. A survey of factor-analytic studies*. Cambridge: Cambridge University Press.

Cockburn, A., & McKenzie, B. (2004). Evaluating spatial memory in two and three dimensions. *International Journal of Human-Computer Studies, 61*(3), 359–373. doi:10.1016/j.ijhcs.2004.01.005

Cohen, C. A., & Hegarty, M. (2007). Individual differences in use of external visualisations to perform an internal visualisation task. *Applied Cognitive Psychology, 21*, 701–711.

Corballis, M. C. (1988). Recognition of disoriented shapes. *Psychological Review, 95*(1), 115–123.

De Koning, B. B., & Jarodzka, H. (2017). Attention guidance strategies for supporting learning from dynamic visualizations. In R. Lowe & R. Ploetzner (Eds.), *Learning from dynamic visualization – Innovations in research and application*. Berlin: Springer (this volume).

Ekstrom, R. B., French, J. W., Harman, H. H., & Dermen, D. (1976). *Manual for kit of factor-referenced cognitive tests* (Vol. 102, p. 117). Princeton, NJ: Educational Testing Service.

Garg, A. X., Norman, G. R., Eva, K. W., Spero, L., & Sharan, S. (2002). Is there any real virtue of virtual reality?: The minor role of multiple orientations in learning anatomy from computers. *Academic Medicine: Journal of the Association of American Medical Colleges, 77*(10 Suppl), 97–99.

Garg, A. X., Norman, G. R., Spero, L., & Taylor, I. (1999). Learning anatomy: Do new computer models improve spatial understanding? *Medical Teacher, 21*, 519–522.

Garg, A. X., Norman, G. R., & Sperotable, L. (2001). How medical students learn spatial anatomy. *Lancet, 357*(9253), 363–364.

Guillot, A., Champely, S., Batier, C., Thiriet, P., & Collet, C. (2007). Relationship between spatial abilities, mental rotation and functional anatomy learning. *Advances in Health Sciences Education: Theory and Practice, 12*, 491–507.

Hays, T. A. (1996). Spatial abilities and the effects of computer animation on short-term and long-term comprehension. *Journal of Educational Computing Research, 14*, 139–155.

Hegarty, M. (2004). Diagrams in the mind and in the world: Relations betweens internal and external visualizations. In A. F. Blackwell, K. Mariott, & A. Shimojima (Eds.), *Diagrammatic representation and inference, Lecture notes in artificial intelligence 2980* (pp. 1–13). Berlin/Heidelberg: Springer.

Hegarty, M. (2005). Multimedia learning about physical systems. In R. E. Mayer (Ed.), *The Cambridge handbook of multimedia learning* (1st ed., pp. 447–465). Cambridge: Cambridge University Press.

Hegarty, M. (2010). Components of spatial intelligence. In B. H. Ross (Ed.), *The Psychology of Learning and Motivation, 52*(10), 265–297. doi:10.1016/S0079-7421(10)52007-3

Hegarty, M., Canham, M. S., & Fabrikant, S. I. (2010). Thinking about the weather: How display salience and knowledge affect performance in a graphic inference task. *Journal of Experimental Psychology: Learning, Memory, and Cognition, 36*, 37–53.

Hegarty, M., Keehner, M., Cohen, C. A., Montello, D. R., & Lippa, Y. (2007). The role of spatial cognition in medicine: Applications for selecting and training professionals. In G. L. Allen (Ed.), *Applied spatial cognition* (pp. 285–315). Mahwah, NJ: Lawrence Erlbaum Associates, Inc.

Hegarty, M., Kozhevnikov, M., & Waller, D. (2004). Perspective taking/spatial orientation test. *Intelligence, 32*, 175–191.

Hegarty, M., & Kriz, S. (2008). Effect of knowledge and spatial ability on learning from animation. In R. K. Lowe & W. Schnotz (Eds.), *Learning with animation: Research implications for design* (1st ed., pp. 3–29). New York: Cambridge University Press.

Hegarty, M., & Sims, V. K. (1994). Individual differences in mental animation during mechanical reasoning. *Memory & Cognition, 22*, 411–430.
Hegarty, M., & Waller, D. (2004). A dissociation between mental rotation and perspective-taking spatial abilities. *Intelligence, 32*, 175–191.
Hinton, G. E., & Parsons, L. M. (1988). Scene-based and viewer-centered representations for comparing shapes. *Cognition, 30*, 1–35.
Höffler, T. N. (2010). Spatial ability: Its influence on learning with visualizations – A meta-analytic review. *Educational Psychology Review, 22*, 245–269.
Höffler, T. N., & Leutner, D. (2011). The role of spatial ability in learning from instructional animations – Evidence for an ability-as-compensator hypothesis. *Computers in Human Behavior, 27*(1), 209–216. doi:10.1016/j.chb.2010.07.042
Hoyek, N., Collet, C., Rastello, O., Fargier, P., Thiriet, P., & Guillot, A. (2009). Enhancement of mental rotation abilities and its effect on anatomy learning. *Teaching and Learning in Medicine, 21*, 201–206.
Huk, T. (2006). Who benefits from learning with 3D models? The case of spatial ability. *Journal of Computer Assisted Learning, 22*, 392–404.
Ionta, S., Fourkas, A. D., Fiorio, M., & Aglioti, S. M. (2007). The influence of hands posture on mental rotation of hands and feet. *Experimental Brain Research, 183*, 1–7.
Jenkinson, J. (2017). The role of craft-based knowledge in the design of dynamic visualizations. In R. Lowe, & R. Ploetzner (Eds.), *Learning from dynamic visualization – Innovations in research and application*. Berlin: Springer (this volume).
Kapandji, A. I. (2009). *Anatomie fonctionnelle* (Maloine, Ed.; 6th ed.). Paris.
Keehner, M., & Khooshabeh, P. (2002). Computerized representations of 3D structure: How spatial comprehension and patterns of interactivity differ among learners. In *Proceedings of the AAAI spring symposium series, reasoning with mental and external diagrams* (pp. 12–17). Menlo Park, CA: AAAI Press.
Kessler, K., & Thomson, L. A. (2010). The embodied nature of spatial perspective taking: Embodied transformation versus sensorimotor interference. *Cognition, 114*, 72–88.
Kessler, K., & Wang, H. (2012). Spatial perspective taking is an embodied process, but not for everyone in the same way: Differences predicted by sex and social skills score. *Spatial Cognition Computation: An Interdisciplinary Journal, 12*, 133–158.
Khacharem, A., Zoudji, B., Kalyuga, S., & Ripoll, H. (2013). Developing tactical skills through the use of static and dynamic soccer visualizations: An expert–nonexpert differences investigation. *Journal of Applied Sport Psychology, 25*, 326–340.
Khooshabeh, P., & Hegarty, M. (2010). Inferring cross-sections: When internal visualizations are more important than properties of external visualizations. *Human-Computer Interaction, 25*, 119–147.
Klatzky, R. L. (1998). Allocentric and egocentric spatial representations: Definitions, distinctions, and inteconnections. In C. Freska, C. Habel, & K. F. Wender (Eds.), *Spatial cognition. Lecture notes in computer science* (Vol. 1404, pp. 1–17). Berlin: Springer.
Kosslyn, S. M. (1980). *Image and mind*. Cambridge, MA: Harvard University Press.
Kosslyn, S. M., Brunn, J., Cave, K. R., & Wallach, R. W. (1984). Individual differences in mental imagery ability: A computational analysis. *Cognition, 18*, 195–243.
Kozhevnikov, M., & Hegarty, M. (2001). A dissociation between object manipulation spatial ability and spatial orientation ability. *Memory & Cognition, 29*, 745–756.
Kozhevnikov, M., Royan, J., Blazhenkova, O., & Gorbunov, A. (2008). The role of immersivity in three-dimensional mental rotation. In J. S. Gero & A. K. Goel (Eds.), *Proceedings of the third international conference on design computing and cognition* (Vol. 3, pp. 143–157). Dordrecht: SAGE Publications.
Levinson, S. C. (1996). Frames of reference and Molyneux' s question: Crosslinguistic evidence. In P. Bloom, M. A. Peterson, L. Nadel, M. F. Garrett, & S. C. Levinson (Eds.), *Language and space* (pp. 109–169). Cambridge, MA: MIT Press.

Linn, M. C., & Peterson, A. C. (1985). Emergence and characterization of sex differences in spatial ability: A meta-analysis. *Child Development, 5*, 1479–1498.
Lowe, R. K. (1999). Extracting information from an animation during complex visual learning. *European Journal of Psychology of Education, 14*, 225–244.
Lowe, R. K., & Schnotz, W. (2004). Reasons for using animation. Unpublished Article.
Luursema, J.-M., & Verwey, W. B. (2011). The contribution of dynamic exploration to virtual anatomical learning. *Advances in Human-Computer Interaction, 2011*, 1–6.
Marks, S. C. (2000). The role of three-dimensional information in health care and medical education: The implications for anatomy and dissection. *Clinical Anatomy, 13*, 448–452.
Marr, D., & Nishihara, H. (1978). Representation and recognition of the spatial organization of three-dimensional shapes. In *Proceedings of the Royal Society of London B: Biological Sciences, 200*, 269–294.
Mayer, R. E. (2002). Cognitive theory and the design of multimedia instruction: An example of the two-way street between cognition and instruction. *New Directions for Teaching and Learning, 89*, 55–71.
Mayer, R. E., & Sims, V. K. (1994). For whom is a picture worth a thousand words? Extensions of a dual-coding theory of multimedia learning. *Journal of Educational Psychology, 86*, 389–401.
Michelon, P., & Zacks, J. M. (2006). Two kinds of visual perspective taking. *Perception & Psychophysics, 68*, 327–337.
Miyake, A., Friedman, N. P., Rettinger, D. A., Shah, P., & Hegarty, M. (2001). How are visuospatial working memory, executive functioning, and spatial abilities related? A latent-variable analysis. *Journal of Experimental Psychology: General, 130*, 621–640.
Mou, W., & McNamara, T. P. (2002). Intrinsic frames of reference in spatial memory. *Journal of Experimental Psychology: Learning, Memory, and Cognition, 28*, 162–170.
Nguyen, N., Nelson, A. J., & Wilson, T. D. (2012). Computer visualizations: Factors that influence spatial anatomy comprehension. *Anatomical Sciences Education, 5*, 98–108.
Pani, J. R. (1993). Limits on the comprehension of rotational motion: Mental imagery of rotations with oblique components. *Perception, 22*, 785–808.
Pani, J. R., William, C. T., & Shippey, G. T. (1995). Determinants of the perception of rotational motion: Orientation of the motion to the object and to the environment. *Journal of Experimental Psychology: Human Perception and Performance, 21*, 1441–1456.
Parsons, L. M. (1994). Temporal and kinematic properties of motor behavior reflected in mentally simulated action. *Journal of Experimental Psychology: Human Perception and Performance, 20*, 709–730.
Parsons, L. M. (1995). Inability to reason about an object's orientation using an axis and angle of rotation. *Journal of Experimental Psychology: Human Perception and Performance, 21*, 1259–1277.
Peters, M., Laeng, B., Latham, K., Jackson, M., Zaiyouna, R., & Richardson, C. (1995). A redrawn Vandenberg and Kuse mental rotations test: Different versions and factors that affect performance. *Brain and Cognition, 29*, 39–58.
Petit, L., Pegna, A. J., Mayer, E., & Hauert, C.-A. (2003). Representation of anatomical constraints in motor imagery: Mental rotation of a body segment. *Brain and Cognition, 51*, 95–101.
Rochford, K. (1985). Spatial learning disabilities and underachievement among university anatomy students. *Medical Education, 19*, 13–26.
Sanchez, C. A., & Wiley, J. (2017). Dynamic visuospatial ability and learning from dynamic visualizations. In R. Lowe, & R. Ploetzner (Eds.), *Learning from dynamic visualization – Innovations in research and application*. Berlin: Springer (this volume).
Schnotz, W., & Lowe, R. K. (2008). A unified view of learning from animated and static graphics. In R. K. Lowe & W. Schnotz (Eds.), *Learning with animation. Research implications for design* (pp. 304–356). New York: Cambridge University Press.

Schwan, S., & Papenmeier, F. (2017). Learning from animations: From 2d to 3d? In R. Lowe, & R. Ploetzner (Eds.), *Learning from dynamic visualization – Innovations in research and application*. Berlin: Springer (this volume).
Shelton, A. L., & McNamara, T. P. (2001). Systems of spatial reference in human memory. *Cognitive Psychology, 43*, 274–310.
Shepard, R. N. (1984). Ecological constraints on intrnal representation: Resonant kinematics of perceiving, imagining, thinking, and dreaming. *Psychological Review, 91*, 417–447.
Shepard, R. N., & Metzler, J. (1971). Mental rotation of three-dimensional objects. *Science, 171*, 701–703.
Shepard, S., & Metzler, D. (1988). Mental rotation: Effects of dimensionality of objects and type of task. *Journal of Experimental Psychology: Human Perception and Performance, 14*, 3–11.
Simons, D. J., & Wang, R. F. (1998). Perceiving real-world viewpoint changes. *Psychological Science, 9*, 315–320.
Stieff, M. (2007). Mental rotation and diagrammatic reasoning in science. *Learning and Instruction, 17*, 219–234.
Stull, A. T., Hegarty, M., & Mayer, R. E. (2009). Getting a handle on learning anatomy with interactive three-dimensional graphics. *Journal of Educational Psychology, 101*, 803–816.
Stull, A. T., Hegarty, M., & Mayer, R. E. (2010). Anatomy learning with virtual objects. In *Proceedings of AAAI spring symposium: Cognitive shape processing* (pp. 39–44). Menlo Park, CA: Association for the Advancement of Artificial Intelligence.
Tamborello, F. P., Sun, Y., & Wang, H. (2012). Spatial reasoning with multiple intrinsic frames of reference. *Experimental Psychology, 59*, 3–10.
Thiriet, P. (1982). *La formation scientifique des professeurs africans d'éducation physique: Contribution à une didactique de l'anatomie et de la physiologie* [Scientific formation of African physical education professors: Contribution to anatomy and physiology didactics] (Doctoral dissertation). Université Lyon II, Lyon.
Vandenberg, S. G., & Kuse, A. R. (1978). Mental rotations, a group test of three-dimensional spatial visualization. *Perceptual and Motor Skills, 47*, 599–604.
Waller, D. (2013). *Human perception & cognition of space I*. Lecture conducted at the International Spatial Cognition Summer Institute, ISCSI 2013, UCSB, Santa Barbara, CA.
Wang, R. F., & Spelke, E. S. (2000). Updating egocentric representations in human navigation. *Cognition, 77*, 215–250.
Wraga, M., Creem, S. H., & Proffitt, D. R. (2000). Updating displays after imagined object and viewer rotations. *Journal of Experimental Psychology: Learning, Memory, and Cognition, 26*, 151–168.
Yang, E., Andre, T., & Greenbowe, T. J. (2003). Spatial ability and the impact of visualization. *International Journal of Science Education, 25*, 329–350.
Zacks, J. M., & Michelon, P. (2005). Transformations of visuospatial images. *Behavioral and Cognitive Neuroscience Reviews, 4*, 96–118.
Zacks, J. M., Mires, J. O. N., Tversky, B., & Hazeltine, E. (2000). Mental spatial transformations of objects and perspective. *Spatial Cognition and Computation, 2*, 315–332.
Zacks, J. M., & Tversky, B. (2005). Multiple systems for spatial imagery: Transformations of objects and bodies. *Spatial Cognition and Computation, 5*, 271–306.
Ziemek, T. R. (2010). *Evaluating the effectiveness of orientation indicators with an awareness of individual differences* (Doctoral dissertation). University of Utah, Salt Lake City, UT.

Part IV
Innovations in Learner Engagement

Two important research foci in the field of learning from dynamic visualizations are the design of the graphical representations themselves (see Part I of this volume) and the design of ancillary augmentations that accompany those representations (see Part III of this volume). A typical aim of these design-oriented approaches is to apply theories and models of human perception and cognition to the way dynamic visualizations are conceived and executed in order to facilitate processes such as the identification, selection, mental organization and integration of key aspects of the presented information. However, on their own, even the very best-designed graphical representations cannot guarantee effective learning. Conversely, effective learning is also possible from poorly designed representations. Good design is therefore neither always necessary nor sufficient. More fundamental is appropriate perceptual and cognitive engagement on the part of the learner, i.e., the processing activities that learners actually apply to the presented visual material.

Research on learner engagement has been stimulated by the frequent observation that many learners – especially in school contexts – follow the prescribed procedures and routines and yet fail to develop the required understandings. Early conceptualizations described learner engagement as a "… psychological investment in and effort directed toward learning, understanding, or mastering the knowledge, skills, or crafts that academic work is intended to promote" (Newmann, Wehlage, & Lamborn, 1992, p. 12). Over the past 20 years, educational and psychological research has considerably broadened how student engagement is now conceptualized (for recent reviews see Lawson & Lawson, 2013; Reschly & Christenson, 2012). Based on an analysis of such conceptualizations, Fredricks, Blumenfeld, and Paris (2004) distinguish three major forms of learner engagement: behavioral, emotional, and cognitive engagement. Although successful learning and understanding typically involves all three forms of engagement, this part focusses on *cognitive learner engagement.* It involves the individual investment in learning and understanding, the self-regulation of cognitive processes, and the use of cognitive learning strategies.

Researchers who study learning from dynamic visualizations have only recently begun to broaden the scope of their investigations to include the effect of cognitive engagement on educational effectiveness. This contrasts markedly with the situation for research on learning from text where there is a long tradition of interrelating theories and models of text processing, principles of text design, and methods for improving the learners' reading competencies. An important thread in research on cognitive engagement is concerned with how learners can be stimulated to carry out specific learning activities that – among other things – induce, support, and sustain effective perceptual and cognitive processes. Applying activities of this type to educational graphics rather than text has the potential to help learners deal more competently with dynamic visualizations that present particularly challenging processing demands.

This part includes three chapters on how cognitive learner engagement can be fostered such that dynamic visualizations are perceptually and cognitively processed in a more comprehensive and accurate way. The first two chapters focus on a specific learning strategy, namely, ***self-generated drawings***. Because drawing is a highly successful strategy in learning from text (for a review see Fiorella & Mayer, 2015), it has recently been investigated whether drawing can also improve learning from dynamic visualizations. While Lowe and Mason (2017, this volume) provide a detailed theoretical analysis of the potentials as well as of the demands associated with drawing for learning from animation, Stieff (2017, this volume) puts forward various principles of how drawing practices can successfully be implemented in science education. In the third chapter, Ploetzner and Breyer (2017, this volume) investigate how more comprehensive learning strategies that combine various specific learning techniques can facilitate learning from animations that either include or do not include verbal explanations.

References

Fiorella, L., & Mayer, R. E. (2015). *Learning as a generative activity – Eight learning strategies that promote understanding*. New York: Cambridge University Press.

Fredricks, J. A., Blumenfeld, P. C., & Paris, A. H. (2004). School engagement: Potential of the concept, state of the evidence. *Review of Educational Research, 74*, 59–109.

Lawson, M. A., & Lawson, H. A. (2013). New conceptual frameworks for student engagement research, policy, and practice. *Review of Educational Research, 83*, 432–479.

Lowe, R., & Mason, L. (2017). Self-generated drawing: A help or hindrance to learning from animation? In R. Lowe & R. Ploetzner (Eds.), *Learning from dynamic visualization – Innovations in research and application*. Berlin: Springer Publishers (this volume).

Newmann, F. M., Wehlage, G. G., & Lamborn, S. D. (1992). The significance and sources of student engagement. In F. M. Newmann (Ed.), *Student engagement and achievement in American secondary schools* (pp. 11–39). New York: Teachers College Press.

Ploetzner, R., & Breyer, B. (2017). Strategies for learning from animation with and without narration. In R. Lowe & R. Ploetzner (Eds.), *Learning from dynamic visualization – Innovations in research and application*. Berlin: Springer Publishers (this volume).

Reschly, A. L., & Christenson, S. L. (2012). Jingle, jangle, and conceptual haziness: Evolution and future directions of the engagement construct. In S. L. Christenson, A. L. Reschly, & C. Wylie (Eds.), *Handbook of research on student engagement* (pp. 3–20). New York: Springer Publishers.

Stieff, M. (2017). Drawing for promoting learning and engagement with dynamic visualizations. In R. Lowe & R. Ploetzner (Eds.), *Learning from dynamic visualization – Innovations in research and application*. Berlin: Springer Publishers (this volume).

Chapter 13
Self-generated Drawing: A Help or Hindrance to Learning from Animation?

Richard Lowe and Lucia Mason

13.1 Introduction

The high expectations that were once held for educational animations have been somewhat moderated in recent years. When technological advances first allowed animated graphics to become a mainstream educational resource, they were expected to offer far more effective ways of learning about dynamic subject matter (Mayer & Anderson, 1992). Rather than having to mentally animate the static depictions used in traditional textbooks, learners instead had access to explicit depictions of the referent dynamics. However, research has shown animation to be a two-edged educational sword (Lowe, 2014), especially when it presents complex subject matter that is unfamiliar to the target audience. The potential benefits that such animations undoubtedly have by virtue of their capacity to represent dynamics directly must be balanced against the information processing costs they may impose on learners. These costs can negatively affect learning by compromising the quality of the mental models that viewers are able to construct of the referent subject matter. Major contributions to these costs come from mismatches between the capacities of the human information processing system and the particularities of how animated graphics present their information. This chapter considers the potential of self-generated drawing to reduce some of these costs and so improve learning from animation. Its theoretical context is the Animation Processing Model (APM; Lowe & Boucheix, 2008, 2011, 2017, this volume).

R. Lowe (✉)
Curtin University, Perth, Australia

University of Burgundy, Dijon, France
e-mail: r.k.lowe@curtin.edu.au

L. Mason
University of Padua, Padua, Italy
e-mail: lucia.mason@unipd.it

R. Lowe, R. Ploetzner (eds.), *Learning from Dynamic Visualization*,
DOI 10.1007/978-3-319-56204-9_13

According to the APM, a significant barrier to learning occurs in Phase 1 of the five processing phases that are posited to be involved in learning from animation. This concerns the challenge of selectively extracting appropriate pieces of information from the animation's continuous dynamic flux. Such information extraction is piecemeal and cumulative because limitations on the human information processing system do not allow us to deal with an animation 'all at once'. To ensure that the information to be internalized fits within these processing constraints, the learner first needs to *decompose* the animated presentation into appropriate pieces. This decomposition is the source of *event units* (where an event unit is an entity plus its associated behavior) that are the raw material for mental model construction. However, proper decomposition of animations can be seriously derailed due to the dominant perceptual effects of their dynamics on which information learners attend to in a display. Learners who are novices with respect to the depicted content tend to select event units on the basis of how visually salient they are, rather than according to their relevance to the topic at hand (Lowe, 2003, 2004). As a result, the raw material learners internalize is inadequate for constructing a high quality mental model. A key step in helping animations fulfill their educational potential is therefore to improve learner extraction of this raw material. In the next section, we briefly consider research on various interventions intended to help learners process animations more effectively.

13.2 Supporting Animation Processing

An inevitable consequence of animations' direct depiction of dynamics is that the information they provide is transient rather than persistent. This characteristic of animations can have negative effects on learning. With system-controlled animations (in which the delivery of information is determined by the presentation system rather than by the user), this defining characteristic severely limits learners' opportunities for extracting and internalizing key information (Lowe, 1999). Transience means that learners are unable to repeatedly and intensively interrogate the available information (as they can do with static depictions). Instead, there can be a tendency merely to 'follow' the animation, behavior that can lead to shallow processing ('underwhelming') and the mere illusion of understanding (Lowe, 2004). In contrast, static graphics impose no intrinsic time constraints on learner interrogation of the presented information. Further, learners are obliged to study static graphics quite intensively if they are to work out the dynamics because these spatiotemporal changes are not presented directly and explicitly. However, the negative effects of animation's transience can be ameliorated (in principle at least) by giving learners control over aspects of the animated presentation such as playing speed, continuity, and direction. User control can be implemented indirectly via a set of 'buttons' or, in tablet computers, by way of more direct touch-based interactions. The many different interrogation possibilities offered by user control of dynamic visualizations have the potential to considerably improve learner processing of the available information. In certain circumstances, learning can indeed be enhanced by providing user control (Schwan & Riempp, 2004) but in others, the additional demands involved in exercising that control may be prejudicial to learning (Boucheix, 2008). Further, user control may not be beneficial if learners' lack of domain-specific background knowledge prevents them from directing their interrogation to the most relevant spatial and temporal locations within the animation (Lowe, 2008).

Failures to benefit from an animation can also be the result of more general processing deficits. For example, a learner may simply lack the strategies required to take proper advantage of the information an animation provides (Ploetzner, Lowe, Schlag, & Hauß, 2012; Ploetzner, Lowe, & Schlag, 2013). In the realm of text comprehension, strategy training has been found beneficial with respect to both recall and inference. A similar approach has been implemented with respect to learning from animations (Kombartzky, Ploetzner, Schlag, & Metz, 2010; Ploetzner & Schlag, 2013). However, the subject matter presented in these studies was relatively simple and the animation was accompanied by text. It is unclear how effective this form of general strategy training would be for animations of more demanding subject matter that did not have this text support (cf. Ploetzner & Breyer, 2017, this volume). Further, domain-*general* strategies do not (by definition) address domain-*specific* challenges that can arise when learners are faced with the particularities of complex, unfamiliar subject matter.

User control and strategy training are both somewhat open and generic types of intervention. Researchers have also investigated the utility of more focused approaches that single out specific features of an animation. When animations present a behaviorally realistic portrayal of relatively complex subject matter, different aspects of the display can compete for the learner's attention. Because our perceptual system tends to privilege information that has a high level of dynamic contrast with its surroundings, learners tend to neglect aspects of an animation that are relatively inconspicuous (Lowe, 2003). However, changing the speed of an animation can change the relative conspicuity of its constituent entities and so increase the likelihood of learners noticing low salience, high relevance aspects that would otherwise remain neglected (Fischer, Lowe, & Schwann, 2008; Meyer, Rasch, & Schnotz, 2010). Unfortunately, this may also result in aspects that were formerly quite salient becoming less so.

Segmentation (cutting an animation into smaller pieces along its time course) is a more targeted approach that is credited with having two main beneficial effects on learner processing of animations (Spanjers, van Gog, & van Merriënboer, 2010; Spanjers, Wouters, van Gog, & van Merriënboer, 2011; Wong, Leahy, Marcus, & Sweller, 2012). First, it is supposed to act as a form of signaling to the learner that indicates boundaries of the main episodes within the animation. This requires the segments to constitute what has been termed a 'meaningful subdivision' of the total animation as a whole. Second, segmentation reduces the amount of information that learners must deal with per unit time by inserting pauses between these episodes. Cognitive load theorists argue that this reduction in processing demands should be beneficial for learning. However, segmentation can be problematic for animations that depict complex subject matter in which there is extensive temporal overlap because multiple events occur simultaneously or in a cascading fashion. In such cases, defining clean inter-episode boundaries becomes problematic because of the overlapping relationships involved. The benefits of segmentation according to one set of relationships may well come at the cost of disrupting other important relationships that take place at the same time. Further, *between*-episode segmentation does nothing to reduce the *within*-episode processing demands associated with simultaneous presentation of multiple events.

The final form of intervention considered here is visual cueing, a technique that has traditionally been based on visuospatial contrast. Compared with the other approaches discussed so far, visual cueing can be considered as far more targeted with respect to the aspects of the animation being addressed. Visual cues are most commonly implemented either by rendering the target entity (or entities) in a color that strongly contrasts with the colors of other entities in the display or by using differences in illumination to produce the required contrast. This approach is intended to direct the learner's selective attention towards high relevance aspects of the display that may otherwise be neglected by raising their perceptual salience. It has essentially been 'borrowed' from a tried-and-true technique that has long been used with static graphics. However, in an animation, the capacity of visual cueing to direct attention tends to be severely compromised because of the way the human perceptual system privileges dynamic contrast. In an animation, it appears that cueing based solely on visuospatial contrast is much less able to command the learner's attention than the natural cueing that is present due to the dynamic contrast that exists within the display itself (Lowe & Boucheix, 2011). Although the effectiveness of visual cueing can be somewhat enhanced by making it dynamic rather than static (Boucheix & Lowe, 2010), the benefits obtained may come at the expense of other more relational aspects of processing.

Each of the approaches discussed above can, under some circumstances, help improve learning from animation. However, the learning gains achieved tend to be modest rather than dramatic and relatively limited in their scope. They may also have undesirable side effects on other aspects of learning. Whether or not worthwhile improvements are made depends on a variety of factors including the characteristics of the learning materials, the type of content addressed in those materials, the nature of the learning task, and the capacities of the target learners. It seems that none of these individual techniques is capable of making substantial across-the-board improvements in learning from animation. Obtaining improvements in the overall quality of the mental model that learners develop of the depicted subject matter remains particularly elusive. In the next section, we consider self-generated drawing as a relatively unexplored approach that has been suggested as a way to improve the quality of mental models that learners construct from their study of animations.

13.3 Drawing as an Aid to Learning

Our consideration of the potential of self-generated drawing to improve learning from animation is motivated by the promising results that have emerged from using this approach to enhance learning from text (e.g., Van Meter & Garner, 2005; Van

Meter, Aleksic, Schwartz, & Garner, 2006; Van Meter & Firetto, 2013). The recent upsurge of interest in drawing to learn (e.g., Quillin & Thomas, 2015) can be seen as part of a broader movement within education in which more active approaches to learning are advocated, including writing to learn (e.g., Bangert-Downs, Hurley, & Wilkinson, 2004) and talking to learn (e.g., Mercer & Littleton, 2007). Various explanations have been advanced about why drawing may have a positive effect on learning. These range from the motivating influence that drawing has on learners to its capacity to help them overcome limitations in their learning materials (Ainsworth, Prain & Tytler, 2011). In the latter case, the educational potential of self-generated drawing approaches to learning from text has been argued for in terms of the Generative Theory of Drawing Construction (GTDC; Van Meter & Garner, 2005) and its later revision, the Cognitive Model of Drawing Construction (CMDC; Van Meter & Firetto, 2013). These two theoretical frameworks owe a considerable legacy to Mayer's Cognitive Theory of Multimedia Learning (CTML; Mayer, 2009). Most of the research in this area has focused on drawing tasks in which learners generate their own depictions as they study text-based learning materials. In general, the findings from such research indicate that such self-generated drawing activities are beneficial, especially with respect to more challenging types of learning outcomes such as problem solving.

At the heart of current accounts for the educational benefits of drawing is the assumption that the process of producing such an external representation involves the student in generative learning. During this activity, verbal information from the original text representation is used to construct a visual representation. The learner's conversion of the text's propositional representation into an image-based representation is credited with producing a more robust mental model of the to-be-learned content. An important consequence attributed to this activity is that information in the source material is processed more deeply than it would be if there was no associated drawing requirement. The implicit assumption here is that the processing activity involved in generating the drawing is not so dominated by the demands associated with producing a drawing *per se* that learning-related processing is effectively sidelined. This is an important consideration in light of Leutner, Leopold and Sumfleth (2009) observation that the positive effects of a drawing for learning strategy may be eliminated if the drawing is too effortful.

Recent theorizing on drawing for learning has also attributed some of the benefits obtained to its self-regulation function. In particular, drawing is thought to increase learner self-monitoring. For example, when the learner reaches a point in developing the drawing beyond which s/he cannot continue, this hiatus acts as an alert that the source instructional material is not sufficiently understood (Van Meter & Firetto, 2013). The learner can presumably then respond by processing the material more deeply in order to overcome any impasse and resume the drawing. However, these beneficial forms of processing are not necessarily as evident during 'free' self-generated drawing as they are when drawing experiences are more structured. Van Meter and colleagues note that self-generated drawing activities are likely to be most effective when accompanied by appropriate instructional support.

13.4 Drawing and Learning from Animation

Is it possible that the existing theoretical frameworks on drawing for learning originally intended for application to text-based resources could also apply to learning media more generally? (see Van Meter & Firetto, 2013). If so, drawing may also be able to benefit learning from animation. To examine if the positive effects of self-generated drawing extended beyond learning from text, Mason, Lowe and Tornatora (2013) investigated the effect of self-generated drawing on learning from animation. In contrast to other research in this area, the source materials that were to be used as the basis for generating drawings contained no text-based information. This was to avoid the possibility of confounding so that if any benefits were found, they could unambiguously be attributed to the effect of drawing on the comprehension of animation (and not to possible contributions from text). This lack of text has important ramifications with respect to two key processes posited by the frameworks of Van Meter and colleagues – (i) the translation of propositional representation into an image-like representation, and (ii) the integration of verbal and non-verbal representations. Because neither of these processes was possible due to the absence of text, it may be that if there were any benefits, other somewhat analogous mechanisms were involved instead. For example, perhaps the learners still engaged in translation but between animated and static representations instead of between propositional and image-like representations. Further, rather than integrating verbal and non-verbal representations, perhaps integration was carried out between dynamic and non-dynamic representations.

The Mason et al.'s (2013) study found that 12-year old participants who self-generated drawings while studying an animation of a Newton's Cradle device (cf. Fig. 13.1a) had scores on immediate and delayed tests of understanding that were superior to those of participants who traced over dotted picture outlines or did no drawing at all. Test performance of those in the self-generation group was also positively correlated with the quality of drawings produced (cf. prognostic effect; Schwamborn, Mayer, Thillmann, Leopold, & Leutner, 2010). Further, higher test scores were associated with portrayal of low salience, high relevance aspects of the Newton's Cradle dynamics. These positive effects of drawing for learning from animation were explained in terms of processing activities posited by the Animation Processing Model (APM; Lowe & Boucheix, 2008, 2011, 2017, this volume) and the Generative Theory of Drawing Construction (GTDC; Van Meter & Garner, 2005).

Although encouraging, the results of this study need to be treated with some caution. As will be discussed below, there are some specific features of the subject matter depicted in this animation that may limit the generalizability of these findings. Further, there are likely to be some fundamental differences in the processes that need to be undertaken in drawing from text versus drawing from animation because of the distinctive way each of these media represents information. Chief amongst these differences is that perceptual considerations play a far more central role in successful processing of animations than they normally do in text processing.

This is important because since both the GTDC and the CMDC assume that static text (not an existing visualization, and certainly not a dynamic one) is the starting material, neither foregrounds perceptual aspects of processing in their accounts of how self-generated drawing is able to improve learning.

There is also the question of which information we wish learners to acquire from studying an external representation. As noted in the introduction to this chapter, the key advantage offered by animated graphics over their static counterparts is their provision of detailed, explicit and comprehensive information about dynamics. Any approach such as self-generated drawing that is intended to improve learning from animation should therefore above all help learners to understand and remember the spatiotemporal changes that are presented. In other words, it should help build high quality runnable mental models of the subject matter that allows the dynamic aspects to be properly represented in the mind of the learner.

The next two sections focus on the possible roles of drawing by considering two key aspects: (i) a drawing as an artefact (product) and (ii) the drawing as an activity (process).

13.5 Drawing *Products* as an Aid to Learning

The previously mentioned transitory nature of animations makes it difficult to characterize the presented information definitively in any detail because it is continually changing. When an animation depicts complex and unfamiliar subject matter, establishing relationships that are key to building high quality mental models can be especially problematic for learners. However, this situation may perhaps be ameliorated if a learner has generated one or more static drawings while studying the animation. One possible function such drawing products could serve is to act as on-going visual checklists that learners can use to monitor how well they are extracting information from the ever-changing animation. This could help prevent the neglect of less conspicuous but important aspects of the display that may otherwise be overlooked due to the dynamic environment. Further, when graphic entities are 'frozen' in the form of a static drawing, it is far easier to ascertain the visuospatial relationships that exist between them than when they are undergoing continual changes. For example, static depictions are far more amenable to the detailed analysis required to establish the visuospatial properties of each entity, how sets of entities are configured, and how configurations change over time (cf. Lowe, Schnotz, & Rasch, 2011). However, not all self-generated drawings will be equally effective in this supportive role.

The usefulness of such learner generated drawing products will very much depend on how well they portray key information available in the animation. Although drawings have much in common with other types of visual images, one of their most distinctive features is that they are created not captured. This makes them very different from representations such as photographs (or even fingerprints) that

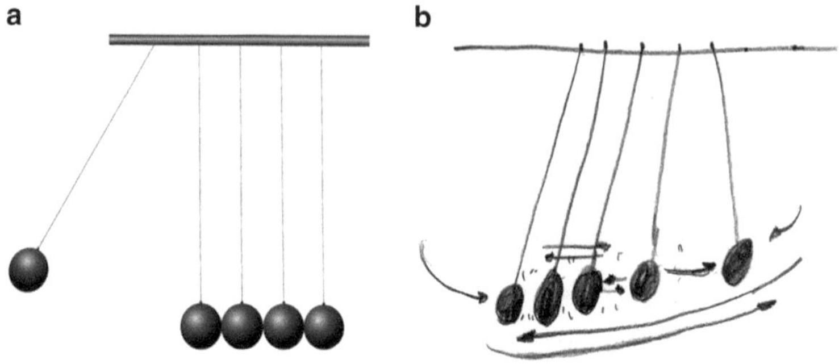

Fig. 13.1 (**a**) Frame from Newton's Cradle animation; (**b**) example student drawing of Newton's Cradle

require minimal input from the 'author' in terms of generating the depiction. Drawings tend to be highly selective rather than comprehensive depictions of the referent subject matter. They are selective both in which aspects of the subject matter are depicted and in how those aspects are portrayed. The types of depictions that participants in drawing-for-learning studies generate tend to be relatively simple, typically consisting of lines of various types (straight, curved, etc.) and shapes defined by enclosing lines (e.g., Schmeck, Mayer, Opfermann, Pfeiffer, & Leutner, 2014). In the main, these drawings are two-dimensional depictions that lack shading to indicate volume. Given that the target learners do not typically possess the advanced drawing skills of professionally trained artists, it is unsurprising that the depictions they produce are relatively unsophisticated.

Figure 13.1 compares a frame from the Newton's Cradle animation with a representative drawing of the type that participants self-generated in the Mason et al.'s (2013) study. A key reason for choosing the Newton's Cradle as the subject matter for this study was that it would be easy to draw with no more than a few simple lines and circles. As exemplified in Fig. 13.1b, the 12-year old participants in this study could indeed generate satisfactory drawings of the Newton's Cradle using such graphic entities. Consistent with the criteria proposed by Van Meter and colleagues in their studies, the basic resemblance between this drawing and its referent is reasonably close. This applies both to the characteristics of the graphic entities from which the drawing is composed and to the way those entities are arranged in space. In principle, the drawing products generated by participants in the Newton's Cradle experiment could serve the types of facilitative functions described above.

However, not all to-be-learned subject matter is equally simple with respect to its visuospatial characteristics. Indeed, animations are often the representation of choice for presenting topics that are complex in both appearance and behavior. Figure 13.2 shows another type of widely used educational content that is nowadays typically presented in animated format – YouTube® abounds with examples of such heart-functioning animations. Due to the nature of this subject matter, an adequate drawing of a heart cross-section that shows its functional aspects would inevitably

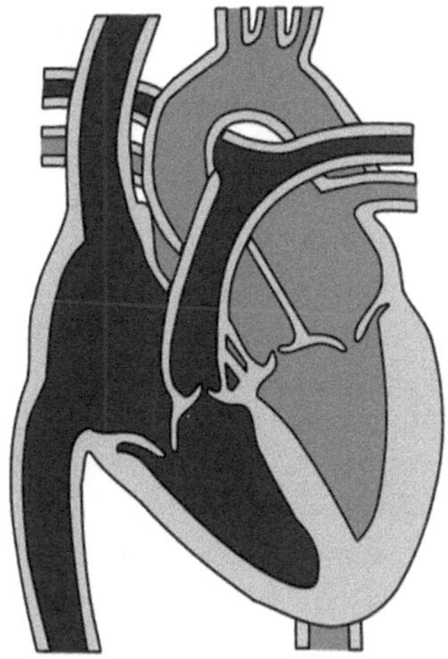

Fig. 13.2 Frame from pumping heart animation. Considerably more drawing skill is required than for Newton's Cradle

need to be considerably more complex than for the previous example. A Newton's Cradle drawing is perfectly adequate if it consists of the same combination of basic graphic elements (a circle on the end of a line) repeated five times in a linear series. However, the situation is very different for even a relatively simplified drawing of the heart (given the resemblance requirement mentioned above). Not only is there far more variety amongst its individual constituent graphic elements, but these elements are also set out in a far more sophisticated arrangement. Further, there is considerable change in the form of certain elements of the heart during its functioning. As products, learner self-generated portrayals of the heart are therefore likely to contain numerous deficiencies because few learners possess the artistic skills necessary to do a good job of drawing this intrinsically difficult subject matter. These deficiencies would limit the utility of a self-generated drawing for purposes such as enhancing engagement with the learning task (see Stieff, 2017, this volume), monitoring one's understanding of the animation or performing a detailed subject matter analysis. As we shall see in the next section, this issue of subject matter complexity also has important processing implications for the learners who might be required to self-generate a drawing of such subject matter.

13.6 Drawing *Processes* as an Aid to Learning

The previous section considered whether the final product of self-generated drawing activity could play a role in aiding learning from animation. However, we also need to consider the extent to which the very act of generating a drawing could itself benefit learning by fostering more effective processing of the animation. It could be argued that generating such drawings involves learners in an activity likely to make them to scrutinize the animation more intensely than they would otherwise do. In the absence of a drawing task, learners may instead merely be swept along by the animation's overall dynamics and so process it holistically and superficially rather than more analytically (cf. Lowe et al., 2011; Stieff, 2017, this volume).

With respect to the processing that will occur, the nature of the reference information provided to the learner as the basis for self-generating a drawing is an important consideration. If learners are given only a text (and no pictures) as their starting point, they are obliged to try and visualize what the subject matter would look like by themselves. For subject matter of even moderate complexity, it can be very challenging to translate text's linear strings of arbitrary symbols into satisfactory two dimensional depictions of the referent entities' forms and arrangements. In practice, even the most skillfully written text can be far too imprecise to provide a sufficiently detailed specification of the many visuospatial subtleties present in a pictorial representation. This is particularly the case for complex subject matter that is unfamiliar to the target learner group. Consider for example the impossibility of writing a text that could adequately represent all the crucial details that are provided in a diagram of the heart (as in Fig. 13.2). It is unsurprising that with respect to content such as this, educators have long appreciated the need to provide learners with complementary text and pictorial representations.

A key advantage that drawing from an animation therefore appears to have over drawing from a purely textual representation is that animation can provide an explicit visual model upon which to base the drawing. For the moment, we will consider just the *visuospatial* information that an animation provides by virtue of its being a pictorial rather a word-based representation. The subsequent section will consider the issue of the animation's dynamic information. In a case such as the heart animation, rather than having to visualize for themselves what the subject matter looks like, learners have the opportunity to directly observe all the crucial visuospatial subtleties that are present. Key aspects ranging from the pathways of the various blood vessels to important differences in heart wall thickness are available and explicit. However, it is one thing to provide this information but quite another for the learner to make effective use of it in self-generated drawing to aid learning. A fundamental assumption underpinning drawing for learning approaches in general is that this approach involves the learner in active processing of the source material (Van Meter & Firetto, 2013). However, it seems there is no guarantee that the type of activity learners engage in during self-generation of such artefacts will necessarily be beneficial (Stull & Mayer, 2007). The process of executing a drawing is of itself one that typically involves significant perceptual, cognitive and psycho-motor activities (Van Sommers, 1984). Further, if the drawing being generated is based on another external representation (such as a text or an animation), complementary processing of that source material in order that a reasonable resemblance can be produced is also required. When the main purpose of drawing is to facilitate learning, these activities would need to be very much subservient to processing related to learning-related tasks such as comprehending the target subject matter.

Across the range of topics that students are required to learn, there are enormous differences in how difficult it would be to draw the to-be-learned subject matter. As already observed, most students would find the task of self-generating a drawing of a heart cross-section to be far more demanding than drawing a Newton's Cradle. At the most fundamental level, very young students may simply lack the level of fine motor control that is required to generate a sufficiently accurate depiction. However, older students are also likely to have considerable difficulty in drawing the heart because they lack the specialized observational and production skills that proficient artists would have developed over many years of drawing practice. As a consequence, it is unlikely that most such students would be able to reproduce the complex, subtle variations in line and form that are required for an accurate portrayal of the heart's functionality. Even if they could produce an acceptable rendition of the heart, they would probably be so consumed by the perceptual and cognitive demands of executing the drawing itself that any consideration of how the heart actually functions would effectively be sidelined. Activity that is largely dominated by the mechanics of drawing process would leave precious few processing resources for the intended task of learning about the subject matter. Under these circumstances, the main task would inevitably be learning how to draw the heart rather than learning how the heart works.

So, there is a danger that with hard-to-draw subject matter, the drawing activity would primarily involve copying visuospatial features rather than understanding how the heart works. However, the likelihood of slavish copying is reduced when the source material is an animation (as opposed to static graphics) because the information is transitory. This means that the learner is required to convert a dynamic portrayal of the subject matter into one or more static depictions. On the surface, this could be regarded as beneficial because it requires deeper processing. The processes learners engage in when re-representing a source to the target representation are supposed to be a major contributor to the benefits of drawing for learning. Presumably, there is a degree of required re-representation that is optimal for achieving maximum benefit. There are two key issues that impinge on the effectiveness of this conversion. One concerns the extent to which the learner is able to extract the *visuospatial* information required for generating one or more static drawings from the continuously changing information flux of the animation. The other concerns the capacity of the learner to incorporate in those drawings an adequate portrayal of the *spatiotemporal* information presented by the animation.

13.7 Demands of Drawing

The skills required to become proficient in reading and writing text are rightly regarded by educators as very demanding to acquire and therefore as needing many years of concentrated, explicit instruction for their development. However, pictorial representation enjoys no such privileged status within the educational enterprise. In contrast to the high proportion of classroom time that is devoted to text-oriented tuition, very little time is allocated to developing students' graphicacy skills. In most cases, the capacities to interpret and generate pictures tend to be treated as an optional (and recreational) extra rather than an essential component of mainstream education. An unsurprising result of this comparative neglect is that most students do not become adept at drawing even relatively simple everyday subject matter (let alone more complex, unfamiliar content).

Studies of the nature of drawing skill suggests that the difference between expert and novice drawers has little or nothing to do with the level of their basic motor abilities, but rather is due to how well they process the to-be-drawn information. For example, Tchalenko (2009) found that experts in drawing were more accurate than novices only when they were copying more complex material. Tchalenko's research also showed that such experts engaged in a more systematic, analytical approach to processing of the original material (Tchalenko, Naim, Moshe, & Miall, 2014). Further, those who have higher levels of drawing skill are more able to report local shape when required while disregarding global shape (Chamberalin, McManus, Riley, Rankin, & Brunswick, 2013). Perdreau and Cavanagh (2013) have associated drawing skill with the ability to construct a robust mental representation of object structure and maintain it relatively intact despite the many potentially disruptive back and forth eye movements that occur between the original object and the drawing. This is important because experienced artists are distinguished from beginners by their greater tendency to shift their gaze back and forth more frequently (Cohen, 2005).

If the drawing skills that most students develop as a result of their schooling are relatively rudimentary, it is also unlikely that their processing of complex visual stimuli will be particularly sophisticated or effective. However, even skilled processing of such a stimulus will not necessarily foster learning. In this context, it is important to consider just what it is that expert drawers target when they process a to-be-drawn stimulus. The over-riding goal in such cases is to produce a drawing that has a high degree of resemblance to the stimulus material. In other words, accuracy of the visuospatial information is the main concern. Recent research confirms that the expert's whole processing approach is finely tuned in order to be optimized for this central purpose (Perdreau & Cavanagh, 2014). There is no doubt that expert drawing involves intensive activity and that such activity is highly generative. Further, the processing carried out during this activity is deep and analytical. Nevertheless, it is not conducted with the aim of comprehending the subject matter (other than is absolutely necessary to ensure good visuospatial resemblance). Expectations that high quality drawing will lead to learning benefits simply because it requires deep, generative, active processing are seriously misguided because they ignore the overriding effect that the drawer's purpose has on the outcomes. However, this is not of course to imply that drawing can never be effective as a tool for learning. Rather, it highlights the importance of considering the nature and purpose of the processing that is associated with such drawing activity.

If the subject matter is not complex and so is relatively easy to draw, it may well be that the drawer's processing could be far more oriented towards learning-related activity than to a quest for resemblance. This would help to explain the positive findings of various drawing for learning studies that have appeared in the literature (e.g., Schmeck et al., 2014). However, it is important not to characterize subject matter complexity solely in terms of its visuospatial attributes (what it looks like). It is perfectly possible that the stimulus material for a drawing is simple in a visuospatial sense but spatiotemporally complex. In other words, it looks easy to draw until it starts to move and then proves to have dynamics that are very challenging to characterize and depict. In the next section, we consider some of the additional demands involved in the drawing of dynamic subject matter and their possible implications for drawing for learning.

13.8 Drawing Dynamic Subject Matter

Extracting accurate information about dynamic subject matter can be very challenging, even for professional artists (let alone for students in the average classroom). For example, the eminent 18th century English horse painter George Stubbs famously failed to depict the configuration of a galloping horse's legs accurately (Calderon, 2011), despite his extensive study of equine anatomy (Stubbs, 1766). This failing has been attributed not to any deficit in his artistic abilities or knowledge of horse structure but to the intrinsic difficulty of observing exactly what happens when a horse moves. The human visual system is simply not up to the task of perceiving such information. It was not until Muybridge (1899) was able to capture horse galloping photographically that the issue was resolved. It may therefore be unreasonable to expect students to make accurate observations of the all the dynamics that occur in an animation, especially if its subject matter is complex and unfamiliar.

Even if the presented information is perceptually accessible, the spatiotemporal variation inherent in dynamic subject matter tends to disrupt efforts to definitively characterize its visuospatial attributes. Such characterization requires analysis of the form of entities depicted in the animation and the relationships amongst them. With complex subject matter, this is hard enough to do when the stimulus material is static (hence artists' traditional preference to use posed models when drawing dynamic scenes). However, the task becomes extremely challenging if drawing directly from a dynamic stimulus is required. It is no surprise that that artists today instead rely on photographic references to support their drawing of action (e.g., Brodatz & Watson, 1968). If the situation is difficult when the subject matter is undergoing translation (moving relative to its context), it is even worse if transformations are involved (i.e., changes in the intrinsic attributes of entities, such as shape, size, etc.). In these circumstances, there is no one visuospatial characterization that can be regarded as inherently definitive. The most challenging situation is when translations and transformations co-occur in an animation, something that is especially likely with biological subject matter.

Generating (static) drawings that provide an effective depiction of dynamic information also requires skills over and above those required to draw static subject matter. Even if the drawer manages to produce a reasonable depiction of the subject matter's visuospatial attributes, it is quite another matter to portray its dynamics effectively in a static picture. Two key challenges in depicting these dynamic aspects are (i) accurately observing the changes that take place over time, and (ii) using suitable graphic techniques to represent those changes in a static drawing. In the Stubbs horse example, the artist attempted to convey the dynamics by way of a single image. He relied solely on the horse's overall pose to indicate how the horse was moving. However, in many cases depictions intended to indicate how things change over time introduce specialized graphic techniques and conventions that provide more elaborated information about dynamics (Cutting, 2002). Such depictions can involve either the subject matter itself (as with composite pictures that use techniques such as dotted images or ghosting to indicate previous positions during a movement) or the addition of ancillary graphic material such as arrows that has the express function of indicating the subject matter's dynamics (see also Jenkinson, 2017, this volume).

Although it may be a relatively straightforward matter to enrich static pictures with dynamic indicators if the changes involved are simple, it can be very challenging to represent spatiotemporal information that is more complex and subtle. Attempts to do so inevitably result in visual clutter that can hinder rather than help comprehension. It is particularly difficult to produce a satisfactory explanatory static depiction of dynamics that involve extensive simultaneity or overlapping spatiotemporal changes. Using graphic techniques and conventions to indicate dynamics effectively on a static picture is not a trivial matter. It requires not only considerable familiarity with these approaches but also the capacity to apply them successfully to the subject matter at hand. Even professional illustrators may struggle to produce entirely satisfactory ways of representing complex dynamic subject matter via static pictures. For example, otherwise capable people may be daunted by the professionally designed static pictorial instructions that are supposed to help us carry out the process of assembling a piece of flat pack furniture. Because drawing, especially if it is from dynamic visualizations of complex unfamiliar subject matter, imposes a number of substantial challenges on learners, is it reasonable to expect them to use it effectively as learning tool without additional support?

13.9 Easy or Hard to Draw?

The discussion so far has identified a number of possible impediments to drawing improving learning from animation. However, as shown by the Mason et al.'s (2013) study, there may be certain circumstances where this strategy would be effective. A proper evaluation of how challenging a specific type of animated subject matter will be for learners to draw seems to be important for deciding on the likely effectiveness of a drawing for learning strategy. As indicated earlier, if the subject matter is too complex, the demands of achieving a reasonable degree of resemblance between the

Fig. 13.3 Static frame from an earthworm locomotion animation

drawing and the stimulus material would be expected to leave little capacity available for learning-related processing. The complexity that confronts the learner is not solely due to characteristics such as how many entities are contained in the display, how varied and involved those entities are with respect to their forms, and how they are set out across the display space. For example, from the perspective of such visuospatial attributes alone, the frame extracted from an animation of a worm's body during its locomotion (Fig. 13.3) would be relatively easy to draw.

This portrayal of a worm was in fact specially designed for ease of drawing while at the same time allowing the key principles of worm locomotion to be presented in the animation. Although it shows the main parts of a worm that are responsible for its movement (i.e., flexible body segments and their tiny subtending retractable anchoring setae), these are reduced in number and simplified. It therefore contains a modest number of entities and these can be drawn using just a few variations on a basic shape (i.e., rounded rectangles with different proportions) and several small pointed lines. The entities are also arranged in a simple linear structure. In this respect, it is very similar to the Newton's Cradle in terms of drawing requirements (although the worm does have 12 segments as opposed to 5 balls in the Cradle). Pilot testing showed that even 10 year old children have no trouble in self-generating drawings of this static worm frame and their drawings closely resemble to the original. However, the situation is very different for the worm *animation*. It introduces some complexities that are not present in the Newton's Cradle. One of these complexities involves the temporally coordinated presentation of several very different sets of events. At the broadest scale, the whole body of the worm moves across the display screen from left to right. Embedded within this overall motion are two other types of spatiotemporal change (i) the protraction and retraction of the worm's setae (depicted by the small lines located at the base of each segment), and (ii) the expansion and contraction of the segments that constitute the worm's body. The worm's forward movement results from the coordinated execution of these two activities and is accompanied by a wave-like pattern that moves along its body in the opposite direction. The net effect of this dynamic activity is that using static drawings to capture a satisfactory representation of what is happening in the worm animation is considerably more challenging than it is for the Newton's Cradle. Work in progress (Boucheix, Lowe, Breyer, & Ploetzner, 2015) suggests that learner activity is dominated by attempts to capture information about the worm's body shape with efforts to include the dynamics being very much a second order issue. Similar results were obtained by Fillisch and Ploetzner (2015).

13.10 Drawing Quality and Learning from Animation

Some research has found an association between the quality of the drawing that individuals produce during a drawing-to-learn intervention and the learning gains made (e.g., Schwamborn et al., 2010; Mason et al., 2013). Accordingly, those who generated higher quality products would presumably have been more successful in translating information provided in the original format into static drawings. However, it is currently difficult to say definitively whether it was the drawing activity itself that caused the learning gains or whether the better drawings were produced as a result of learning gains that had other causes. For example, it may be that those who learned more effectively already had relevant prior knowledge, higher spatial ability, or more successful processing strategies. In such cases, superior drawings would be produced as a result of their more effective learning rather than being responsible for it. However, if it turned out that drawing quality is causally related to learning gains, the possibility is raised that training learners to produce better drawings could improve their learning. This raises two important questions; (i) how should drawing quality be assessed, and (ii) what type of drawing training should learners be given?

In assessing the quality of drawings produced for the purposes of learning, there are several attributes that need to be taken into account and the relative weighting given to each could well need to vary according to the particularities concerned. With respect to visuospatial features, a certain level of accuracy will be required in depicting the form and arrangement of the entities involved. However, the closeness of this mapping between the drawing and its referent that is required will very much depend on the specific content involved. For example, because the dynamics of the Newton's Cradle involve the interaction of discrete entities that undergo no intrinsic changes during the process, it matters little how well the balls and strings that support them are depicted. Almost any sort of a circle (the ball) attached to a line (the string), even if quite rough, will probably suffice. What really matters in Newton's Cradle drawings is the various *arrangements* that the set of ball-string elements are shown as adopting. The drawings of the ball-string units serve essentially as tokens by which the different configurations of the Newton's Cradle device as a whole can be shown. However, the situation would be very different for learner drawings made from a heart animation (Fig. 13.2). In this case, the respective shapes and sizes of the various regions of the heart are of crucial importance in demonstrating an understanding of how the heart works, as are the intrinsic changes that occur in these aspects during its operation. The criteria for assessing learner drawings of the heart would therefore need to be much stricter with respect to these visuospatial characteristics than for the Newton's Cradle.

With learner drawings made from animations, it is not sufficient to assess the quality of the products with respect to their portrayal of visuospatial features alone. It is also important (usually even more so) to assess how well the dynamic aspects of the subject matter have been depicted in learners' drawings. This aspect of assessment can be more challenging than assessing drawings' visuospatial attributes because in static depictions, dynamics can be indicated only indirectly and via a variety of techniques (including the use of arrows to show movement and the use of dotted lines or multiple drawings to show successive states). There is thus no standard criterion that can be used as the basis for assessing how well the animation's spatiotemporal information has been depicted. This contrasts with the situation for visuospatial information where notions such as the presence or absence of key entities, relationships and properties are far less ambiguous in terms of interpretation by an assessor. Issues such as what is an appropriate grain size for representing the dynamics also need to be considered. In complex animations, multiple levels of hierarchically related spatiotemporal relationships may exist and these are extremely difficult to depict clearly, especially for learners who lack advanced graphicacy capacities. A further complicating factor for complex subject matter is that the inclusion of such dynamics can greatly add to the visual clutter of the depiction and make it difficult to distinguish between visuospatial and spatiotemporal aspects.

Another even more fundamental issue to be addressed is whether or not the quality of drawing (even if it can be assessed relatively unambiguously) is necessarily an indicator of learning benefit. If learners are not specifically told to make the drawing 'explanatory' (that is, capable of providing a comprehensive explanation of the subject matter to another person), they may well not bother to include all aspects that they have observed in the animation. Indeed, they may deliberately abstract the presented information both to make their drawing task easier and to help clarify fundamental relationships that underlie what the animation shows. This notion of abstracting a realistic depiction in order to make the subject matter more tractable to the learner is a standard procedure taught to students to help them solve physics problems (as well as being used by domain experts as a way of cutting through the surface information to expose the crux of a problem). Under such circumstances, the self-generated drawing is not a representative indicator of the individual's understanding of the presented subject matter. Rather, it is only a complement to learner's internal representation. To have a proper appreciation of the learner's mental model of the subject matter, we would need to know what aspects of the subject matter have been deliberately omitted from the external drawing. It could even be that lower quality drawings (in terms of comprehensiveness) is an indicator of a superior (rather than inferior) mental model because it reflects a higher level, more generalizable understanding of the subject matter that goes beyond the specifics of a particular animation.

If drawing quality is a casually related indicator of learning success, interventions that aim to help learners produced better drawings should perhaps be considered. Two possibilities here would be to raise students' general capacity to generate drawings that capture key aspects of the target subject matter, or to provide specific help with the drawing of the particular content shown in a target animation. Currently, most educational systems place relatively low value on developing graphicacy skills. Despite the burgeoning use of visualizations in teaching and learning in recent years, the traditional dominance of literacy and numeracy within classrooms remains essentially unchanged. Typically, any graphicacy acquisition that does occur during a student's school career is largely incidental rather than the result of the type of deliberate, intensive instruction that we see with respect to words and numbers. This means that even very basic aspects of graphicacy such as the capacity to distinguish between and interpret fundamental graphic conventions develops relatively slowly (Boucheix, Lowe, & Thibaut, 2015). Further, although the capacity to express ideas via the written word is a highly regarded aspect of literacy development, there is no comparable emphasis given to developing students' fluency in expressing ideas in the form of drawings or other visual representations. Specialized techniques such as how to portray dynamic information effectively via static drawings receives scant if any attention in most classrooms. It may be that if school systems implemented rigorous programs of graphicacy development through the years, students would be better equipped to self-generate more effective drawings from animations and perhaps use those skills to improve their learning.

It seems rather unlikely that graphicacy development will become a number one educational priority any time soon. Perhaps it would therefore be more practical to provide students with more immediate support for drawing the specific type of subject matter shown in a particular animation. This approach has been used to some extent by researchers who have been investigating drawing for learning from text and has produced promising results. In those studies, one technique has been to provide learners with pre-drawn entities which they can either copy or assemble in order to generate their drawing. Something similar could also be useful when the starting representation is an animation rather than a text. Perhaps this provision of ready-made entities would help shift the learner emphasis from trying to produce drawings with an acceptable resemblance to the original to a focus on how they are arranged in space and vary over time.

13.11 Conclusion

The use of self-generated drawing to enhance learning is both strongly advocated (e.g., Ainsworth et al., 2011; Quillin & Thomas, 2015) and supported by findings from empirical research (e.g., Van Meter et al., 2006; Schmeck et al., 2014). However, to date, the main focus of this research has been upon improving learning from text rather than other forms of representation. With the increased reliance on non-text ways of presenting to-be-learned content in today's educational resources, it is appropriate to consider if the benefits of self-generated drawing extend to media such as dynamic visualizations (cf. Van Meter & Firetto, 2013). This chapter considered the potential of self-generated drawing to help improve learning from animations of complex, unfamiliar subject matter. It was prompted by the limited effectiveness of a variety of other interventions that have been examined (user control, strategy training, speed alternations, segmentation and cueing), particularly their relative lack of impact on mental model quality. A key rationale for considering self-generated drawing as a way of supporting learning is that it could lead to an animation being processed more deeply.

Given that animations are most widely used for presenting subject matter that features change over time, the primary aim of interventions intended to make animations more educationally effective should presumably be to improve learning about spatiotemporal (rather than visuospatial) aspects of these representations. But to what extent can self-generation of *static* graphics be expected to improve learner processing of *dynamic* visualizations? This chapter has suggested that for subject matter of even modest complexity, the requirement to generate a static drawing from an animation is likely to lead to learners' finite processing resources being misdirected. Instead of targeting comprehension of the referent subject matter's dynamics, learners' efforts tend to be dominated by the demands of achieving visuospatial resemblance. We do not question the likelihood that the generative aspect of a drawing task makes learners considerably more active in their processing of an animation. However, we do challenge the assumption that this activity is necessarily beneficial for learning, especially when complex, difficult-to-draw content is involved. If the nature of the processing that learners engage in is more concerned with reproducing the visuospatial attributes of the material shown in the animation than understanding the dynamics, the capacity to build a high quality mental model will be compromised because crucial spatiotemporal information will be neglected. Requiring learners to portray *dynamic* information in *static* form essentially emphasizes the primacy of visuospatial over spatial temporal information. The processing of the animation when drawing is required may well be deeper, but if that depth of processing is concerned with visuospatial rather than spatiotemporal information, it will not foster better mental models.

An alternative argument is that instead of hindering learning, this dynamic-to-static conversion activity could actually be of benefit because it requires learners to analyze the depiction spatiotemporal information more deeply in order to depict it in their drawings. However, this assumes that learners are equipped with the levels of artistic capacity and graphicacy needed to devise effective ways of showing dynamics by means of static graphics. Given that this type of task is challenging for professional graphic designers, it seems unrealistic to expect it to be performed adequately by most learners. Failures in such performance would have implications for both product and process aspects of how learning might be expected to benefit from the self-generation of drawings from animation. These implications could be particularly negative for animations depicting complex, difficult-to-draw subject matter. With respect to products, this is because the inadequate drawings generated by learners would provide a poor basis for monitoring their developing understanding of the animation (because they fail to capture key aspects). As such, the possible benefits of comparative processing amongst the source, emerging drawing and developing mental representation would be severely restricted. With respect to process, this is because the learners' available processing resources would be dominated by the challenges of trying to provide a drawing which had a reasonable resemblance to what is shown in the animation and because they are simply not equipped to convert the animation's dynamics into adequate static representations. When the subject matter involves many and varied spatiotemporal changes that take place simultaneously (which is often the case when it is complex), the task of capturing such dynamics in a static graphic is just far too demanding. The positive results from the Mason et al.'s (2013) study should therefore be treated with some caution, given that they were for subject matter that was very easy to draw. It would be a serious mistake to assume they generalize to other more complex and harder to draw types of content. If requiring learners to self-generate static drawings from animations can effectively sabotage their processing of the all-important dynamic information, perhaps such an intervention is misconceived. One possible way around this problem is to put spatiotemporal information (rather than visuospatial information) front and center in designing learning interventions. For example, instead of requiring learners to self-generate static drawings from animations, why not have them generate their own dynamic visualizations in which visuospatial considerations are down-played rather than foregrounded?

Such an approach has been pioneered at the Centre of Education and Animation, VIA University College in Viborg, Denmark (http://animatedscience.dk). It involves students using techniques such as stop motion photography to create their own dynamic visualizations from cardboard cut-outs or models. Although not conceptualized by its originators as an approach for improving learning from existing educational animations, it could be readily adapted for this purpose. By photographing pre-made cut-outs and models, learners can focus their generative activity on coming to grips with the dynamics rather than being side-tracked by other competing but essentially secondary demands. In a sense, this approach could be seen as capable of providing the benefits attributed to drawing for learning from text in that learners are engaged in generative activity. However, the emphasis is upon generating dynamics rather than generating visuospatial resemblance because learners are freed from the responsibility to produce their own portrayals of the entities that participate in the animation's action. Future research is needed to investigate the potential utility of such interventions.

References

Ainsworth, S., Prain, V., & Tytler, R. (2011). Drawing to learn in science. *Science, 333*, 1096–1097.

Bangert-Downs, R. L., Hurley, M. M., & Wilkinson, B. (2004). The effects of school-based writing-to-learn interventions on academic achievement: A meta-analysis. *Review of Educational Research, 74*, 29–58.

Boucheix, J.-M. (2008). Young learners' control of technical animations. In R. K. Lowe & W. Schnotz (Eds.), *Learning with animation: Research implications for design* (pp. 208–234). New York: Cambridge University Press.

Boucheix, J.-M., & Lowe, R. K. (2010). An eye tracking comparison of external pointing cues and internal continuous cues in learning with complex animations. *Learning and Instruction, 20*, 123–135.

Boucheix, J.-M., Lowe, R. K., Breyer, B., & Ploetzner, R. (2015, August). *How self-generated drawing may impact learning from animation.* Paper presented at the 16th Biennial EARLI Conference for Research on Learning and Instruction, Limassol, Cyprus.

Boucheix, J.-M., Lowe, R. K., & Thibaut, J.-P. (2015, August). *Graphicacy in children: An experimental study of the comprehension of paired graphics conventions.* Paper presented at the 16th Biennial EARLI Conference for Research on Learning and Instruction, Limassol, Cyprus.

Brodatz, P., & Watson, D. (1968). *The human form in action and repose: A photographic handbook for artists.* New York: Reinhold.

Calderon, W. F. (2011). *Animal painting and anatomy.* New York: Dover.

Chamberlain, R., McManus, I. C., Riley, H., Rankin, Q., & Brunswick, N. (2013). Local processing enhancements associated with superior observational drawing are due to enhanced perceptual functioning, not weak central coherence. *The Quarterly Journal of Experimental Psychology, 66*, 1448–1466.

Cohen, D. J. (2005). Look little, look often: The influence of gaze frequency on drawing accuracy. *Perception & Psychophysics, 67*, 997–1009.

Cutting, J. E. (2002). Representing motion in a static image: Constraints and parallels in art, science, and popular culture. *Perception, 31*, 1165–1193.

Fillisch, B., & Ploetzner, R. (2015, August). *The role of support in drawing for learning from complex animation.* Paper presented at the 16th Biennial European Conference for Research on Learning and Instruction, Limassol, Cyprus.

Fischer, S., Lowe, R. K., & Schwan, S. (2008). Effects of presentation speed of a dynamic visualization on the understanding of a mechanical system. *Applied Cognitive Psychology, 22*, 1126–1141.

Jenkinson, J. (2017). The role of craft-based knowledge in the design of dynamic visualizations. In R. Lowe & R. Ploetzner (Eds.), *Learning from dynamic visualization – Innovations in research and application.* Berlin: Springer (this volume).

Kombartzky, U., Ploetzner, R., Schlag, S., & Metz, B. (2010). Developing and evaluating a strategy for learning from animations. *Learning and Instruction, 20*, 424–433.

Leutner, D., Leopold, C., & Sumfleth, E. (2009). Cognitive load and science text comprehension: Effects of drawing and mentally imagining text content. *Computers in Human Behavior, 25*, 284–289.

Lowe, R. K. (1999). Extrating information from an animation during complex visual learning. *European Journal of Psychology of Education, 14*, 225–244.

Lowe, R. K. (2003). Animation and learning: Selective processing of information in dynamic graphics. *Learning and Instruction, 13*, 157–176.

Lowe, R. K. (2004). Interrogation of a dynamic visualization during learning. *Learning and Instruction, 14*, 257–274.

Lowe, R. K. (2008). Learning from animation: Where to look, when to look. In R. K. Lowe & W. Schnotz (Eds.), *Learning with animation: Research implications for design* (pp. 49–68). New York: Cambridge University Press.

Lowe, R. K. (2014). Dynamic visualizations: A two-edged sword? In W. Huang (Ed.), *Handbook of human centric visualization* (pp. 581–604). New York: Springer.

Lowe, R., & Boucheix, J.-M. (2008). Learning from animated diagrams: How are mental models built? In G. Stapleton, J. Howse, & J. Lee (Eds.), *Diagrammatic representations and inference – Proceedings of the fifth international conference, diagrams 2008* (pp. 266–281). Berlin: Springer.

Lowe, R., & Boucheix, J.-M. (2011). Cueing complex animations: Does direction of attention foster learning processes? *Learning and Instruction, 21*, 650–663.

Lowe, R., & Boucheix, J.-M. (2017). A composition approach to design of educational animations. In R. Lowe & R. Ploetzner (Eds.), *Learning from dynamic visualization – Innovations in research and application.* Berlin: Springer (this volume).

Lowe, R., Schnotz, W., & Rasch, T. (2011). Aligning affordances of graphics with learning task requirements. *Applied Cognitive Psychology, 25*, 452–459.

Mason, L., Lowe, R., & Tornatora, M. C. (2013). Self-generated drawings for supporting comprehension of a complex animation. *Contemporary Educational Psychology, 38*, 211–224.

Mayer, R. E. (2009). *Multimedia learning* (2nd ed.). New York: Cambridge University Press.

Mayer, R. E., & Anderson, R. B. (1992). The instructive animation: Helping students build connections between words and pictures in multimedia learning. *Journal of Educational Psychology, 84*, 444–452.
Mercer, N., & Littleton, K. (2007). *Dialogue and the development of children's thinking*. London: Routledge.
Meyer, K., Rasch, T., & Schnotz, W. (2010). Effects of animation's speed of presentation on perceptual processing and learning. *Learning and Instruction, 20*, 136–145.
Muybridge, E. (1899). *Animals in motion*. London: Chapman and Hall.
Perdreau, F., & Cavanagh, P. (2013). The artist's advantage: Better integration of object information across eye movements. *Perception, 4*, 380–395.
Perdreau, F., & Cavanagh, P. (2014). Drawing skill is related to the efficiency of encoding object structure. *i-perception, 5,* 101–119.
Ploetzner, R., & Breyer, B. (2017). Strategies for learning from animation with and without narration. In R. Lowe & R. Ploetzner (Eds.), *Learning from dynamic visualization – Innovations in research and application*. Berlin: Springer (this volume).
Ploetzner, R., Lowe, R., & Schlag, S. (2013). A systematic characterization of cognitive techniques for learning from textual and pictorial representations. *Journal of Education and Learning, 2*, 78–95.
Ploetzner, R., Lowe, R., Schlag, S., & Hauß, C. (2012). Lernen mit statischen und dynamischen Bildern – Welche kognitiven Techniken kennen Lehramtsstudierende? [Learning from static and dynamic pictures – Which cognitive techniques are known by pre-service teachers?] In I. Oomen-Welke & M. Staiger (Hrsg.), *Bilder in Medien, Kunst, Literatur, Sprache, Didaktik* [Pictures in media, art, literature, language, pedagogy] (S. 249–259). Freiburg, Germany: Fillibach Verlag.
Ploetzner, R., & Schlag, S. (2013). Strategic learning from expository animations: Short- and mid-term effects. *Computers & Education, 69*, 159–168.
Quillin, K., & Thomas, S. (2015). Drawing to learn: A framework for using drawings to promote model-based reasoning in biology. *CBE-Life Sciences Education, 14*, 1–16.
Schmeck, A., Mayer, R. E., Opfermann, M., Pfeiffer, V., & Leutner, D. (2014). Drawing pictures during learning from scientific text: Testing the generative drawing effect and the prognostic drawing effect. *Contemporary Educational Psychology, 39*, 275–286.
Schwamborn, A., Mayer, R. E., Thillmann, H., Leopold, C., & Leutner, D. (2010). Drawing as a generative activity and drawing as a prognostic activity. *Journal of Educational Psychology, 102*, 872–879.
Schwan, S., & Riempp, R. (2004). The cognitive benefits of interactive videos: learning to tie nautical knot. *Learning and Instruction, 14*, 293–305.
Spanjers, I. A. E., van Gog, T., & van Merrienboer, J. J. G. (2010). A theoretical analysis of how segmentation of dynamic visualizations optimizes students' learning. *Educational Psychology Review, 22*, 411–423.
Spanjers, I. A. E., Wouters, P., van Gog, T., & van Merrienboer, J. J. G. (2011). An expertise reversal effect of segmentation in learning from animated worked-out examples. *Computers in Human Behavior., 27*, 46–52.
Stieff, M. (2017). Drawing for promoting learning and engagement with dynamic visualizations. In R. Lowe & R. Ploetzner (Eds.), *Learning from dynamic visualization – Innovations in research and application*. Berlin: Springer (this volume).
Stubbs, G. (1766). *The anatomy of the horse*. London: J. Purser.
Stull, A. T., & Mayer, R. E. (2007). Learning by doing versus learning by viewing: Three experimental comparisons of learner-generated versus author provided graphic organizers. *Journal of Educational Psychology, 99*, 808–820.
Tchalenko, J. (2009). Segmentation and accuracy in copying and drawing: Experts and beginners. *Vision Research, 49*, 791–800.
Tchalenko, J., Nam, S.-H., Moshe, L., & Miall, R. C. (2014). The gazeshift strategy in drawing. *Psychology of Aesthetics, Creativity, and the Arts, 8*, 330–339.

Van Meter, P., Aleksic, M., Schwartz, A., & Garner, J. (2006). Learner-generated drawing as a strategy for learning from content area text. *Contemporary Educational Psychology, 31*, 142–166.

Van Meter, P., & Firetto, C. M. (2013). Cognitive model of drawing construction: Learning through the construction of drawings. In G. Schraw, M. McCrudden, & D. Robinson (Eds.), *Learning through visual displays* (pp. 247–280). Scottsdale, AZ: Information Age Publishing.

Van Meter, P., & Garner, J. (2005). The promise and practice of learner-generated drawing: Literature review and synthesis. *Educational Psychology Review, 17*, 285–325.

Van Sommers, P. (1984). *Drawing and cognition: Descriptive and experimental studies of graphic production processes*. New York: Cambridge University Press.

Wong, A., Leahy, W., Marcus, N., & Sweller, J. (2012). Cognitive load theory, the transient information effect and e-learning. *Learning and Instruction, 22*, 449–457.

Chapter 14
Drawing for Promoting Learning and Engagement with Dynamic Visualizations

Mike Stieff

14.1 Introduction

Modern curriculum innovations for teaching science at the elementary and secondary levels advocate the use of dynamic computer visualizations, specifically animations and simulations, as a means of improving student achievement (Linn, Lee, Tinker, Husic, & Chiu, 2006; Plass et al., 2012; Roschelle et al., 2010). The benefits attributed to using such tools are varied and include providing opportunities for students to view otherwise imperceptible phenomena (Stieff, Bateman, & Uttal, 2005), allowing students to make perceptual inferences from large datasets (Edelson, Gordin, & Pea, 1999), and permitting students to experience by proxy experiments that would be dangerous or unethical in practice (Hofstein & Lunetta, 2003). In the past two decades, there has been a burgeoning in the number and type of dynamic visualizations used for teaching science at all levels of instruction. There is increasing evidence that dynamic visualizations can help improve science learning (e.g., Edelson et al., 1999; Linn et al., 2006; Reiser et al., 2001; Stieff, 2011a; Wu, Krajcik, & Soloway, 2001); however, the observed improvements in learning outcomes are often marginal, which challenges the wisdom of their widespread adoption.

The marginal improvements seen from learning with dynamic visualizations are the subject of much debate in the learning sciences and science education communities. Despite the clarity of animations and simulations to expert scientists, students can struggle to assimilate information displayed in dynamic visualizations without explicit scaffolds to guide their attention (De Koning & Jarodzka, 2017, this volume; Kozma & Russell, 1997; Lowe, 2003; Lowe & Boucheix, 2017, this volume; Quintana et al., 2004). Even in instances where sufficient scaffolds are provided, animations and simulations too often fail to facilitate learning. Such failures are

M. Stieff (✉)
Department of Chemistry, Learning Sciences Research Institute,
University of Illinois, Chicago, Chicago, IL, USA
e-mail: mstieff@uic.edu

R. Lowe, R. Ploetzner (eds.), *Learning from Dynamic Visualization*,
DOI 10.1007/978-3-319-56204-9_14

particularly prevalent with simulations that permit students to design hypotheses and test them by running virtual experiments. Although the impressive features of simulations would seem ideally suited to supporting science learning, students actually experience the same challenges in making predictions and designing experiments with these tools as they do in the laboratory (de Jong & van Joolingen, 1998).

The efficacy of animations and simulations is known to be heavily dependent on the design of the activity in which these tools are embedded. The way in which dynamic visualizations are embedded in educational activities impacts not only the quality of *science learning* that occurs with a dynamic visualization (Harris, Mishra, & Koehler, 2009; Lowe & Schnotz, 2008) but also *student engagement* with those representations (Wu & Huang, 2007). Further, learning and engagement can be affected in very different ways. Design-based researchers have thus made a concerted effort to develop and test a variety of activities with the intention of improving both learning and engagement with animations and simulations. *Drawing* has recently been identified as one such activity that has considerable potential for the science classroom (Ainsworth, Prain, & Tytler, 2011). Although it has received relatively little research attention to date, several empirical studies indicate that drawing can both significantly improve learning of science content (Leutner & Schmeck, 2014; Van Meter & Firetto, 2013) and foster students' engagement in science learning (Prain & Tytler, 2012).

The extant research on the effectiveness of drawing activities has focused primarily either on learning from text-based learning materials or on student engagement with scientific practices. Studies of the effect of drawing activities on learning with animations or simulations are relatively rare (e.g., Ploetzner & Fillisch, 2017; Mason, Lowe, & Tornatora, 2013). Researchers have devoted even less attention to the potential of drawing activities for improving engagement in the science classroom (either with texts or with dynamic visualizations). Although empirical evidence is scant, it seems possible that theoretical arguments for the efficacy of drawing in promoting learning and engagement in general could be seen as encompassing animation and simulations. To that end, this chapter reviews the research on drawing to promote learning and engagement with science and examines the potential of drawing to achieve similar outcomes when coupled with dynamic visualizations in inquiry-based curricula. This potential is illustrated with examples from *The Connected Chemistry Curriculum* (Stieff, Nighelli, Yip, Ryan, & Berry, 2012), a technology-infused secondary chemistry curriculum that emphasizes drawing while working with molecular-level simulations.

14.2 Review of Research on Drawing for Promoting Science Learning and Engagement

14.2.1 Drawing for Promoting Science Learning

The role of drawing for promoting science learning has been of interest to the science education community for some time (Ainsworth et al., 2011; Van Meter & Firetto, 2013). Drawing activities appear to offer unique benefits to the science learner because drawing is a fundamental act of scientific practice that permits experts to achieve near instant perceptual understanding for many spatial and temporal phenomena that may otherwise go misunderstood (Latour, 1990). Accordingly, drawing is credited not only with being able to support a student's learning of science content, but also with being able to support the training of observational practices that are useful in the laboratory (Ainsworth et al., 2011; Hayes, Symington, & Martin, 1994; Van Meter & Garner, 2005). Thus, there is reason to suspect that novel curricula that couple animations and simulations with drawing activities have the potential to improve students' understanding of dynamic visualizations.

According to the Generative Theory of Drawing Construction (Van Meter & Garner, 2005), drawing promotes learning by fundamentally altering a student's approach to understanding an idea or concept presented in instructional texts. Based on information processing models of memory (Miller, Galanter, & Pribram, 1960) and dual coding theory (Paivio, 1991), the Generative Theory of Drawing Construction (GTDC) asserts that drawing helps learners to better comprehend ideas presented in text by constraining attention because the learner must select and organize information in order to construct a pictorial representation. A key assumption of the theory is that the act of drawing compels the learner to integrate verbal and nonverbal representations when planning a drawing. Propositional knowledge is thus translated into a perceptual image as the learner prepares to draw, which produces a more robust mental model of the concept and facilitates learning through structural analogy.

Ainsworth and Iacovides (2005), Ainsworth and Loizou (2003) and Cromley et al. (2013) have argued independently that drawing of itself may not be directly responsible for improving student learning. These researchers contend that the observed benefits of drawing are actually due to the self-explanation that is prompted by the constructive act of drawing (Chi, 2009). Self-explanation prompts of various kinds have been shown to greatly enhance student understanding of texts, diagrams, and visualizations, such as animations and simulations (Berthold, Eysink, & Renkl, 2009; Berthold, Röder, Knörzer, Kessler, & Renkl, 2011; Roy & Chi, 2005). According to these researchers, when students are required to generate a drawing, they are effectively engaged in a process where they are likely to self-explain their understanding of a concept in order to represent it via the drawing. Moreover, the drawing leaves an external representation of that understanding with which students can self-assess their learning.

In a revision to the GTDC, Van Meter and Firetto (2013) acknowledged that drawing can indeed support self-explanation and additionally argued that drawing supports learning more generally by stimulating students to engage in multiple self-regulation strategies. The requirement to generate a drawing forces the learner to set goals for the quality of their drawing, and to monitor the quality of their own understanding as they create a drawing. With respect to learning from text-based instructional materials, the authors argue that drawing prompts a cyclic process of revision wherein the learner must continually interrogate the text to identify whether the generated picture accurately reflects information represented by the text. When the learner notes a discrepancy between the picture and the text, the learner recognizes an incomplete understanding, which prompts the application of alternative learning strategies or help seeking behavior. Thus, the revised theory, the Cognitive Model of Drawing Construction, posits two casual mechanisms by which drawing promotes learning. First, drawing facilitates information uptake by constraining attention and forcing knowledge integration. Second, drawing improves the use of self-regulation strategies.

Results from empirical investigations of the effectiveness of drawing for supporting learning in science have been mixed. Some studies indicate that drawing significantly improves student learning (e.g., Chang, Quintana, & Krajcik, 2009; Kelly & Jones, 2007; Schwamborn, Mayer, Thillmann, Leopold, & Leutner; 2010; Linn, 2010; Lowe & Mason, 2017, this volume; Stieff, 2011a; Zhang & Linn, 2011). Other work indicates that drawing is no more effective than simply reading texts or viewing illustrations (Ainsworth & Iacovides, 2005; de Bock, Verschaffel, & Janssens, 1998; Gobert, 2000, 2005; Leutner, Leopold, & Sumfleth, 2009; Rasco, Tennyson, & Boutwell, 1975; Snowman & Cunningham, 1975). Comprehensive reviews (Van Meter & Garner, 2005; Van Meter & Firetto, 2013), offer a partial explanation for the variability of the findings: drawing is not universally effective for promoting learning independent of the context of its use. Specifically, optimal improvements in learning appear to occur when drawing is accompanied by instructional scaffolds that (a) constrain the number of features to be drawn and (b) explicitly prompt the learner to employ self-regulation strategies, such as monitoring understanding of the task and evaluating drawing quality (Van Meter & Firetto, 2013).

The interaction between drawing and instructional scaffolds is evident across various studies that have investigated the efficacy of drawing for promoting learning. Early work by Britton and Wandersee (1997) investigated the effect of drawing activities on students' understanding of complex biological processes. In that study, students viewed several pictures that corresponded to specific steps in the polymerase chain reaction, a complex series of chemical reactions that produce nucleic acid polymers. Importantly, students were not given a picture for every step. The students were then asked to place the illustrations in order and to construct drawings to represent missing steps described in a textual representation. The researchers observed that the act of drawing the missing steps resulted in higher order thinking and deeper content understanding. Similarly, Stein and Power (1996) asked secondary students who were learning about states of matter to construct drawings that represented water, as it would be seen under a powerful microscope. The students in the study constructed multiple drawings that reflected their propositional knowledge regarding the behavior of water molecules in various states. Consistent with the findings of Britton and Wandersee, Stein and Power observed that improvements in the quality of student drawings predicted learning outcomes; they argued this result was due to the drawing activities, which prompt self-reflection while constructing drawings and "force students to think" (Stein & Power, 1996, p. 66).

In contrast, Gobert later demonstrated in two experiments that drawing might not always promote learning. In one experiment (Gobert, 2000), three groups of secondary students learned about plate tectonics in different lesson conditions: (1) reading and summarizing texts, (2) constructing drawings that represented the main ideas in provided texts, and (3) reading texts alone. Achievement measures revealed that students in the summarize group performed better on formative assessments, but students in the drawing group performed better on summative assessments, which showed that drawing did not provide a universal benefit. In a second experiment, Gobert (2005) further investigated the precise role of drawing activities. Specifically, she examined whether drawing facilitates self-explanation during reading and the effectiveness of drawing compared with that of direct self-explanation prompts. Students either were asked to draw diagrams at specific time points while reading a text or were prompted to write short explanations instead of drawing. In this study no differences in achievement between the groups were found on either formative or summative assessments, which suggested that the drawing activity provided no specific support for learning.

However, more recent work by Schwamborn et al. (2010) contradicted this finding by showing drawing can benefit learning as long as adequate instructional supports are provided to the learner. In an experimental design, students were presented with a science text explaining the chemical processes that occur during laundry washing. Five groups were compared: students who read the text alone, students who read the text and received drawing prompts, and three groups of students who read the text and received various scaffolds to support their drawing activities. These scaffolds included prompts to (1) underline important information in the text; (2) underline important information, organize it, and integrate it with prior knowledge; and (3) organize important information and integrate it with prior knowledge. All students in the drawing groups were instructed to use a digital drawing tool that provided important features to include in their final drawings. Students in all four drawing groups scored higher on post-test achievement measures than students in the read-only group. Importantly, the results of this study suggest that drawing activities can indeed foster students' learning from science texts, but this requires a minimum of instructional support to constrain the number of features they attend to in a text while constructing a drawing.

The discrepancy between these findings demonstrates the important influence of instructional materials and activities on the relationship between drawing and learning outcomes. In studies where drawing was 'free' (i.e., with no instructional scaffolds), it was not an effective learning strategy. In contrast, when drawing activities were embedded in more complex tasks that prompted students to reflect on the quality of their drawings and to compare their drawings with provided texts or illustrations or to their own propositional knowledge about the phenomena, drawing improved students' understanding of science concepts and enhanced retention. The differences in these findings are consistent with Van Meter and Firetto's (2013) revised theory: Drawing will not yield improvements in learning without adequate instructional supports that guide attention, constrain the range of features to be depicted in a drawing, and prompt self-regulation strategies.

14.2.2 Drawing for Promoting Student Engagement in Science

Most studies that have investigated the effectiveness of drawing in the science classroom have narrowly focused on how drawing might provide direct support for learning. However, an independent line of research has examined the potential of drawing for supporting *student engagement* with science (which may in turn have an indirect influence on learning). In contrast to the information processing model of memory underlying the GTDC (Van Meter & Garner, 2005), this line of research centers on the assumption that drawing can work to increase one or more types of engagement with science practices and science learning activities rather than (or possibly concurrent with) improving information uptake or conceptual change. Although the GTDC contributes to our understanding of how drawing can help learners to comprehend scientific information presented in instructional texts, it does not inform our understanding of whether drawing activities could be used to improve students' enjoyment in science classrooms, increase their motivation to learn science, or help persuade them to pursue science careers. Studies investigating the relationship between drawing and engagement are relatively scarce. However, they are no less important than those investigating direct learning because student engagement and interest are more predictive of persistence in science than learning outcomes assessed by achievement measures (Tai, Liu, Maltese, & Fan, 2006).

Student engagement has been broadly defined as "[a] student's psychological investment in and effort directed toward learning, understanding, and mastering the knowledge, skills, or crafts that academic work is intended to promote" (Newmann, Wehlage, & Lamborn, 1992, p. 12). More recently, Fredricks, Blumenfeld, and Paris (2004) have argued that student engagement is a multi-faceted construct that includes cognitive engagement, emotional engagement, and behavioral engagement. *Cognitive engagement* refers to engagement with meaningful learning and the effort expended to understand a concept or idea (see also Ploetzner & Breyer, 2017, this volume). *Emotional engagement* refers to affective engagement with the classroom, peers, or the learning activities. *Behavioral engagement* refers to engagement with specific behaviors in a classroom setting, such as completing an assignment or following teacher instructions. In addition to these three engagement types identified by Fredricks et al., Hegedus and Kaput (2004) distinguished *social engagement* as a fourth type of engagement related to an individual's commitment to, or participation in, an interactive learning community. Extant research on engagement suggests that curricular artifacts, including dynamic visualizations, do not intrinsically promote any of these types of engagement (Harris et al., 2009). Rather, the structure of learning activities stimulates student engagement with curricular artifacts that ultimately yields improvements in learning outcomes, self-efficacy, or attitude.

Prain and Tytler (2012) have offered a theoretical account of how drawing may help to promote engagement with science practices and science learning activities through three overlapping affordances. Their theory of Representational Construction Affordances (RCA) posits that the act of creating scientific representations, including diagrams, promotes student engagement by scaffolding student participation in *semiotic*, *epistemic*, and *epistemological practices* of the science community. Semiotic affordances relate to the functional features of science representations and symbols that are used to model natural phenomena. Epistemic affordances relate to the knowledge-building practices that define science inquiry around a phenomenon. Epistemological affordances relate to how representations are leveraged to support reasoning and understanding of a phenomenon. With its focus on these three affordances, the RCA theory accounts for how students come to identify with, and behave as, members of the scientific community and to develop strategies that characterize scientific reasoning and problem solving. Again, while the evidence is limited, a small, but growing number of studies indicate that these affordances of drawing work to promote cognitive and social engagement with science learning and science as a discipline, as predicted by the RCA theory.

Most importantly, drawing can work to promote cognitive engagement with science learning by helping learners to adopt important strategies and skills that characterize scientific reasoning, such model-based reasoning and thought experiments. Within the RCA framework, cognitive engagement is significantly different from cognitive processing as described by Van Meter and Garner (2005) as well as by Van Meter and Firetto (2013). Whereas the GTDC focuses carefully on how drawings can help learners to select, organize, and integrate information represented in texts, Prain and Tytler (2012) argue that "student generated representations, including drawing, …can be understood as enacting science learning and reasoning because this kind of activity is consistent with how knowledge is developed and communicated in the science community" (p. 2757). Drawing is an epistemic practice of scientists in that scientific reasoning frequently involves the generation of diagrams or other external representations to facilitate meaning making and support explanation (Gooding, 2004; Latour, 1990). As a cognitive artifact, a drawing is a product of model-based reasoning that involves the creation of pictorial representations that are structurally analogous to scientists' explanatory models of the physical world. In many cases, the fidelity of a scientific diagram to the model is sufficiently high that it helps the scientist who constructed it and other viewers to animate the diagram mentally in order to make inferences about structural and temporal relationships in the represented model or to make predictions about the outcome of alterations to the model.

Several researchers have documented this role of drawing diagrams in scientific practice (Latour, 1990; Gooding, 2004; Kozma, 2003; Kozma, Chin, Russell, & Marx, 2000). Field observations by Kozma et al. (2000) provide compelling insights about the role of diagram construction as an epistemic practice of scientists. By embedding themselves in a chemistry laboratory, these researchers were able to observe how working scientists generated diagrams to initiate their solving of challenging problems and to justify claims made during the problem solving process. The researchers observed that the laboratory space was designed to promote drawing, with white boards and glass surfaces reserved as dedicated spaces for creating transient inscriptions. When the scientists came together to discuss their experiments and explain their observations, they initiated conversations with the generation of a diagram around which interlocutors engaged in argumentation and group problem solving. Notably, the researchers also observed that expert members in the group routinely engaged novices with techniques for generating and interpreting diagrams as they worked together in the laboratory.

Khishfe and Abd-El-Khalick (2002) observed a similar function of drawing tasks embedded within inquiry activities to support students' cognitive engagement with science learning in a school classroom. These researchers documented how students came to understand the epistemological assumptions underlying different science practices, such as the tentative nature of scientific knowledge and how scientists generate inferences from observations. There were significant improvements in both students' understanding of the material and social affordances (Kozma, 2003) of creating diagrams to reason about imperceptible phenomena and to communicate after they completed drawing activities in the context of the lesson. However, the authors observed improvement only among children who received explicit guidance to reflect on the purpose of their drawings. Students who received guidance exhibited a much better understanding of the nature of science practices and the role of making diagrams for supporting scientific reasoning than students who completed the same drawing activity without guidance. As with the work on drawing for promoting learning, the findings of Khishfe and Abd-El-Khalick suggest that the benefits of drawing for supporting cognitive engagement are not guaranteed by drawing alone; task goals and instructional supports significantly affect the effectiveness of drawing.

Drawing can also work to promote social engagement with science by helping learners to engage in a learning community through shared experiences around diagrams in their role as cultural tools of science (Prain & Tytler, 2012). As Latour (1990) notes, scientists are constantly "drawing things together" in an epistemic practice that not only supports reasoning but also serves as a tool to support social interaction. The field observations of Kozma et al. (2003) and Khishfe and Abd-el-Khalick (2002) show clear evidence that drawing serves as an activity where learners and experts come together around inscriptions that are publically shared to facilitate communication. In both studies, the drawings generated by both students and experts were not produced solely for the purpose of improving their own understanding of a phenomenon. The construction of a drawing also helped to support communication among peer learners, whether in the laboratory or classroom, where individuals produced drawings and shared them to invite critique and dialogue.

Wu and Huang (2007) documented the explicit use of drawing in this manner as a pedagogical tool of science teachers. In the classroom under study, students were tasked with using animations to learn about concepts related to force and motion. After all students had viewed an animation, the teacher initiated a whole class discussion of the underlying concepts represented in the animation. The discussion began with an invitation to one student to draw a free-body diagram that represented all of the forces acting on a falling object on the chalkboard in front of the classroom. The act of producing a drawing led to a debate among the class about how to best use a diagram to represent the forces that were observed in the animation. In this approach, the students publicly question the relationship between their own understanding and the diagram drawn by the volunteer and recommend alterations to the diagram that reflect community-sanctioned conventions. As seen in this study, individuals do not create drawings solely to demonstrate their knowledge or understanding for assessment. Rather, drawings can sometimes also serve to spur participation in a learning community. Importantly, learning activities that couple drawing with structured social interactions appear to have the greatest effect on learning outcomes. This suggests that social engagement while drawing may be necessary to achieve the greatest benefits of drawing for promoting learning.

Finally, there is evidence that drawing can also work to support emotional engagement in a manner distinct from the cognitive and social engagement accounted for by Prain and Tytler's (2012) RCA theory. Drawing is a creative act that permits students to express their own understanding in an imaginative or expressive way, and students, especially young science learners, appear to enjoy drawing (Hayes et al., 1994). Indeed, the act of drawing can trigger positive or negative affective states and help individuals to regulate their emotional states and mood (Dalebroux, Goldstein, & Winner, 2008; De Petrillo & Winner, 2005). To make and share drawings publicly can cue positive emotions in individuals who take pleasure from drawing or prime a relaxed or playful mood among those who perceive drawing as a form of artistic expression or hobby.

There is emerging empirical support for the potential of drawing to support emotional engagement in science learning. Clegg et al. (2012) compared student's experiences using two technology-based learning environments that were created to promote personal meaning making with students and engage them in science learning and practice. One of these learning environments (StoryKit) supported personal sketching and the other (Zydeco) had similar features but lacked the sketching function. Study participants reported that they found StoryKit more emotionally engaging than Zydeco because the ability to make drawings allowed them to be creative in ways that brought more personal meaning to science learning. Further, the researchers also reported that in their collaborations with children on the design of learning environments one of the more frequently requested features by students was the opportunity to draw.

14.3 Drawing with Animations and Simulations

The research reported in the previous section indicates that drawing has considerable potential for improving student learning and engagement in science. According to the Generative Theory of Drawing Construction (Van Meter & Garner, 2005), drawing is an activity that fundamentally alters a student's learning strategy to improve comprehension, retention, and self-regulation. According to the Representation Construction Affordances Theory (Prain & Tytler, 2012), drawing is an activity that reflects a defining epistemic practice of scientists that can foster cognitive and social engagement with science and science learning. These theoretical arguments that drawing can improve learning and student engagement in science are compelling, and empirical support continues to emerge that is consistent with both. Unfortunately, research on the potential of drawing activities and their effectiveness in science classrooms has focused narrowly on the relationship between drawing and learning from science texts or drawing and engagement in scientific practices. It remains unclear whether drawing activities are equally effective at improving learning and also engagement when paired with dynamic visualizations in the science classroom.

Predictions from both the Generative Theory of Drawing Construction and the Representation Construct Affordance are not limited to learning with texts, and the authors of both theories have explicitly stated that benefits of drawing should support learning and engagement in learning contexts that involve representations other than text. Results from the available research on drawing with texts raise the possibility that drawing activities with adequate instructional scaffolds could also improve learning and engagement with animations and simulations. However, there are reasons to question whether such benefits would necessarily extend to such non-text representations. First, animations and simulations present information in the form of a dynamic visualization that often requires the coordination of a significant number of variables that are continually updating. As a result, learners can face significant challenges in making accurate observations from dynamic visualizations, particularly with respect to identifying information of most relevance to the nominated learning task (Lowe, 2003; Lowe & Boucheix, 2017, this volume; Lowe & Schnotz, 2008). These challenges may be exacerbated when the visualization contains multiple representations that must be coordinated in order to understand the referent content (Kozma & Russell, 1997). The extent to which drawing can help overcome such challenges and hence support learning and engagement remains unclear.

Second, animations and simulations that effectively support learning are often designed in such a way to direct a learners' attention to information that is typically not present in texts or illustrations (Gilbert, 2005). Such information includes dynamic process information that learners must infer from texts or static pictures. Extensive studies on alternative designs of learning technologies that include animations have shown that explanatory narrative voiceovers (for a review see Mayer, 2009) or other instructional scaffolds that guide attention and reflection

(Quintana et al., 2004) can help learners to better attend to the critical features of the visualization to achieve robust understanding. Thus, it is not clear that drawing activities support or replace learning from an animation or simulation compared to alternative scaffolds. In fact, recent work suggests that construction a static drawing of a dynamic process may even inhibit learning from an animation of visualization (Ploetzner & Fillisch, 2017).

There is emerging evidence to suggest that drawing can offer unique benefits despite the constraints and affordances of dynamic visualizations themselves. A preliminary study by Zhang and Linn (2011) demonstrated that students who construct drawings to explain their observations of an animation of molecular interaction perform marginally better on learning outcome measures than students who view the same animation but do not construct drawings. Mason et al. (2013; see also Lowe & Mason, 2017, this volume) conducted a more extensive study that also indicated drawing could foster learning from animation. These authors compared students who only viewed a physics animation with two groups of students who completed drawings as well as viewing the animation. One group generated their own drawings and the other group traced over a dotted outline drawing provided by the researchers. Students in the self-generated drawing group performed better on both immediate and delayed tests of comprehension than those in the other two groups. These findings suggest that the benefits similar to those seen with texts can also be found when drawing is used as to support learning from animations and simulations are, but also raise important question about the design of the drawing activity itself. As above, prior research on drawing for supporting learning from texts shows that free drawing activities are not effective for supporting text comprehension; the results of Mason et al. show the opposite effect when students are working with an animation. Whether the conflicting results are due to the design of the drawing activities in each study or a unique interaction between drawing and learning from dynamic visualizations is not clear. An important additional consideration is the level of drawing demands required of the learner in order to generate a satisfactory depiction of the subject matter presented in an animation (see Lowe & Mason, 2017, this volume).

The potential of drawing to improve engagement with science practices or science learning when coupled to animations and simulations is also unclear because no studies have investigated this interaction directly. It is clear, however, that dynamic visualizations in and of themselves do not guarantee increased engagement (Dickey, 2005). The most notable work in this area was conducted by Wu and Huang (2007) who explored how different curriculum designs impacted student engagement and learning from the 'Physlets' physics simulation (Christian & Titus, 1998). The developers of Physlets created the technology from a strictly cognitive perspective with the aim of helping students to link multiple representations, visualize abstract concepts, and perform inquiry experiments. They were not concerned with whether the technology itself improved student engagement. Working from the assumption that the technology would have little impact on student learning independent of curriculum activities, Wu and Huang embedded the technology in both a student-centered and a teacher-centered activity. In the student-centered design, students were able to work with the computer simulations in isolation or in pairs; in the teacher-centered design, the teacher demonstrated the simulations and students were not provided opportunities to manipulate the simulations or work interactively.

The investigators examined how each of these two designs impacted student engagement and subsequent learning outcomes. Although the authors gave no explicit statement about their design goals, the student-centered design was apparently intended to target improvements in social engagement and emotional engagement (because social interaction with peers and individual control and exploration were both promoted). Both designs included activities that emphasized cognitive engagement: students in both conditions were required to record their observations of the simulation and to reflect on those observations with respect to the relevant physics laws. The results showed in that the student-centered classroom, which promoted individual choice and peer interactions, learners had significantly higher levels of emotional engagement than those in the teacher-centered classrooms. However, low-achieving students appeared to have lower cognitive engagement in the student-centered classroom where they struggled to identify important visual elements in the simulation, posed significantly fewer conceptual questions about their experiments, and often sought help from their high-achieving peers to understand the visualization.

These preliminary studies on the relationship between drawing, learning, engagement, and dynamic visualizations suggest that drawing may have some potential for improving learning and engagement. However, empirical studies that investigate precisely how drawing might support learning and engagement when paired with animations or simulations are lacking. New investigations are needed that explore the interactive nature of drawing activities and dynamic visualizations to identify possible affordances and constraints that could apply to the use of drawing intended to support learning with animations and simulations. Additionally, new studies are needed to explore possible additive or interactive benefits of drawing activities and dynamic visualizations to promote various types of student engagement. If any potential that drawing activities have to improve learning and engagement is to be realized, concerted efforts are needed to understand how, when, why, and for whom drawing activities support learning and engagement in technology-infused learning environments.

14.4 Design Principles for Drawing with Animations and Simulations

The studies discussed above do not offer definitive answers regarding the ultimate outcome of coupling drawing and dynamic visualizations. Nevertheless, it seems likely that drawing and dynamic visualizations together may impact learning and engagement in different ways, depending on how they are paired in a learning activity. Although there has been some laboratory investigation in this area, no rigorous classroom-based studies have been conducted to explore the relationship between learning activities that include both drawing and dynamic visualizations tools for promoting learning and engagement. As a result, formal empirically-based design principles for developing learning activities that include both drawing and dynamic visualizations are currently lacking. There are clues, however, in the extant literature that suggest design principles for creating technology-infused learning environments with scaffolds that (1) support learning with drawing and (2) improve engagement in science practices and science learning. Here, I posit four design principles derived from the literature that merit testing in authentic educational settings.

Design Principle 1: Drawing activities should include explicit instructions regarding what features of a dynamic visualization should be represented in a self-generated drawing. The extensive work on the relationship between drawing and texts indicates that learning is most improved when instructional scaffolds are included that constrain students' attention to important features that should be depicted in a drawing (e.g., Van Meter & Firetto, 2013). In contrast, free-drawing activities may selectively impair learning as students struggle to identify important information in a dynamic visualization, given the complexity often found in educational materials that use this medium (Kozma & Russell, 1997). By scaffolding students' observations of a dynamic visualization and directing them to attend to crucial elements of the display, drawing activities could also support cognitive engagement among all learners by capitalizing on some of the effective features in teacher-centered designs (Wu & Huang, 2007).

The specific ways in which attention can be constrained by particular design elements warrants careful investigation by the designed-based research community. Studies are needed to compare different types of scaffolds for constraining attention, the number of scaffolds provided, and the number of features to be drawn. Alternative designs might provide students with a list of key features to note (e.g., Stieff et al., 2012), limited tool palettes (e.g., Chang et al., 2009), prompts to integrate particular features (e.g., Schwamborn et al., 2010), or instructions on how to make a diagram for self-explanation (e.g., Ainsworth & Iacovides, 2005). For example, students may gain no benefit from copying or completing partially constructed drawings because they are not able to select features themselves or to reflect on what they are constructing. As seen in the work by Mason et al. (2013), students who completed partially constructed drawings did not perform better than students who completed free drawings, which leaves open questions about the effectiveness of free-drawing as well. Because of this future studies that compare the effectiveness of various designs that pair drawing with dynamic visualizations should include free-drawing activities as control conditions.

Design Principle 2: Drawing activities should include explicit prompts that ask students to reflect on their drawings as they are generated and explicit prompts to reflect on the utility of their completed drawings. The research on drawing with texts suggests that the design of a drawing activity can promote, but does not guarantee, self-regulation strategies or cognitive engagement. As reported by Van Meter and Firetto (2013), studies that report a beneficial effect of drawing on self-monitoring behaviors often include explicit scaffolds and prompts to help students reflect on the quality of their drawing and its relationship to the information in the text. To help learners engage in self-reflection with dynamic visualizations similar prompts will likely be required. Moreover, such prompts should also include epistemological questions that help the learner to reflect on how drawing is useful for representing scientific models and for communicating in science. The act of creating a drawing is a foundational epistemic practice of science, and learners require significant support to appreciate the significance of making a drawing beyond simply representing their observations (Prain & Tytler, 2012). The form of these self-reflection prompts can be quite simple. As seen in the work of Khishfe and Abd-El-Khalick (2002), students can achieve appreciable understanding of the utility of drawings by receiving direct prompts from a teacher to think about why it is helpful to make a drawing while learning about a science concept.

Design Principle 3: Students should receive feedback on how effectively a self-generated drawing represents important features while they are engaged with dynamic visualizations. Given the evidence that drawing can support learning when students are able to evaluate the quality of their drawing against the information they are required to learn, drawing activities must include opportunities for students to receive feedback on drawing quality. Feedback can be self-guided in the form of prompts that guide the student to compare their drawings with other drawings or with the visualization. As in earlier studies with learning from text, learners can receive substantial feedback simply by comparing their own drawings to exemplar illustrations that represent the same information (e.g., Britton & Wandersee, 1997).

Prior research on the effectiveness of drawings for supporting learning have focused solely on how well individual learners can evaluate the quality of their own drawings. However, in authentic educational settings feedback may be more effective when provided by peers or instructors. There is some limited empirical support for the use of peer feedback for supporting learning from drawing in a science classroom. Chang et al. (2009) compared the achievement of middle school students who were randomly placed into three groups with different activity structures: (1) design, interpret, and peer-evaluate a drawing, (2) design and interpret a drawing, or (3) view and interpret teacher-made drawings. Analysis of post-test achievement scores revealed that students who engaged in evaluating their peers' animations performed much better than the students in the other two groups. The opportunity to give feedback to peers and to receive feedback from them facilitates social engagement in a learning community that can both motivate students to engage with science learning and help students better reflect on their own developing understanding as they interact with a dynamic visualization.

Design Principle 4: Drawing activities should be incorporated into student-centered classrooms that include peer interaction with dynamic visualizations. As with other student-centered designs that promote social or emotional engagement (e.g., Wu & Huang, 2007), students should have the opportunity to interact with dynamic visualizations to discover disciplinary concepts, pose conceptual questions, generate drawings with their peers, and share their individual drawings with each other. Constructing drawings in the public space helps to catalyze productive social interactions among scientists and among learners that goes beyond feedback (Prain & Tytler, 2012). Activities that involve students creating and interpreting drawings from dynamic visualizations should afford students the opportunity to work together in a learning community that fosters their engagement with science learning. Although it is clear that learning can be significantly improved with self-generated drawings, more substantial gains may be possible when students are afforded the opportunity to work together.

In addition to the social engagement, peer interaction can also afford opportunities for learners to help each other select and integrate important features in a dynamic visualization, to prompt self-reflection and other metacognitive behaviors, and to provide feedback. Indeed, Stieff (2011a) demonstrated that peer interactions can significantly improve learning when students are tasked with making drawings while viewing a dynamic visualization. In that study, Stieff analyzed the behavior of students who were required to work in pairs to make inferences about chemical equilibrium using a molecular-level simulation. As they constructed their drawings, each student in a pair attended to different features of the display and did not effectively integrate information across multiple representations. By engaging with one another, however, the two students were collectively able to make correct inferences by evaluating the observations depicted by each individual. In subsequent interviews, individual students working alone were unable to make the correct inferences they had achieved in pairs. This suggests that peer interaction can strongly influence learning from a dynamic visualization.

The design principles offered above have been proposed on the basis of a diverse range of studies that have not specifically examined the interaction between drawing, dynamic visualizations, learning, and engagement in authentic educational settings. Empirical work is therefore needed to determine whether new educational interventions that rely on these design principles can improve learning outcomes or increase interest in science. Moreover, studies of the relative efficacy of alternative designs that embody these design principles will be needed to determine how learning and engagement is affected and best enhanced by specific design elements and whether these benefits can be realized across diverse groups of students.

14.5 Investigating the Relationship Between Drawing, Engagement, and Learning with the Connected Chemistry Curriculum

I conclude the chapter by illustrating how these four design principles shaped the design of drawing activities in an innovative chemistry curriculum in which the use of dynamic visualization is central. *The Connected Chemistry Curriculum* (CCC; Stieff et al., 2012) is a stand-alone secondary chemistry curriculum that covers over 120 days of instruction with activities that feature active student-driven explorations of animated simulations in a "molecular laboratory interface" (Fig. 14.1).

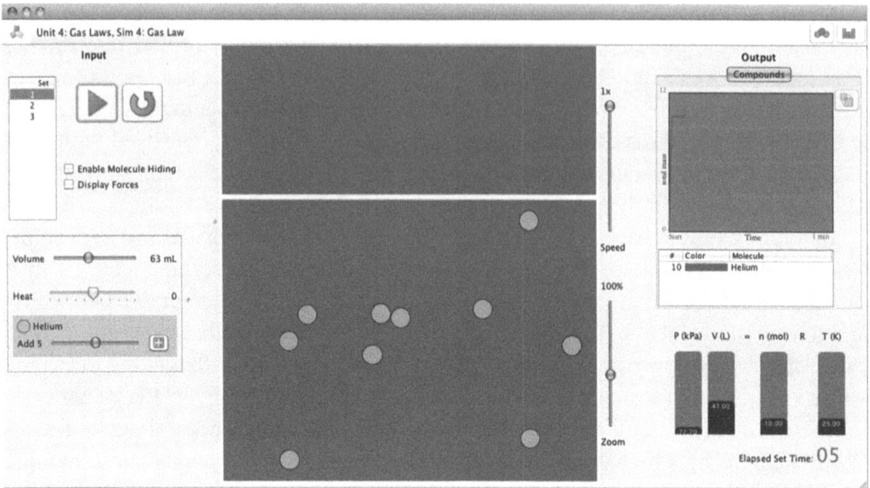

Fig. 14.1 Screenshot of the Java interface from the molecular laboratory interface from the CCC "Gas Law Sim" demonstrates user inputs, molecular visualizations, plotting, and variable outputs

Using CCC computer simulations, students work in pairs to manipulate various parameters of a chemical system simulated in the interface, make predictions based on their understanding of relevant chemistry concepts, and then compare their observations of the simulation with their predictions. Student inquiry with CCC simulations is systematically supported by activities that ask students to draw their predictions about molecular interactions, to record their observations in drawings, and to revise their initial drawings given their observations. CCC couples the interactive 'predict-test-observe' process with representation construction via drawing to help students engage with two fundamental activities that define the practice of chemistry.

Drawing activities in CCC reflect all four drawing design principles proposed above. First, all drawing activities in CCC include explicit scaffolds that help guide students to important features of the molecular laboratory interface to constrain their attention (Design Principle 1). A complete model that describes a chemical phenomenon includes information about the composition and motion of chemical species, their relative location in a system, and how they interact with other species. Students are prompted to identify these four pieces of information in each activity, depict them in their drawings, and write a verbal description to accompany the drawing. Early editions of the curriculum lacked these explicit scaffolds, and the drawing activities were free rather than guided. Figure 14.2 presents side-by-side comparisons of student responses to the same drawing activity with scaffolds (Fig. 14.2a) and without scaffolds (Fig. 14.2b) that illustrates the substantial effect the scaffolds can have on the quality of student drawings. When scaffolds are present, students pay more attention to the important features in the visualization and depict more information in their drawings.

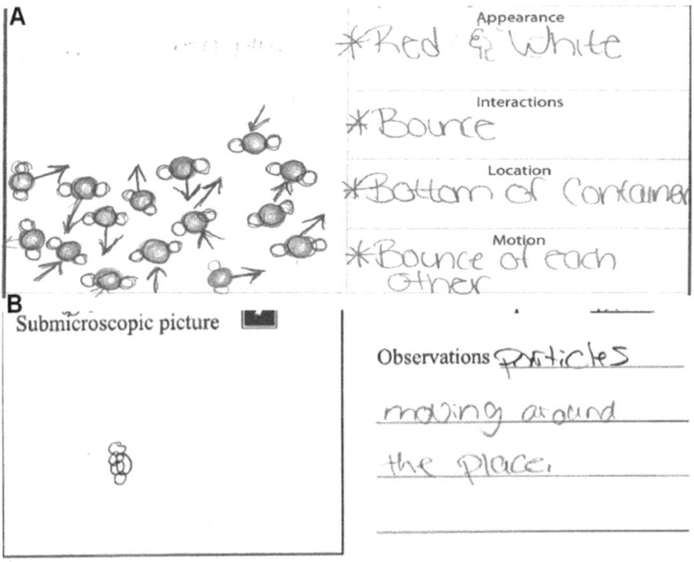

Fig. 14.2 A CCC activity that prompts students to make a drawing of a dynamic visualization that represents water molecules in a liquid state. With explicit prompts that constrain attention (**a**), students better represent important features in their drawing than they do without prompts (**b**)

Second, each CCC guided inquiry activity where students work with simulations includes prompts to construct drawings on at least three occasions and to reflect on these drawings (Design Principle 2). Initially (prior to viewing a simulation), students are tasked with constructing *predictive drawings* of what they think the submicroscopic world would look like for a given phenomenon. Later, while exploring a simulation, students are asked to construct *observational drawings* that capture what they are seeing for later reference in discussion and laboratory activities. Finally, after completing their exploration, students are asked to construct *reflection drawings* of chemical systems in which they reflect on their earlier drawings and how they can use them to support claims about chemical behavior. The systematic use of drawings in this way helps learners to employ self-regulation strategies as well as to engage with an important epistemic practice of the discipline.

Third, CCC drawing activities help students monitor their own learning by observing the results of their manipulations of system variables and providing feedback in real time that students can use to evaluate the quality of their drawings (Design Principle 3). Throughout the curriculum explicit prompts ask students to compare what they have drawn to the simulation and judge whether their peers or instructor would be likely to understand what the simulation represents from the drawing alone. Because CCC activities require students to work in pairs under the supervision of a teacher, the curriculum also helps students receive feedback from others. Teacher materials also include examples of student work from earlier implementations of CCC materials (Fig. 14.3) to help them identify problematic drawings among their own students while they are working with the simulations. Thus, CCC

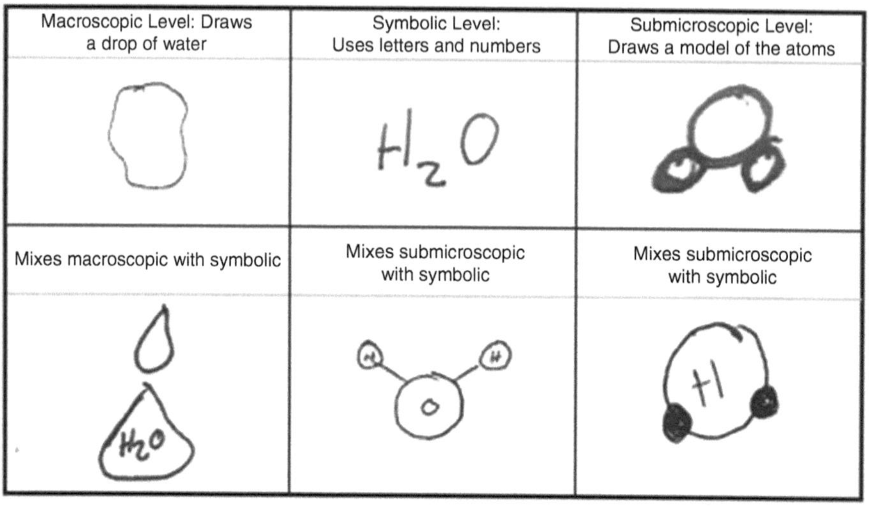

Fig. 14.3 CCC Teacher materials include student work to help teachers identify problematic ideas of phenomena while students are constructing drawings while viewing a simulation

drawing activities serve as an important formative assessment tool with which students and their instructors can monitor understanding.

Fourth, all CCC drawing activities are embedded within guided-inquiry investigations that students complete in pairs or in groups (Design Principle 4). With the explicit scaffolds that constrain attention, prompt reflection, and support self-evaluation, CCC materials encourage students to work collaboratively. Many CCC activities task individual students with making drawings that they must then compare or combine with drawings made by peers in order to successfully complete an investigation. Discussion activities that conclude each lesson prompt instructors to invite students to the board to share their drawings and to foster argumentation around the affordances of a specific drawing to represent what was observed in a simulation activity and foster a learning community that uses drawings to represent ideas.

These various drawing activities of CCC illustrate how prior research on the use of drawing to improve learning from (mainly) text can be leveraged to couple drawing with dynamic visualizations to support science learning in authentic contexts. Previous studies have demonstrated the effectiveness of these curriculum materials for promoting conceptual change and representational competence, increasing engagement with chemistry learning, and improving student achievement on summative assessments (e.g., Stieff, 2005, 2011b; Stieff & McCombs, 2006; Stieff & Wilensky, 2003; Ryan, Yip, Stieff, & Druin, 2013). However, none of these prior studies was specifically directed towards investigating whether or not a causal relationship exists between the drawing activities and learning or engagement. Thus, despite the demonstrated effectiveness of CCC, the extent to which the observed benefits are due to the drawing activities, the dynamic visualizations, the inquiry investigations, or an interaction among all three components of the curriculum remains unclear. Nevertheless, CCC presents an innovative learning environment for studying how drawing with dynamic visualizations may influence learning and engagement.

14.6 Summary

This chapter reviewed research on the effectiveness of drawing for supporting learning and engagement in science learning, and considered the potential of drawing to improve learning and engagement when coupled together with dynamic visualizations. Preliminary investigations suggest that learning outcomes may be improved by interventions that couple drawing with dynamic visualizations (Linn, 2010; Lowe & Mason, 2017, this volume; Mason et al., 2013; Stieff, 2011a). However, uncertainty remains regarding the mechanisms that could be involved in producing any such improvements. Because the effects of dynamic visualizations on the information to which learners direct their attention are very different from those occurring in texts or illustrations (Gilbert, 2005), the role of drawing for supporting learning from a dynamic visualization may differ from the role it plays in supporting learning from text. However, research in this area has been very limited so the extent to which drawing activities can best enhance learning from visualizations requires further investigation. New investigations that explore how drawing activities moderate learning from dynamic visualizations are needed as educators place an increasing emphasis on incorporating both visualizations (Linn et al., 2006) and drawing in the classroom (Ainsworth et al., 2011).

Some recent investigations have also indicated that drawing can increase student engagement in science learning (Clegg et al., 2012; Wu & Huang, 2007). In particular, drawing appears to promote both cognitive and social engagement in science learning. As an epistemic practice of science, drawing facilitates model-based reasoning and group problem solving and presents an activity that can help students engage in a science learning community (Prain & Tytler, 2012). Moreover, researchers have observed teachers to employ drawing activities as a way of promoting student participation and peer interaction in science classrooms in order to increase student engagement (Khishfe & Abd-El-Khalick, 2002; Wu & Huang, 2007). Understanding this specific affordance of drawing to promote engagement and how it may interact with the presence of dynamic visualizations in the science classroom is an important emerging research area in science education. Student interest is an important predictor of persistence in science (Tai et al., 2006), and drawing may offer a novel pathway to increase the number of students who pursue science even if it fails to improve learning directly.

By considering research on the use of drawing to promote learning and engagement with science, I have suggested four design principles for coupling drawing activities with dynamic visualizations. First, drawing activities should include explicit instructions regarding what features of a dynamic visualization should be represented in a self-generated drawing. Second, drawing activities should include explicit prompts that ask students to reflect on their drawings as they are generated and explicit prompts to reflect on the utility of their drawings. Third, students should receive feedback on how effectively a self-generated drawing represents important features while they are engaged with dynamic visualizations. Finally, drawing activities should be incorporated into student-centered classrooms that include peer interaction with dynamic visualizations. Empirical investigations are needed to test the utility of these design principles for developing new learning environments such as those found in The Connected Chemistry Curriculum (Stieff et al., 2012).

Although the proposed affordances of drawing in science are only partially supported by empirical data, there is an increased emphasis on supporting science education with drawing activities (e.g., Ainsworth et al., 2011; Johnson & Reynolds, 2005; McGrath & Brown, 2005), and computer tools that support drawing for learning proliferate (e.g., Clegg et al., 2012; Cooper, Groves, Pargas, Bryfczynski, & Gatlin, 2009; Jee, Gentner, Forbus, Sageman, & Uttal, 2009). Much remains to be learned about the underlying causal mechanisms by which drawing activities might be able to improve student learning and engagement when paired with dynamic visualizations. The emerging findings suggest that drawing may be beneficial for science learners, but empirical studies demonstrating precisely how drawing supports learning and engagement and how to capitalize on the benefits of drawing and dynamic visualizations in the classroom are lacking. Increasing the persistence and achievement of science students remain elusive goals for the education research community, and drawing presents interesting new opportunities that may help to foster these important outcomes.

Acknowledgements This research was supported in part by a grant from the National Science Foundation (DRL-1102349). Any opinions, findings, or conclusions expressed in this article are those of the authors and do not necessarily represent the views of these agencies.

References

Ainsworth, S. E., & Iacovides, I. (2005). *Learning by constructing self-explanation diagrams*. Paper presented at the 11th EARLI Conference, Munich, Germany. Retrieved from http://www.psychology.nottingham.ac.uk/staff/sea/earli2005/ainsworth_abstract.pdf

Ainsworth, S., & Loizou, A. T. (2003). The effects of self-explaining when learning with text or diagrams. *Cognitive Science, 27*, 609–681.

Ainsworth, S., Prain, V., & Tytler, R. (2011). Drawing to learn in science. *Science, 333*, 1096–1097.

Berthold, K., Eysink, T. H. S., & Renkl, A. (2009). Assisting self-explanation prompts are more effective than open prompts when learning with multiple representations. *Instructional Science, 37*, 345–363.

Berthold, K., Röder, H., Knörzer, D., Kessler, W., & Renkl, A. (2011). The double-edged effects of explanation prompts. *Computers in Human Behavior, 27*, 69–75.

Britton, L. A., & Wandersee, J. H. (1997). Cutting up text to make moveable, magnetic diagrams: A way of teaching and assessing biological processes. *The American Biology Teacher, 59*, 288–291.

Chang, H., Quintana, C., & Krajcik, J. S. (2009). The impact of designing and evaluating molecular animations on how well middle school students understand the particulate nature of matter. *Science Education, 94*, 73–94.
Chi, M. T. H. (2009). Active-constructive-interactive: A conceptual framework for differentiating learning activities. *Topics in Cognitive Science, 1*, 73–105.
Christian, W., & Titus, A. (1998). Developing web-based curricula using Java physlets. *Computers in Physics, 12*, 227–232.
Clegg, T.L., Bonsignore, E., Yip, J. C., Gelderblom, H., Kuhn, A., Valenstein, T. & Druin, A. (2012). Technology for promoting scientific practice and personal meaning in life-relevant learning. In *Proceedings of the 11th International Conference on Interaction Design and Children (IDC' 12)* (pp. 152–161). New York: ACM.
Cooper, M. M., Groves, N. P., Pargas, R., Bryfczynski, S. P., & Gatlin, T. (2009). OrganicPad: An interactive freehand drawing application for drawing Lewis structures and the development of skills in organic chemistry. *Chemistry Education Research and Practice, 10*, 296–301.
Cromley, J. G., Bergey, B. W., Fitzhugh, S. L., Newcombe, N., Wills, T. W., Shipley, T. F., & Tanaka, J. C. (2013). Effects of three diagram instruction methods on transfer of diagram comprehension skills: The critical role of inference while learning. *Learning and Instruction, 26*, 45–58.
Dalebroux, A., Goldstein, T. R., & Winner, E. (2008). Short-term mood repair through art-making: Positive emotion is more effective than venting. *Motivation and Emotion, 32*, 288–295.
de Bock, D., Verschaffel, L., & Janssens, D. (1998). The predominance of the linear model in secondary school students' solutions of word problems involving length and area of similar plane figures. *Educational Studies in Mathematics, 35*, 65–83.
de Jong, T., & van Joolingen, W. R. (1998). Scientific discovery learning with computer simulations of conceptual domains. *Review of Educational Research, 68*, 179–201.
De Koning, B. B., & Jarodzka, H. (2017). Attention guidance strategies for supporting learning from dynamic visualizations. In R. Lowe & R. Ploetzner (Eds.), *Learning from dynamic visualization – Innovations in research and application*. Berlin: Springer (this volume).
De Petrillo, L., & Winner, E. (2005). Does art improve mood? A test of a key assumption underlying art therapy. *Art Therapy, 22*, 205–212.
Dickey, M. (2005). Engaging by design: How engagement strategies in popular computer and video games can inform instructional design. *Educational Technology Research and Development, 53*, 67–83.
Edelson, D. C., Gordin, D. N., & Pea, R. D. (1999). Addressing the challenges of inquiry-based learning through technology and curriculum design. *Journal of the Learning Sciences, 8*, 391–450.
Fredricks, J. A., Blumenfeld, P. C., & Paris, A. H. (2004). School engagement: Potential of the concept, state of the evidence. *Review of Educational Research, 74*, 59–109.
Gilbert, J. (2005). *Visualization in science education*. Dordrecht: Springer.
Gobert, J. (2000). A typology of models for plate tectonics: Inferential power and barriers to understanding. *International Journal of Science Education, 22*, 937–977.
Gobert, J. (2005). The effects of different learning tasks on conceptual understanding in science: Teasing out representational modality of diagramming versus explaining. *Journal of Geoscience Education, 53*, 444–455.
Gooding, D. (2004). Visualization, inference and explanation in the sciences. In G. Malcolm (Ed.), *Studies in multidisciplinarity* (Vol. 2, pp. 1–25). Amsterdam: Elsevier.
Harris, J., Mishra, P., & Koehler, M. (2009). Teachers' technological pedagogical content knowledge and learning activity types: Curriculum-based technology integration reframed. *Journal of Research on Technology in Education, 41*, 393–416.
Hayes, D., Symington, D., & Martin, M. (1994). Drawing during science activity in the primary school. *International Journal of Science Education, 16*, 265–277.
Hegedus, S., & Kaput, J. (2004). An introduction to the profound potential of connected algebra activities: Issues of representation, engagement and pedagogy. In *Proceedings of the 28th*

Conference of the International Group for the Psychology of Mathematics Education (pp. 129–136). Toronto: OISE/UT.

Hofstein, A., & Lunetta, V. N. (2003). The laboratory in science education: Foundations for the twenty-first century. *Science Education, 88*, 28–54.

Jee, B. D., Gentner, D., Forbus, K., Sageman, B., & Uttal, D. H. (2009). Drawing on experience: Use of sketching to evaluate knowledge of spatial scientific concepts. In N. A. Taatgen & H. van Rijn (Eds.), *Proceedings of the 31st Annual Conference of the Cognitive Science Society* (pp. 2499–2504). Amsterdam: Cognitive Science Society.

Johnson, J. K., & Reynolds, S. J. (2005). Concept sketches – Using student- and instructor-generated, annotated sketches for learning, teaching, and assessment in geology courses. *Journal of Geoscience Education, 53*, 85–95.

Kelly, R. M., & Jones, L. L. (2007). Exploring how different features of animations of sodium chloride dissolution affect students' explanations. *Journal of Science Education and Technology, 16*, 413–429.

Khishfe, R., & Abd-El-Khalick, F. (2002). Influence of explicit and reflective versus implicit inquiry-oriented instruction. *Journal of Research in Science Teaching, 39*, 551–578.

Kozma, R. (2003). Material and social affordances of multiple representations for science understanding. *Learning and Instruction, 13*, 205–226.

Kozma, R., Chin, E., Russell, J., & Marx, N. (2000). The role of representations and tools in the chemistry laboratory and their implications for chemistry learning. *Journal of the Learning Sciences, 9*, 105–143.

Kozma, R., & Russell, J. (1997). Multimedia and understanding: Expert and novice responses to different representations of chemical phenomena. *Journal of Research in Science Teaching, 34*, 949–968.

Latour, B. (1990). Drawing things together. In M. Lynch & S. Woolgar (Eds.), *Representation in scientific practice* (pp. 19–68). Cambridge, MA: MIT Press.

Leutner, D., Leopold, C., & Sumfleth, E. (2009). Cognitive load and science text comprehension: Effects of drawing and mentally imagining text content. *Computers in Human Behavior, 25*, 284–289.

Leutner, D., & Schmeck, A. (2014). The generative drawing principle in multimedia learning. In R. E. Mayer (Ed.), *The Cambridge handbook of multimedia learning* (pp. 433–448). New York: Cambridge University Press.

Linn, M. C. (2010). How can selection and drawing support learning from dynamic visualizations. In K. Gomez, L. Lyons & J. Radinsky (Eds.), *Proceedings of the Ninth International Conference of the Learning Sciences (ICLS)* (Vol. 1, pp. 165–166). Mahwah, NJ: Erlbaum.

Linn, M. C., Lee, H.-S., Tinker, R., Husic, F., & Chiu, J. L. (2006). Teaching and assessing knowledge integration in science. *Science, 313*, 1049–1050.

Lowe, R. (2003). Animation and learning: Selective processing of information in dynamic graphics. *Learning and Instruction, 13*, 157–176.

Lowe, R., & Boucheix, J.-M. (2017). A composition approach to design of educational animations. In R. Lowe, & R. Ploetzner (Eds.), *Learning from dynamic visualization – Innovations in research and application*. Berlin: Springer (this volume).

Lowe, R., & Mason, L. (2017). Self-generated drawing: A help or hindrance to learning from animation? In R. Lowe & R. Ploetzner (Eds.), *Learning from dynamic visualization – Innovations in research and application*. Berlin: Springer (this volume).

Lowe, R., & Schnotz, W. (Eds.). (2008). *Learning with animation: Research implications for design*. New York: Cambridge University Press.

Mason, L., Lowe, R., & Tornatora, M. C. (2013). Self-generated drawings for supporting comprehension of a complex animation. *Contemporary Educational Psychology, 38*, 211–224.

Mayer, R. E. (2009). *Multimedia learning* (2nd ed.). New York: Cambridge University Press.

McGrath, M. B., & Brown, J. R. (2005). Visual learning for science and engineering. *IEEE Computer Graphics and Applications, 25*, 56–63.

Miller, G. A., Galanter, E., & Pribram, K. H. (1960). *Plans and the structure of behavior*. New York: Holt, Rinehart, & Winston.

Newmann, F. M., Wehlage, G. G., & Lamborn, S. D. (1992). The significance and sources of student engagement. In F. M. Newman (Ed.), *Student engagement and achievement in American secondary schools* (pp. 11–39). New York: Teachers College Press.

Paivio, A. (1991). Dual coding theory: Retrospect and current status. *Canadian Journal of Psychology, 45*, 255–287.

Plass, J. L., Milne, C., Homer, B. D., Jordan, T., Schwartz, R. N., Hayward, E. O., & Barrientos, J. (2012). Investigating the effectiveness of computer simulations for chemistry learning. *Journal of Research in Science Teaching, 49*, 394–419.

Ploetzner, R., & Breyer, B. (2017). Strategies for learning from animation with and without narration. In R. Lowe & R. Ploetzner (Eds.), *Learning from dynamic visualization – Innovations in research and application*. Berlin: Springer (this volume).

Ploetzner, R., & Fillisch, B. (2017). Not the silver bullet: Learner-generated drawings make it difficulty to understand broader spatiotemporal structures in complex animations. *Learning and Instruction, 47*, 13–24.

Prain, V., & Tytler, R. (2012). Learning through constructing representations in science: A framework of representational construction affordances. *International Journal of Science Education, 34*, 2751–2773.

Quintana, C., Reiser, B. J., Davis, E. A., Krajcik, J. S., Fretz, E., Duncan, R. G., & Soloway, E. (2004). A scaffolding design framework for software to support science inquiry. *Journal of the Learning Sciences, 13*, 337–386.

Rasco, R. W., Tennyson, R. D., & Boutwell, R. C. (1975). Imagery instructions and drawings in learning prose. *Journal of Educational Psychology, 67*, 188–192.

Reiser, B. J., Tabak, I., Sandoval, W. A., Smith, B., Steinmuller, F., & Leone, T. J. (2001). BGuILE: Strategic and conceptual scaffolds for scientific inquiry in biology classrooms. In S. M. Carver & D. Klahr (Eds.), *Cognition and instruction: Twenty-five years of progress* (pp. 263–305). Mahwah, NJ: Erlbaum.

Roschelle, J., Schechtman, N., Tatar, D., Hegedus, S., Hopkins, B., Empson, S., & Gallagher, L. (2010). Integration of technology, curriculum, and professional development for advancing middle school mathematics: Three large-scale studies. *American Educational Research Journal, 47*, 833–878.

Roy, M., & Chi, M. T. H. (2005). The self-explanation principle in multimedia learning. In R. E. Mayer (Ed.), *The Cambridge handbook of multimedia learning* (pp. 271–286). New York: Cambridge University Press.

Ryan, S., Yip, J., Stieff, M., & Druin, A. (2013). Cooperative inquiry as a community of practice. In N. Rummel, M. Kapur, M. Nathan, & S. Puntambekar (Eds.), *Proceedings of the 10th International Conference on Computer-Supported Collaborative Learning* (pp. 145–148). Madison, WI: International Society for the Learning Sciences.

Schwamborn, A., Mayer, R. E., Thillmann, H., Leopold, C., & Leutner, D. (2010). Drawing as a generative activity and drawing as a prognostic activity. *Journal of Educational Psychology, 102*, 872–879.

Snowman, J., & Cunningham, D. J. (1975). A comparison of pictorial and written adjunct aids in learning from text. *Journal of Educational Psychology, 67*, 307–311.

Stein, M., & Power, B. (1996). Putting art on the scientist's palette. In R. S. Hubbard & K. Ernst (Eds.), *New entries: Learning by writing and drawing* (pp. 59–68). Portsmouth, NH: Heinemann.

Stieff, M. (2005). Connected chemistry – A novel modeling environment for the chemistry classroom. *Journal of Chemical Education, 82*, 489–493.

Stieff, M. (2011a). Fostering representational competence through argumentation with multi-representational displays. *Proceedings of the 9th international conference on computer-supported collaborative learning* (Vol. 1, pp. 288–295). Mahwah, NJ: Erlbaum.

Stieff, M. (2011b). Improving representational competence using multi-representational learning environments. *Journal of Research in Science Teaching, 48*, 1137–1158.

Stieff, M., Bateman, R., & Uttal, D. H. (2005). Teaching and learning with three-dimensional representations. In J. Gilbert (Ed.), *Visualization in science education* (pp. 93–120). Oxford: Oxford University Press.

Stieff, M., & McCombs, M. (2006). Increasing representational fluency with visualization tools. In *Proceedings of the Seventh International Conference of the Learning Sciences (ICLS)* (Vol.1, pp. 730–736). Mahwah, NJ: Erlbaum.

Stieff, M., Nighelli, T., Yip, J., Ryan, S., & Berry, A. (2012). *The connected chemistry curriculum (Vols. 1–9)*. Chicago, IL: University of Illinois.

Stieff, M., & Wilensky, U. (2003). Connected Chemistry – incorporating interactive simulations into the chemistry classroom. *Journal of Science Education & Technology, 12*, 285–302.

Tai, R., Liu, C. Q., Maltese, A. V., & Fan, X. (2006). Planning early for careers in science. *Science, 312*, 1143–1144.

Van Meter, P., & Firetto, C. M. (2013). Cognitive model of drawing construction: Learning through the construction of drawings. In G. Schraw, M. McCrudden, & D. Robinson (Eds.), *Learning through visual displays* (pp. 247–280). Scottsdale, AZ: Information Age Publishing.

Van Meter, P., & Garner, J. (2005). The promise and practice of learner-generated drawing: Literature review and synthesis. *Educational Psychology Review, 17*, 285–325.

Wu, H.-k., & Huang, Y.-L. (2007). Ninth-grade student engagement in teacher-centered and student-centered technology-enhanced learning environments. *Science Education, 91*, 727–749.

Wu, H.-k., Krajcik, J. S., & Soloway, E. (2001). Promoting conceptual understanding of chemical representations: Students' use of a visualization tool in the classroom. *Journal of Research in Science Teaching, 38*, 821–842.

Zhang, H. Z., & Linn, M. C. (2011). Can generating representations enhance learning with dynamic visualizations? *Journal of Research in Science Teaching, 48*, 1177–1198.

Chapter 15
Strategies for Learning from Animation With and Without Narration

Rolf Ploetzner and Bianka Breyer

15.1 Learning Strategies

"Don't make me think!" is Krug's (2014) famous first law of Web usability. It means that "... when I look at a Web page it should be self-evident. Obvious. Self-explanatory." (Krug, 2014, p. 11). While the mere operation of user interfaces should not require us to think, learning material, on the other hand, almost always should and does. Quite the contrary, educators are very often faced with the challenge of how to design the learning material in such a manner that makes their students sufficiently think (e.g., Jonassen, Howland, Marra, & Crismond, 2011; Linn & Eylon, 2011). One approach to encourage and support students' information processing is the principled design of the learning material (cf. Clark & Mayer, 2011; Mayer, 2014a). Based on theories and models of human memory and learning, this approach essentially aims at designing the learning material in such a way that the students attend to and adequately process the relevant information. Over the past 15 years, educational research has focused on this approach and it has been demonstrated in numerous studies that the principled design of instructional material fosters learning (for overviews see Mayer, 2009, 2014a). With respect to learning from animation, an example of such an important design principle is to combine an animation with spoken text rather than written text (cf. Ginns, 2005; Low & Sweller, 2014; Mayer, 2009).

The design of the learning material alone, however, cannot guarantee effective learning; it is only one side of the coin. The other side involves the students' perceptual, cognitive, and metacognitive engagement during learning (see Stieff, 2017,

R. Ploetzner (✉)
University of Education at Freiburg, Freiburg, Germany
e-mail: rolf.ploetzner@ph-freiburg.de

B. Breyer
Leibniz Institute for the Social Sciences, Mannheim, Germany
e-mail: bianka.breyer@gesis.org

R. Lowe, R. Ploetzner (eds.), *Learning from Dynamic Visualization*,
DOI 10.1007/978-3-319-56204-9_15

this volume). This engagement is manifested in the perceptual, cognitive, and metacognitive techniques and strategies that the students make use of. Therefore, another approach to encourage and support the students' information processing is the principled design of learning strategies. Also based on theories and models of human memory and learning, this approach aims at empowering students to competently initiate, plan, organize, monitor, and regulate their own learning, especially when dealing with challenging learning material (cf. Pressley & Harris, 2006; Winne & Hadwin, 1998; Wittrock, 1974, 1989; Zimmerman, 2002). With respect to learning from animation, this approach has been largely neglected up until now.

According to Streblow and Schiefele (2006), a learning strategy can be perceived as "... a sequence of efficient learning techniques, which are used in a goal-oriented and flexible way, are increasingly automatically processed, but can be called into awareness when needed" (p. 353, translation by the authors). Learning techniques are hence the individual components that are collectively employed as an overall learning strategy. Examples of learning techniques include selecting and highlighting specific material as well as enriching the material with annotations and notes. Learning techniques constitute a learning strategy only when a number of them are coordinated together in a goal-oriented way. The aim of applying a learning strategy is to induce, support, and sustain effective learning processes.

In educational research, three broad classes of learning strategies are typically distinguished (e.g., Streblow & Schiefele, 2006): cognitive strategies, metacognitive strategies, and resource strategies. The aim of cognitive strategies is to foster effective primary information processing, whereas metacognitive strategies focus on the super-ordinate level of planning, monitoring and regulating cognitive processes. In contrast to these internally focused strategies, resource strategies focus on establishing supportive external learning conditions such as ensuring access to learning resources and avoiding distractions. This chapter focuses on cognitive learning strategies with an emphasis on strategies for the selection, organization, elaboration, and memorization of information (cf. Weinstein & Mayer, 1986; Wittrock, 1974, 1989). Perceptual strategies, in contrast, have been almost entirely neglected in educational research (cf. Ploetzner, Lowe, & Schlag, 2013). That is, although educational research has produced a range of strategies that make students think, to our knowledge, it has yet to produce strategies that make students look.

Research on learning strategies has a long and extensive tradition with respect to learning from text which reaches back to the influential work of Robinson (1946). In stark contrast, strategies for learning from pictorial representations remain relatively neglected by researchers. This might explain why strategies for learning from pictorial representations are rarely taught to students (cf. Kremling, 2008). Educators who mistakenly assume that pictorial representations are not only intrinsically beneficial but also self-explanatory might exacerbate the potentially harmful consequences of this neglect (cf. Lieber, 2008). Further, students may experience "illusions of understanding" when asked to learn from pictorial representations (cf. Salomon, 1983, 1984), but in truth, they do not actually comprehend them. Research on learning from animation has repeatedly demonstrated that students can have severe difficulties in understanding the animated subject matter (e.g., Lowe, 2003,

2004, 2008). However, with respect to the question of how learning from animation can be enhanced beyond optimizing the animation design, research is still in its infancy.

15.2 Previous Research on Learning Strategies

Early research on learning strategies was carried out at a time when written text was the dominant form of information representation. Not surprisingly, it focused on how more successful and less successful learners differed in their strategic behavior during learning from text. Marton and Säljö (1984), for example, identified two different approaches to learning from text. They distinguished between a surface level approach, in which the text may be repeatedly recited to remember it, and a deep level approach, in which the text is elaborated in order to comprehend it. Successful learners were found to favor deep level approaches. Similar observations with respect to the depth of text processing have been made in other studies such as those by Dornisch, Sperling, and Zeruth (2011), Pask (1976), and Svensson (1977).

Consistent with these findings, strategies that aim to induce and promote elaboration processes during learning from text were devised. For instance, building on the early work of Robinson (1946), Thomas and Robinson (1972) devised the well-known PQ4R-Method (Preview, Question, Read, Reflect, Recite, Review) that structures learning from text by means of six steps: (1) survey the material to get a general overview (Preview), (2) formulate questions about the text (Question), (3) read the text thoroughly while keeping the formulated questions in mind (Read), (4) reflect on the text by relating the information to prior knowledge and formulating examples (Reflect), (5) answer the questions by giving an account of the text in one's own words (Recite), and (6) try to recall or summarize the information that has been read without looking at the text (Review). Further examples of strategies for learning from text are MURDER (Mood, Understanding, Recall, Digest, Expanding, Review) from Dansereau et al. (1979) and REDUTEX (Reduce Text) from Friedrich (1995).

Current research on learning strategies is still geared mainly to learning from text (for overviews see Artelt, 2000; Gambrell, Morrow, & Pressley, 2007; Mandl & Friedrich, 2006; Ploetzner et al., 2013; Pressley & Harris, 2006). This is despite the fact that today's learning materials commonly contain a substantial proportion of pictures, and quite often, even more than the amount of text. In order to provide a form of support paralleling that available for text-only materials, researchers such as Larson et al. (1986), Schlag and Ploetzner (2009, 2011), Seufert (2009), as well as Stalbovs, Scheiter, and Gerjets (2015) have proposed strategies for learning from illustrated texts. For instance, the strategy developed by Schlag and Ploetzner (2009, 2011) is made up of a sequence of learning techniques that aims to stimulate not only the processing of textual and pictorial information, but also the construction of relations between both types of representation. Two experimental studies showed that students who took advantage of the strategy exhibited significantly larger learning gains than students who wrote a summary of the learning material. Metz and Wichert (2009) observed similar results when they investigated the educational effectiveness of a strategy for learning from illustrated text.

With respect to learning from static pictures, only single learning techniques – as opposed to more comprehensive learning strategies – have been proposed and empirically evaluated. For example, in order to encourage learners to process pictures more deeply, Salomon (1984) and Weidenmann (1989) asked learners to pay special attention to the pictures while orienting themselves with the learning material. Weidenmann (1988) prompted learners to compare different pictures, whereas Peeck (1994) and Weidenmann (1994) required learners to answer questions about the target pictures, and Ainsworth and Loizou (2003) encouraged learners to explain the pictures to themselves. Although each of these techniques has its own value in helping students to comprehend static pictures, individually they do not constitute a comprehensive learning strategy.

In respect to dynamic pictures such as animations, there is even less research on learning techniques and strategies available (cf. Ploetzner, 2016). De Koning, Tabbers, Rikers, and Paas (2010), for instance, asked learners to explain an animation to themselves. Hegarty, Kriz, and Cate (2003) asked learners to describe relations that hold between the verbal and pictorial information presented in an animation. Mason, Lowe, and Tornatora (2013) asked students to produce drawings in order to depict the animated processes (see also Lowe & Mason, 2017, this volume). Moreno and Valdez (2005) supported learning from animation by requesting less successful learners to organize segments of an animation into an appropriate sequence. Successful learners, in contrast, benefited the most from splitting an animation into meaningful segments. Schmidt-Weigand (2005) encouraged learners to stop an animation when the display showed important information and to then analyze the components on the display as well as their relations. As for learning from static pictures, each of these techniques has its own value in supporting learning from animation; on their own, however, they do not constitute a comprehensive learning strategy. Up until now, only Kombartzky, Ploetzner, Schlag, and Metz (2010) have proposed and empirically evaluated a comprehensive strategy for learning from narrated animation.

15.3 Two Types of Animation and Two Theories of Learning from Animation

Two important types of expository animation can be distinguished: animations that contain verbal explanations and animations that do not include verbal explanations. Both types of animation place different processing demands on the students. Unsurprisingly, learning from each type of animation is accounted for by a different theory: while the Animation Processing Model (Lowe & Boucheix, 2008, 2011; see also Lowe & Schnotz, 2014) focuses on learning from animations without verbal enhancement, the Cognitive Theory of Multimedia Learning (Mayer, 2009, 2014b) describes learning from animations containing verbal explanations.

The explanatory focus of expository animations can be on structural aspects (i.e., the entities that constitute a process), behavioral aspects (i.e., the events that take place in a process), and/or functional aspects (i.e., the – possibly causal – relations that hold between different entities and events in a process) of the animated subject matter (cf. Ploetzner & Lowe, 2012). Due to their different representational characteristics, pictorial and verbal representations are differently suited to express these aspects (cf. Larkin & Simon, 1987; Oestermeier & Hesse, 2000; Prain & Tytler, 2013; Schnotz, 2002; Stenning & Oberlander, 1995). Pictorial representations limit abstraction but aid in processibility. They can depict certain classes of information almost entirely, for instance, geometric properties and spatial relations. In many cases, pictorial representations encompass information that can be easily read off. Pictorial representations are therefore especially suited to depict structural and behavioral aspects of the animated subject matter.

Verbal representations, in contrast, are less limited with respect to abstraction. They allow for the expression of abstract relations such as causal dependencies, for example. The attempt to entirely describe certain classes of information such as geometric properties and spatial relations, nevertheless, can result in extensive, but still incomplete descriptions. Furthermore, the inference of new information from verbal representations can be challenging and error prone. Verbal representations are therefore particularly suited to describe functional aspects of the animated subject matter. Verbal information in an animation, however, does not merely complement the information included in the pictorial display. Rather, it may also influence how the pictorial display itself is processed. For instance, the students' processing of the verbal information may result in the activation of specific concepts and schemata relevant to the animated subject matter. As a consequence, these concepts and schemata may top-down impact what is being perceived in the display. Furthermore, verbal information might indirectly (e.g., by being synchronized with the pictorial information) or directly (e.g., by spelling out hints where to look) direct the students' attention to specific regions in the display. Therefore, verbal representations might support the students not only at the cognitive level, but also at the perceptual level. In contrast, when verbal information is not present, the students rely exclusively on the pictorial display and their prior knowledge.

The Animation Processing Model (APM; Lowe & Boucheix, 2008, 2011, 2017, this volume; Lowe & Schnotz, 2014) describes how learning from animations without verbal explanations progresses as a cumulative activity in which bottom-up and top-down processes interact in order to construct an increasingly comprehensive mental model of the animated subject matter. If a student possesses only little prior knowledge about the animated subject matter that could influence her or his assessment of the pictorial display, then the student's analyses are mostly limited to bottom-up processes. That is, the student starts by perceptually identifying local event units (Phase 1 of the APM) and then successively combines them into broader regional structures (Phase 2 of the APM). Next, the student may proceed to link regional structures by establishing higher-order relations that may cover the animation's entire spatial and temporal scope (Phase 3 of the APM). Without having access to additional information, however, it is rather unlikely that she or he will further progress to Phases 4 and 5 of the Animation Processing Model (cf. Lowe, 2004). That is, based solely on the pictorial display, the student will hardly be able to establish functional relationships or construct a comprehensive and coherent mental model of the animated subject matter.

Derived from the Animation Processing Model, various principles for the design of animations have been proposed and empirically evaluated: the presentation of dynamic cues to direct the students' attention to relevant regions in the pictorial display (Boucheix, Lowe, Putri, & Groff, 2013), the incremental presentation of animations to facilitate the students' mental model building (Lowe & Boucheix, 2012, 2017, this volume), and the simultaneous presentation of multiple animation episodes to support the students in identifying higher-order relationships (Ploetzner & Lowe, 2014, 2017, this volume).

Animations that encompass verbal explanations require students not only to process the pictorial as well as verbal information, but also to mentally relate and integrate both sources of information. Thus, while the focus of the Animation Processing Model is on perceptual and cognitive intra-representational processes that are applied to the pictorial display, the focus of the Cognitive Theory of Multimedia Learning (Mayer, 2009, 2014b) is on cognitive inter-representational processes that aim to relate – previously selected and organized – pictorial and verbal information. In accord with Atkinson and Shiffrin (1968), as well as Baddeley (1986), it is assumed that the human memory is made up of three main components: the sensory registers, the working memory, and the long-term memory. The working memory plays a pivotal role in processing information. Mayer (2009, 2014b) formulates three basic assumptions concerning the working memory. First, the working memory consists of an auditory-verbal and a visual-pictorial channel (cf. Baddeley, 1999; Paivio, 1986). Second, the capacity of the working memory is limited, i.e., only a limited amount of information can be processed simultaneously (cf. Atkinson & Shiffrin, 1968; Baddeley, 1999). Third, successful learning from different representations requires (1) the selection of relevant pictorial and verbal information, (2) the organization of the selected information to construct a pictorial and a verbal model, and (3) the integration of the pictorial and verbal model into one coherent mental structure by taking advantage of prior knowledge. Furthermore, it is assumed that pictorial information can be mentally transformed into a verbal representation (e.g., the mental verbalization of information presented in a picture) and vice versa (e.g., the construction of a mental image that depicts information from a verbal description). Because each representation has its own strengths and weaknesses with respect to expressiveness and processability, these transformations can support the students to utilize the presented information more effectively and efficiently.

Various principles for the design of multimedia learning material have been devised on the basis of the Cognitive Theory of Multimedia Learning (for an overview see Mayer, 2009, 2014a). With respect to learning from expository animation, an important design principle – the so-called modality principle – is to combine an animation with spoken explanations rather than written explanations (cf. Ginns, 2005;

Low & Sweller, 2014). When an animation is combined with written explanations, then both the pictorial as well as the verbal information need to be taken up by the eyes resulting in a split-attention effect (cf. Ayres & Sweller, 2014) that impedes learning. In contrast, when an animation is combined with spoken explanations, the eyes can attend to the pictorially presented information and – simultaneously – the ears can attend to the verbally presented information allowing the students to process both representations more completely. Throughout the rest of this chapter, animations that are combined with spoken explanations will be denoted as narrated animations.

15.4 A Strategy for Learning from Narrated Animation

To support learning from narrated animation, Kombartzky et al. (2010) conceptualized a learning strategy. In accord with Mayer's (2009, 2014b) Cognitive Theory of Multimedia Learning, the strategy aims to systematically induce and sustain the cognitive processes of selecting, organizing, and transforming verbal and pictorial information as well as integrating both types of information. The strategy is made up of eight learning techniques:

- Overview, activation of prior knowledge, and formulation of expectations: (1) watch the animation, listen to the narration, and note what can be learned from the animation.
- Selection and organization of information: (2) identify important frames in the animation and sketch them, (3) identify important statements in the narration and take notes on them, (4) identify important regions in the selected frames and mark them, (5) identify important assertions in the selected statements and mark them, (6) label regions in the selected frames by taking advantage of the selected statements.
- Transformation and integration of information: (7) express relations between the selected frames and the selected statements in your own words, (8) summarize the overall process in your own words.

Kombartzky et al. (2010) conducted two experimental studies in order to evaluate the educational effectiveness of the proposed strategy. In both studies, students learned from a narrated animation about the dances of honeybees. The animation was taken from Microsoft Encarta Professional (Microsoft Corporation, 2002). It consists of realistic pictures, schematic pictures, symbols, labels, and narration. In 2 min and 20 s, the animation visualizes and explains how honeybees dance in order to communicate to other bees where resources in the environment are located. While one type of dance is used to communicate the location of resources that can be found within 100 m of the honeycomb, a second type of dance is used to communicate the location of resources that are more than 100 m away from the honeycomb. In these dances, honeybees take into account the relative position of the honeycomb, the sun, and the resources.

In the first study, a group of 21 sixth-graders were encouraged to make use of the proposed strategy during learning from the animation. The actual use of the strategy, however, was not monitored. A second group of 22 sixth-graders learned from the animation without the strategy. Instead, these students were requested to write a summary of the animated processes. The students who were encouraged to take advantage of the strategy showed significantly more conceptual understanding and better transfer than the students who were asked to write a summary. The effect sizes were medium to large.

Because an analysis of the students' worksheets revealed that only 19% of the students fully took advantage of the learning strategy, a second study with three groups of sixth graders was conducted. One group consisted of 53 students who were encouraged to make use of the strategy during learning from the animation; it was additionally monitored whether the students actually used the strategy. The second group was comprised of 52 students who were also encouraged to make use of the strategy. In this case, however, the actual use of the strategy was not monitored. The third group consisted of 49 students who learned from the animation without the strategy. Again, these students were requested to write a summary of the animated processes. The results of the second study replicated the findings of the first study. Furthermore, learning was most successful when the students' use of the learning strategy was monitored. The effect sizes were again medium to large.

In a more recent study, Ploetzner and Schlag (2013) investigated whether the learning strategy proposed by Kombartzky et al. (2010) also improves learning from different narrated animations, leads to an acquisition of knowledge which is available beyond the learning period, and/or equally benefits students with low and high cognitive ability alike. A total of 152 sixth graders participated in the study: 69 students learned from the same animation about the dances of honeybees that was employed by Kombartzky et al. (2010) and 83 students learned from a narrated animation about sailing. The animation about sailing was produced by the authors based on a textbook about sailing (Bark, 2009). Figure 15.1 shows a picture that was included in the animation. The animation is made up of schematic pictures, symbols, labels, and narration. In 3 min and 27 s, it visualizes and explains four courses that a yacht can sail in relation to the wind direction: running (i.e., directly downwind), broad reach, close hauled, and tacking. An explanation of how the sail of the yacht needs to be adjusted, how the main forces act on the yacht, and how the forces affect the yacht's speed are provided with respect to each course.

A treatment group and a control group were formed for each animation respectively. While the students in the treatment groups were encouraged to apply the learning strategy proposed by Kombartzky et al. (2010), the students in the control groups were requested to write a summary. Before learning took place, the students' prior knowledge and cognitive ability were assessed. The students worked on a posttest immediately after learning. Furthermore, the students completed a follow-up test that was identical to the posttest 1 week after the learning period.

The study yielded four main results. First, with respect to the animation about the dances of honeybees, it fully replicated the findings of Kombartzky et al. (2010). Second, regarding the animation about sailing, the study validates that taking advantage of the strategy enhances learning from other narrated animations as well. With respect to both animations, the observed effect sizes are medium to large. Third, with respect to both animations, the students who took advantage of the learning strategy showed superior performance not only directly after the learning period, but also 1 week later. Again, the effect sizes are medium to large. Fourth, non-significant interactions between the experimental treatments and the students' cognitive ability do not support the assumption of an aptitude-treatment interaction. The results of the study suggest rather that the learning strategy benefits students with low and high cognitive ability alike.

The strategy proposed by Kombartzky et al. (2010) successfully supported learning from different narrated animations. The strategy thereby focused on inducing cognitive processes such as the selection, organization, transformation, and integration of verbal and pictorial information. Is it possible that a strategic approach that focuses on cognitive processes could successfully support learning from animation without narration as well?

15.5 A Strategy for Learning from Animation Without Narration

The sailing animation employed by Ploetzner and Schlag (2013; see Fig. 15.1) consists of four sailing episodes. Each episode depicts one of four courses that a yacht can sail in relation to the wind direction: running, broad reach, close hauled, and tacking. Any one of these episodes depicts a specific set of local relationships

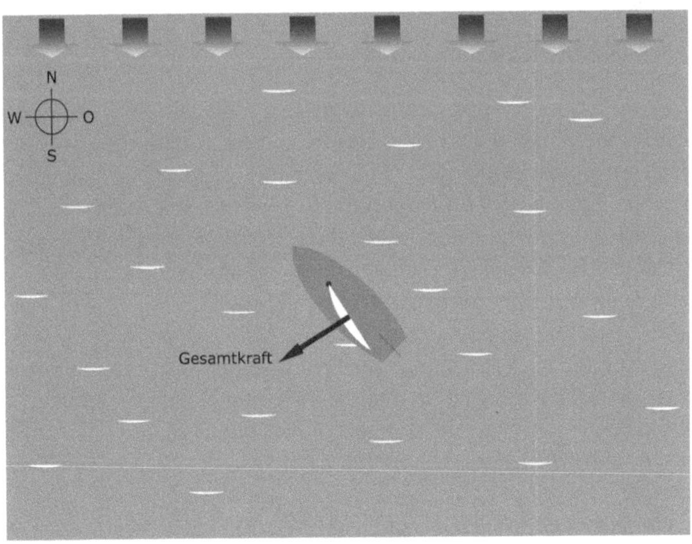

Fig. 15.1 A screen shot taken from the animation about sailing depicting a yacht that sails close hauled

amongst entities such as the wind direction, the yacht's hull and sail, and the forces that act on the yacht. However, there are also higher-order relationships connecting the individual episodes at a more general level. These relationships link the wind direction to the sail's orientation, the sail's orientation to the directions and magnitudes of the different forces, and the directions and magnitudes of the forces to the yacht's speed. For instance, the relationship that links the wind direction to the sail's orientation could be expressed as "the closer the yacht sails towards the wind direction, the closer the sail is oriented towards the hull".

If such a multi-episode animation lacks the verbal explanations that – indirectly or directly – direct the students' attention to relevant entities and events in the display as well as explicitly express relations between these entities and events, then the students have to identify them on their own. According to the Animation Processing Model (Lowe & Boucheix, 2008, 2011, 2017, this volume; Lowe & Schnotz, 2014), in order to develop a comprehensive and hierarchically structured mental model, the students need to internally represent not only local relationships from within individual episodes, but also higher-order, between-episode relationships that encompass information from the animation as a whole. In this situation, constructing a satisfactory mental model requires learners to compare and contrast relevant entities and events as well as their local relationships across individual episodes in order to identify and internalize higher-order relationships. That is, the students need to engage first in perceptual processes as described in the Animation Processing Model, and then in inductive processes which are also known to place high demands on learners (cf. Holland, Holyoak, Nisbett, & Thagard, 1986; Klauer & Phye, 2008; see also Ploetzner & Lowe, 2017, this volume).

Accordingly, a learning strategy was developed that aims to systematically induce – rather indirectly – the perceptual processes of the first two phases and – more directly – the cognitive processes of the third phase of the Animation Processing Model. The strategy is made up of nine learning techniques:

- Overview, activation of prior knowledge, and formulation of expectations: (1) watch the animation and note what can be learned from the animation.
- Localized perceptual exploration (Phase 1 of the APM): (2) watch the animation again; select five out of the eleven provided pictures that show the most important states in the animation; (3) arrange the selected pictures in the same sequence as they appear in the animation; (4) watch the animation again and give each selected picture a caption so that you can refer to it.
- Regional structure formation (Phase 2 of the APM): (5) look at the selected pictures and note whether something changes from one picture to another picture; (6) describe the changes that take place from one picture to another; watch the animation again in order to check whether your descriptions match what actually happens in the animation; (7) try to express each change by means of an if-then-rule.
- Global characterization (Phase 3 of the APM): (8) look across all of your if-then-rules and try to summarize them as best as possible into statements with the form

"the more/the less …, the more/the less …"; watch the animation again in order to check whether your statements match what actually happens in the animation; (9) summarize the overall process in your own words.

In an experimental study, we investigated whether use of the proposed strategy improves learning from the sailing animation without narration more than writing a summary does.

15.5.1 Design and Hypotheses

Two groups of students were investigated. Both groups learned from the same animation about sailing without narration. While one group of students was encouraged to make use of the learning strategy described above (strategy group), the second group of students was requested to write a summary (summary group). Before and after learning, the students had to process pre- and posttests.

We assumed that the learning strategy would support the students in processing the first three phases of the Animation Processing Model. Without taking advantage of the strategy, the students might spontaneously, but not systematically, engage in the perceptual and cognitive processes required for effective learning from animations without narration (cf. Lowe, 2003, 2004, 2008). Accordingly, we expected the strategy group to learn more successfully than the summary group.

15.5.2 Participants, Material, and Procedure

A total of 58 eighth-graders volunteered for the study. They received financial compensation for their participation in the study. None of the students who participated in the study had experience in sailing. The students were randomly assigned to the strategy group (19 female and 13 male students, mean age = 13.67 years, $SD = 0.51$) and the summary group (17 female and 9 male students, mean age = 13.65 years, $SD = 0.30$).

The students were investigated in groups. Each student was individually seated in front of a computer. To begin, the students worked on a pretest. The test consisted of six items that assessed the students' prior knowledge of the principles that apply to sailing. Next, the students studied three printed pages that depicted and explained the graphic entities shown in the animation: a compass, arrows indicating the wind direction, an arrow representing the magnitude and direction of a force, a parallelogram with a description how it is used to resolve a force into its component forces, and a buoy that the yacht is to reach by tacking. Thereafter, the strategy group was given an introduction to the learning strategy. A demonstration of the learning strategy was provided using an animation of the different moon phases. The summary group also watched the animation of the moon phases, however, the learning strategy was not demonstrated.

Next, printed instructions informed the students that they could study the animation for up to 50 min, make use of a media player to start, stop, forward and rewind the animation, and watch the animation as often as they wished within the limits of the learning time. During the subsequent learning phase, each student watched the animation by taking advantage of a standard media player on the computer. The animation used a bird's eye view to present a schematic and idealized depiction of four different courses that a yacht can sail in relation to the wind direction: running, broad reach, close hauled, and tacking. For the first three of these courses – all downwind – the animations show how the yacht's hull and sail are oriented, how different forces act on the yacht, and how the yacht moves. For the fourth course, the animations show how the yacht needs to sail in order to approach a goal that is located straight into the wind, i.e., upwind. The animation included labels naming the sailing courses as well as the forces that act on the yacht. However, it did not include explanatory text or narration.

The students in the strategy group received two printed pages that described each learning technique and the sequence in which the different techniques had to be applied. Furthermore, with respect to each learning technique, the students were requested to keep track of their achievements in three different worksheets. The results of Technique 1 (overview and learning expectation) were written on the first worksheet. The results of Techniques 2–8 were documented on the second worksheet. This worksheet was printed on paper in DIN A3 landscape format. It was presented as a table in such a manner that the students were able to look over their achievements at any time. For instance, the pictures the students selected in Step 2 of the strategy were glued into the top of the worksheet. Figure 15.2 illustrates the layout of this worksheet. The results of Technique 9 (summary) were written on the third worksheet. The students in the summary group received one worksheet in order to write down their summaries.

After learning, the students completed a posttest. The test contained 12 multiple-choice items that were provided in a verbal format (for sample items, see Table 15.1) as well as 12 multiple-choice items that were presented in a mixed verbal-graphic format (for sample items, see Fig. 15.3).

15.5.3 Results

Both groups of students possessed very little prior knowledge about the mechanisms underlying sailing (strategy group M = 0.42 (2.71%), SD = 0.79; summary group M = 0.44 (2.84%), SD = 0.90). The difference between the strategy group and the summary group in prior knowledge was not significant ($t(56)$ = 0.70, *n.s.*). Contrary to our hypothesis, there was no significant difference between the strategy group and the summary group on the posttest ($t(56)$ = −1.16, *n.s.*). Descriptively, the summary group (M = 21.38 (44.55%), SD = 5.45) exhibited even better results on the posttest than the strategy group (M = 19.69 (41.02%), SD = 5.67). Internal consistency of the posttest was acceptable (*Cronbach's* α = 0.70).

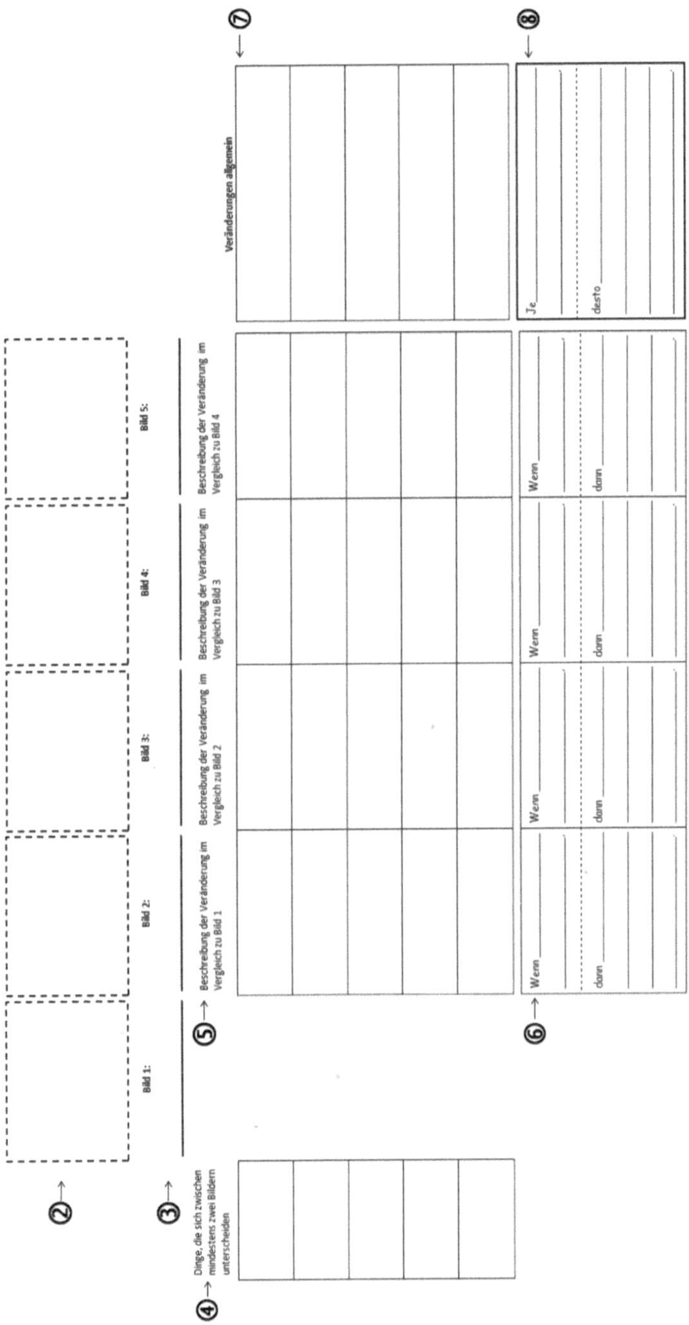

Fig. 15.2 The layout of the worksheet for Techniques 2–8

Table 15.1 Two examples of items presented in a verbal format (translation by the authors)

The yacht sails as close hauled as possible. What would happen if the yacht turned 10 degrees further upwind? ☐ The sail would swing to the other side of the yacht's hull ☐ The sail would flutter ☐ The sail would move closer to the yacht's hull ☐ The sail would move further away from the yacht's hull ☐ The sail would remain where it is
For which course is the distance between the sail and the hull of the yacht largest? ☐ Running ☐ Broad reach ☐ Close hauled ☐ Tacking ☐ The distance is the same for all courses

To gain further insight into the students' learning, the worksheets were analyzed. In particular, it was analyzed to which extent the students processed Phase 3 of the APM, i.e., to which extent the students were able to characterize the animated processes globally. By means of a task analysis, six global rules which could be learned from the animation were identified (cf. Table 15.2). Two independent raters judged whether a student expressed any of these rules on her or his worksheet. For each rule expressed, a score of one point was assigned. Inter-rater reliability ($ICC(3,1) = 0.72$) was moderate.

Table 15.3 shows how often each rule was expressed on the students' worksheets. A multivariate analysis of variance (MANOVA) revealed at the multivariate level, i.e., across all six rules, that the strategy group expressed significantly more rules than the summary group (*Wilk's Lambda* = 0.48, $F(6,51) = 9.06$, $p = 0.000$). The results at the univariate level, i.e., with respect to the individual rules, are provided in Table 15.3.

15.5.4 Discussion

For more than half a century, strategies for learning from text have been developed and empirically evaluated on the basis of theories of text comprehension. Accordingly, text comprehension strategies are commonly taught to and exercised by students at school as well as pre-service teachers at university. Although animations are being more and more frequently employed in digital learning material, research on strategies for learning from animation is still in its infancy. Correspondingly, strategies for learning from animation are rarely taught to students at school or pre-service teachers at university (cf. Ploetzner, Lowe, Schlag, & Hauß, 2012). As a consequence, as previous research has demonstrated (e.g., Lowe, 2003, 2004; Marton & Saljö, 1984; McKeachie, 1995), students quite often develop individual learning approaches during their school career that are not optimal from a psychological point of view. For instance, many students develop an approach to

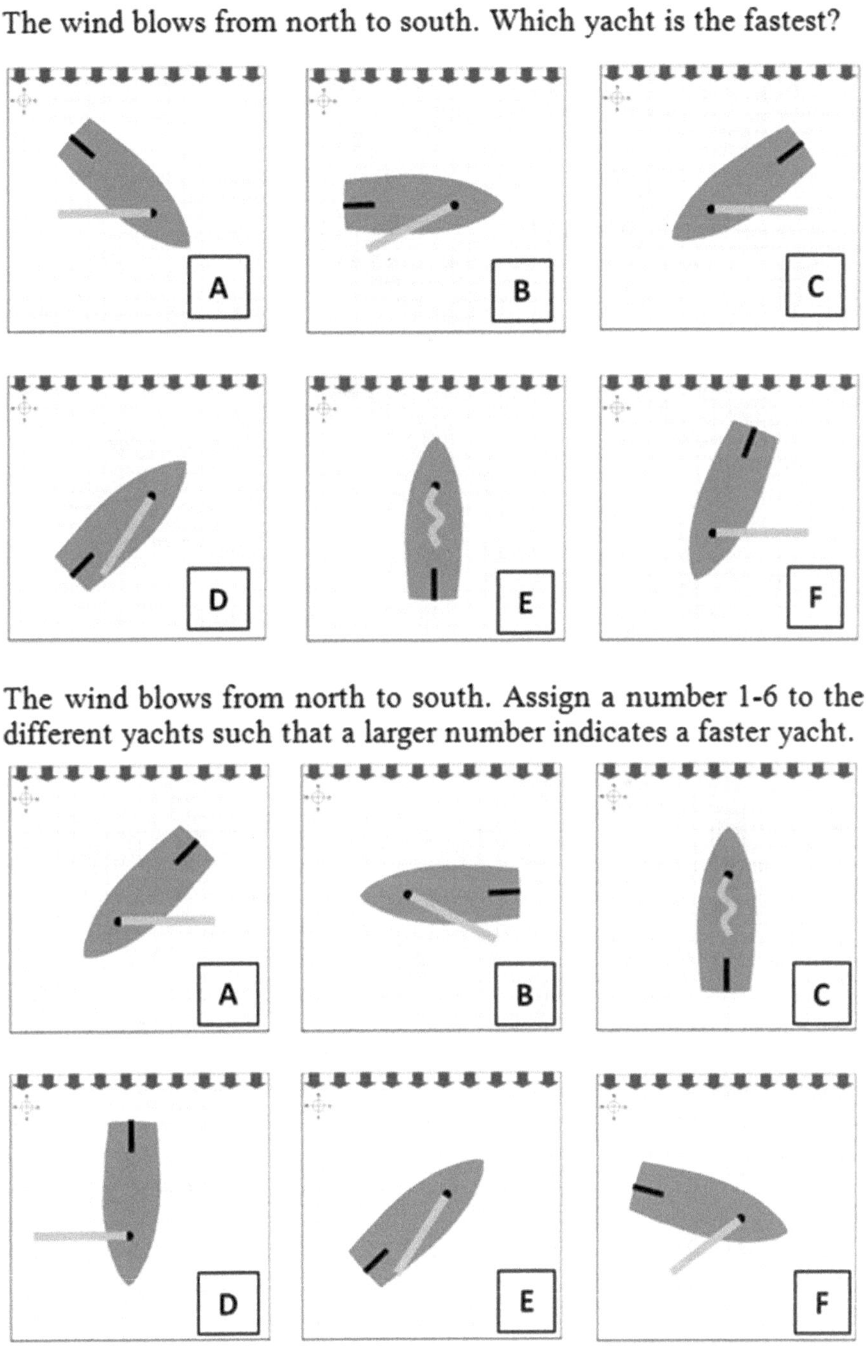

Fig. 15.3 Two examples of items presented in a verbal-graphic format (translation by the authors)

Table 15.2 Six global rules that can be learned from the animation about sailing

1. The closer the yacht sails towards the wind direction, the closer the sail is oriented towards the yacht's hull (and vice versa)
2. The closer the yacht sails towards the wind direction, the larger is the drifting force (and vice versa)
3. The more the yacht sails towards the wind direction, the smaller is the driving force (and vice versa)
4. The smaller the driving force, the less the yacht's speed (and vice versa)
5. If the yachts turns too much towards the wind direction, the yacht stops and the sail flutters
6. If a goal is located upwind, the yacht needs to sail a tacking course

Table 15.3 The mean frequencies of how often each rule was expressed in the students' worksheets

Rule		Strategy group		Summary group		F-value	Probability
1	*M*	0.59	59.4%	0.12	11.5%	$F(1,56) = 17.72$	$p = 0.000$
	SD	0.50		0.32			
2	*M*	0.56	56.3%	0.27	26.9%	$F(1,56) = 5.32$	$p = 0.025$
	SD	0.50		0.45			
3	*M*	0.72	71.9%	0.35	34.6%	$F(1,56) = 9.03$	$p = 0.004$
	SD	0.46		0.49			
4	*M*	0.25	25.0%	0.27	26.9%	$F(1,56) = 0.03$	*n.s.*
	SD	0.44		0.45			
5	*M*	0.78	78.1%	0.65	65.4%	$F(1,56) = 1.15$	*n.s.*
	SD	0.42		0.49			
6	*M*	0.00	0.0%	0.42	42.3%	$F(1,56) = 22.66$	$p = 0.000$
	SD	0.00		0.50			

learning that is rather surface oriented and emphasizes memorization instead of elaboration (Marton & Saljö, 1984). Unfortunately, the use of such approaches can be highly automatized and therefore compete with learning strategies that later are taught to the students (cf. mathemathantic effects; Clark, 1990).

In order to provide a form of support paralleling that available for learning from text, two strategies for learning from animation were proposed: one for learning from narrated animation and one for learning from animation without narration. Both types of animation place different processing demands on the students. When learning from narrated animation, one important processing demand – beyond the need to process the pictorial display – is to mentally relate the verbal and pictorial information provided. For instance, the students need to map linguistic terms and relations expressed in the narration to perceptual entities and events in the pictorial display and vice versa. Because the Cognitive Theory of Multimedia Learning (Mayer, 2009, 2014b) describes the cognitive processes that are relevant to the successful integration of verbal and pictorial representations, it formed the conceptual basis for the strategy for learning from narrated animations. In four experimental studies overall, it was demonstrated that the proposed strategy substantially improves learning from different narrated animations (Kombartzky, Ploetzner, Schlag, & Metz, 2010; Ploetzner & Schlag, 2013). Furthermore, the strategy had beneficial effects beyond the immediate learning phase and for students with lower and higher cognitive ability alike (Ploetzner & Schlag, 2013).

However, because the strategy is made up of eight different learning techniques, the question as to how each technique contributes to the strategy's overall learning effectiveness remains open. Perhaps it is the technique of identifying and sketching important frames of the animation that mainly supports learning (cf. Mason, Lowe, & Tornatora, 2013; see also Lowe & Mason, 2017, this volume). Or perhaps it is the technique of articulating relations between the pictorially and verbally presented information that predominantly facilitates learning (cf. Ainsworth, 2006; Schlag & Ploetzner, 2011). If we could better understand the function of each of the individual learning techniques, perhaps the strategy could be simplified without compromising its learning effectiveness.

A further perspective is to conceptualize and implement a training program that aims at teaching and exercising the strategy in the classroom so that students would begin to internalize the strategy and process it more and more automatically (cf. Streblow & Schiefele, 2006). That is, the learning strategy would become an integral part of the students' self-regulatory skills (cf. Winne & Hadwin, 1998; Zimmerman, 2002). However, from research on strategies for text comprehension, it is well known that long lasting training programs are required to achieve this goal (for overviews see Gambrell et al., 2007; Mandl & Friedrich, 2006; Pressley & Harris, 2006).

In learning from animation without narration, the students receive neither complementary verbal descriptions and explanations nor verbal guidance that tells them "where to look and when to look" (Lowe, 2008; Lowe & Boucheix, 2017, this volume). Instead, the students need to find out themselves what constitutes local entities and events in the pictorial display. Thereafter, they might have to identify complex compositions of local entities and events that might be distributed in space and time. Finally, they need to determine how the identified entities and events are related to each other at multiple hierarchical levels. Because the Animation Processing Model (Lowe & Boucheix, 2008, 2011, 2017, this volume; Lowe & Schnotz, 2014) gives an account of how these processes interplay, it forms the basis for the strategy for learning from animation without narration.

In an experimental study, however, the strategy did not improve learning more than did just watching the animation and writing a summary. This is true despite the fact that the worksheet for guiding the students' processing was even tailored to the structure of the episodes in the animation about sailing (cf. Fig. 15.2). On the one hand, the students "successfully" processed the worksheet in that they expressed a large proportion of the global rules that could be learned from the animation. On the other hand, the students were not able to take advantage of their achievements on the posttest. It might well be that the students in the strategy group followed the affordances offered by the worksheet without actually understanding what they were doing and why they were doing it. Furthermore, the worksheet emphasizes compare-and-contrast processes between the first three sailing courses shown in the animation, namely running, broad reach, and close hauled. The fourth sailing course, tacking, fits only partially into this schema of compare-and-contrast processes.

The worksheet only affords the expression of local changes to the sail of the yacht. More global considerations of the overall purpose of tacking are not offered by the worksheet. As a consequence, the majority of students in the strategy group expressed the rule concerning the yacht's sail but none of the students in the strategy group expressed the rule concerning the purpose of tacking.

The results of our study generally underline the findings reported by Lowe (2004). He found that students who were novices to the animated subject matter rarely engaged in the perceptual and cognitive interrogation processes required for identifying higher-order relationships in animations without narration. According to our results, even the systematic elicitation of the required perceptual processes by means of a learning strategy poses a challenge. Strategies that merely address perceptual processes indirectly and focus mainly on cognitive processes do not seem to effectively support the perceptual processing of animations without narration. However, if an animation is insufficiently processed at the perceptual level, the informational basis for the required cognitive processes is missing. In this case, supporting the cognitive processes is doomed to be ineffective as well.

Perhaps it is unrealistic to support perceptual processes by means of learning strategies, especially if the strategies are to remain applicable to diverse animations without narration. If learning strategies have their strengths in supporting cognitive processes, what could be beneficial in supporting perceptual processes? It is plausible to assume that perceptual processes are strongly influenced by the design of animations, for instance, the visuospatial and dynamic contrasts realized within the animations (cf. Lowe & Schnotz, 2014; Schnotz & Lowe, 2008). Therefore, instead of attempting to support perceptual processes by means of learning strategies, it might be more promising to support perceptual processes by means of effective design strategies. Examples of such strategies are attention guidance strategies, e.g., cueing (cf. Boucheix et al., 2013; De Koning, Tabbers, Rikers, & Paas, 2009; De Koning & Jarodzka, 2017, this volume) and the stepwise presentation of animations (Lowe & Boucheix, 2012, 2017, this volume). Perhaps design strategies that support animation processing at the perceptual level and learning strategies that support animation processing at the cognitive level could be concurrently applied to synergistically complement each other.

Acknowledgements This research was supported by the Deutsche Forschungsgemeinschaft under contract PL 224/17-1. We thank Benjamin Fillisch and Patrick Gewald for supporting the data collection.

References

Ainsworth, S. (2006). DeFT: A conceptual framework for considering learning with multiple representations. *Learning and Instruction, 16*, 183–198.

Ainsworth, S., & Loizou, A. (2003). The effects of self-explaining when learning with text or diagrams. *Cognitive Science, 27*, 669–681.

Artelt, C. (2000). *Strategisches Lernen* [Strategic learning]. Münster: Waxmann Verlag.

Atkinson, R. C., & Shiffrin, R. M. (1968). Human memory: A proposed system and its control processes. In K. W. Spence & J. T. Spence (Eds.), *The psychology of learning and motivation* (Vol. 2, pp. 89–195). New York: Academic Press.

Ayres, P., & Sweller, J. (2014). The split-attention principle in multimedia learning. In R. E. Mayer (Ed.), *The Cambridge handbook of multimedia learning* (2nd ed., pp. 206–226). New York: Cambridge University Press.

Baddeley, A. D. (1986). *Working memory*. Oxford: Clarendon Press.

Baddeley, A. D. (1999). *Essentials of human memory*. Hove: Psychology Press.

Bark, A. (2009). *Sportküstenschifferschein* [License for coastal sport skippers] (9 Aufl.). Bielefeld: Delius Klasing Verlag.

Boucheix, J.-M., Lowe, R., Putri, D. K., & Groff, J. (2013). Cueing animations: Dynamic signaling aids information extraction and comprehension. *Learning and Instruction, 25*, 71–84.

Clark, R. E. (1990). When teaching kills learning: Studies of mathematantics. In H. Mandl, E. de Corte, N. S. Bennet, & H. F. Friedrich (Eds.), *Learning and instruction: European research in an international context* (pp. 1–22). Oxford: Pergamon Press.

Clark, R. C., & Mayer, R. E. (2011). *E-learning and the science of instruction* (3rd ed.). San Francisco, CA: Pfeiffer.

Dansereau, D. F., Collins, K. W., McDonald, B. A., Holley, C. D., Garland, J. C., Diekhoff, G., & Evans, S. H. (1979). Development and evaluation of a learning strategy training program. *Journal of Educational Psychology, 71*, 64–73.

De Koning, B. B., & Jarodzka, H. (2017). Attention guidance strategies for supporting learning from dynamic visualizations. In R. Lowe & R. Ploetzner (Eds.), *Learning from dynamic visualization – Innovations in research and application*. Berlin: Springer (this volume).

De Koning, B. B., Tabbers, H. K., Rikers, R. M., & Paas, F. G. W. C. (2009). Towards a framework for attention cueing in instructional animations: Guidelines for research and design. *Educational Psychology Review, 21*, 113–140.

De Koning, B. B., Tabbers, H. K., Rikers, R. M., & Paas, F. G. W. C. (2010). Learning by generating vs. receiving instructional explanations: Two approaches to enhance attention cueing in animations. *Computers & Education, 55*, 681–691.

Dornisch, M., Sperling, R. A., & Zeruth, J. A. (2011). The effects of levels of elaboration on learners' strategic processing of text. *Instructional Science, 39*, 1–26.

Friedrich, H. F. (1995). *Training und transfer reduktiver-organisierender Strategien für das Lernen mit Texten* [Training and transfer of reductive and organizational strategies for learning from texts]. Münster: Aschendorff.

Gambrell, L. B., Morrow, L. M., & Pressley, M. (2007). *Best practices in literacy instruction* (3rd ed.). New York: Guilford Press.

Ginns, P. (2005). Meta-analysis of the modality effect. *Learning and Instruction, 15*, 313–331.

Hegarty, M., Kriz, S., & Cate, C. (2003). The roles of mental animations and external animations in understanding mechanical systems. *Cognition and Instruction, 21*, 325–360.

Holland, J. H., Holyoak, K. J., Nisbett, R. E., & Thagard, P. R. (1986). *Induction – Processes of inference, learning, and discovery*. Cambridge, MA: MIT Press.

Jonassen, D. H., Howland, J., Marra, R. M., & Crismond, D. (2011). *Meaningful learning with technology* (4th ed.). Upper Saddle River, NJ: Pearson.

Klauer, K. J., & Phye, G. D. (2008). Inductive reasoning: A training approach. *Review of Educational Research, 78*, 85–123.

Kombartzky, U., Ploetzner, R., Schlag, S., & Metz, B. (2010). Developing and evaluating a strategy for learning from animation. *Learning and Instruction, 20*, 424–433.

Kremling, C. (2008). "Lehren und Lernen mit Bildern" – Erste Ergebnisse der Eingangserhebung des Forschungsprojektes "Bildliteralität und ästhetische Alphabetisierung" ["Teaching and learning with pictures" – First results of a survey within the research project "picture literacy and aesthetical alphabetization"]. In G. Lieber (Hg.), *Lehren und Lernen mit Bildern* [Teaching and learning with pictures] (S. 115–123). Hohengehren: Schneider Verlag.

Krug, S. (2014). *Don't make me think –A common sense approach to web and mobile usability* (3rd ed.). Berkeley, CA: New Riders.
Larkin, J. H., & Simon, H. A. (1987). Why a diagram is (sometimes) worth ten thousand words. *Cognitive Science, 11*, 65–99.
Larson, C. O., Dansereau, D. F., Hythecker, V. I., O'Donnell, A., Young, M. D., Lambiotte, J. G., & Rocklin, T. R. (1986). Technical training: An application of a strategy for learning structural and functional information. *Contemporary Educational Psychology, 11*, 217–228.
Lieber, G. (Hg.). (2008). *Lehren und Lernen mit Bildern* [Teaching and learning with pictures]. Hohengehren: Schneider Verlag.
Linn, M., & Eylon, B.-S. (2011). *Science learning and instruction: Taking advantage of technology to promote knowledge integration*. New York: Routledge.
Low, R., & Sweller, J. (2014). The modality principle in multimedia learning. In R. E. Mayer (Ed.), *The Cambridge handbook of multimedia learning* (2nd ed., pp. 227–246). New York: Cambridge University Press.
Lowe, R. K. (2003). Animation and learning: Selective processing of information in dynamic graphics. *Learning and Instruction, 13*, 157–176.
Lowe, R. K. (2004). Interrogation of a dynamic visualization during learning. *Learning and Instruction, 14*, 257–274.
Lowe, R. K. (2008). Learning from animation: Where to look, when to look. In R. K. Lowe & W. Schnotz (Eds.), *Learning with animation – Research implications for design* (pp. 49–68). New York: Cambridge University Press.
Lowe, R., & Boucheix, J.-M. (2008). Learning from animated diagrams: How are mental models built? In G. Stapleton, J. Howse, & J. Lee (Eds.), *Diagrammatic representations and inference – Proceedings of the fifth international conference, diagrams 2008* (pp. 266–281). Berlin: Springer.
Lowe, R., & Boucheix, J.-M. (2011). Cueing complex animations: Does direction of attention foster learning processes? *Learning and Instruction, 21*, 650–663.
Lowe, R., & Boucheix, J.-M. (2012). Dynamic diagrams: A composition alternative. In P. T. Cox, B. Plimmer, & P. Rodgers (Eds.), *Diagrammatic representations and inference – Lecture notes in computer science* (Vol. 7352, pp. 233–240). Berlin: Springer.
Lowe, R., & Boucheix, J.-M. (2017). A composition approach to design of educational animations. In R. Lowe & R. Ploetzner (Eds.), *Learning from dynamic visualization – Innovations in research and application*. Berlin: Springer (this volume).
Lowe, R., & Mason, L. (2017). Self-generated drawing: A help or hindrance to learning from animation? In R. Lowe & R. Ploetzner (Eds.), *Learning from dynamic visualization – Innovations in research and application*. Berlin: Springer (this volume).
Lowe, R. K., & Schnotz, W. (2014). Animation principles in multimedia learning. In R. E. Mayer (Ed.), *The Cambridge handbook of multimedia learning* (2nd ed., pp. 513–546). New York: Cambridge University Press.
Mandl, H., & Friedrich, H. F. (Hrsg.). (2006). *Handbuch Lernstrategien* [Handbook learning strategies]. Göttingen: Hogrefe Verlag.
Marton, F., & Säljö, R. (1984). Approaches to learning. In F. Marton, D. J. Hounsell, & N. Entwistle (Eds.), *The experience of learning* (pp. 39–58). Edinburgh: Scottish Academic Press.
Mason, L., Lowe, R., & Tornatora, M. C. (2013). Self-generated drawings for supporting comprehension of a complex animation. *Contemporary Educational Psychology, 38*, 211–224.
Mayer, R. E. (2009). *Multimedia learning* (2nd ed.). New York: Cambridge University Press.
Mayer, R. E. (Ed.). (2014a). *The Cambridge handbook of multimedia learning* (2nd ed.). New York: Cambridge University Press.
Mayer, R. E. (2014b). Cognitive theory of multimedia learning. In R. E. Mayer (Ed.), *The Cambridge handbook of multimedia learning* (2nd ed., pp. 43–71). New York: Cambridge University Press.
McKeachie, W. J. (1995). Learning styles can become learning strategies. *The National Teaching and Learning Forum, 4*, 1–3.

Metz, B., & Wichert, A. (2009). Lernen mit illustrierten Texten [Learning with illustrated texts]. In R. Ploetzner, T. Leuders, & A. Wichert (Hrsg.), *Lernchance Computer – Strategien für das Lernen mit digitalen Medienverbünden* [The computer as a learning opportunity – Strategies for learning with digital media] (S. 141–161). Münster: Waxmann Verlag.

Microsoft Corporation. (2002). *Microsoft® Encarta® professional* [software]. Redmond, WA: Microsoft Corporation.

Moreno, R., & Valdez, A. (2005). Cognitive load and learning effects of having students organize pictures and words in multimedia environments: The role of student interactivity and feedback. *Educational Technology Research and Development, 53*, 35–45.

Oestermeier, U., & Hesse, F. W. (2000). Verbal and visual causal arguments. *Cognition, 75*, 65–104.

Paivio, A. (1986). *Mental representations: A dual-coding approach*. New York: Oxford University Press.

Pask, G. (1976). Styles and strategies of learning. *British Journal of Educational Psychology, 46*, 128–148.

Peeck, J. (1994). Wissenserwerb mit darstellenden Bildern [Knowledge acquisition with representative pictures]. In B. Weidenmann (Hg.), *Wissenserwerb mit Bildern* [Knowledge acquisition with pictures] (S. 59–94). Bern: Huber Verlag.

Ploetzner, R. (2016). *Understanding explanatory animation – An introduction to cognitive design and user support*. Freiburg: eXplanatory!media.

Ploetzner, R., & Lowe, R. (2012). A systematic characterisation of expository animations. *Computers in Human Behavior, 28*, 781–794.

Ploetzner, R., & Lowe, R. (2014). Simultaneously presented animations facilitate the learning of higher-order relationships. *Computers in Human Behavior, 34*, 12–22.

Ploetzner, R., & Lowe, R. (2017). Looking across instead of back and forth – How the simultaneous presentation of multiple animation episodes facilitates learning. In R. Lowe & R. Ploetzner (Eds.), *Learning from dynamic visualization – Innovations in research and application*. Berlin: Springer (this volume).

Ploetzner, R., Lowe, R., & Schlag, S. (2013). A systematic characterization of cognitive techniques for learning from textual and pictorial representations. *Journal of Education and Learning, 2*, 78–95.

Ploetzner, R., Lowe, R., Schlag, S., & Hauß, C. (2012). Lernen mit statischen und dynamischen Bildern – Welche kognitiven Techniken kennen Lehramtsstudierende? [Learning from static and dynamic pictures – Which cognitive techniques are known by pre-service teachers?] In I. Oomen-Welke & M. Staiger (Hrsg.), *Bilder in Medien, Kunst, Literatur, Sprache, Didaktik* [Pictures in media, art, literature, language, pedagogy] (S. 249–259). Freiburg: Fillibach Verlag.

Ploetzner, R., & Schlag, S. (2013). Strategic learning from expository animations: Short- and mid-term effects. *Computers & Education, 69*, 159–168.

Prain, V., & Tytler, R. (2013). Learning through the affordances of representation construction. In R. Tytler, V. Prain, P. Hubber, & B. Waldrip (Eds.), *Constructing representations to learn in science* (pp. 67–82). Rotterdam: Sense Publishers.

Pressley, M., & Harris, K. R. (2006). Cognitive strategies instruction: From basic research to classroom instruction. In P. A. Alexander & P. H. Winne (Eds.), *Handbook of educational psychology* (pp. 265–286). Mahwah, NJ: Lawrence Erlbaum Associates.

Robinson, F. P. (1946). *Effective study*. New York: Harper and Row.

Salomon, G. (1983). The differential investment of mental effort in learning from different sources. *Educational Psychologist, 18*, 42–50.

Salomon, G. (1984). Television is easy and print is tough – The differential investment of mental effort in learning as a function of perceptions and attributions. *Journal of Educational Psychology, 76*, 647–658.

Schlag, S., & Ploetzner, R. (2009). Lernen mit informierenden Bildern in Texten [Learning with informational pictures in texts]. In R. Ploetzner, T. Leuders & A. Wichert (Hrsg.), *Lernchance computer – Strategien für das Lernen mit digitalen Medienverbünden* [The computer as a learning opportunity – Strategies for learning with digital media] (S. 123–140). Münster: Waxmann Verlag.

Schlag, S., & Ploetzner, R. (2011). Supporting learning from illustrated texts: Conceptualizing and evaluating a learning strategy. *Instructional Science, 39*, 921–937.

Schmidt-Weigand, F. (2005). *Dynamic visualizations in multimedia learning: The influence of verbal explanations on visual attention, cognitive load and learning outcome* (PhD Thesis). Institut für Psychologie der Universität Gießen, Gießen.

Schnotz, W. (2002). Towards an integrated view of learning from text and visual displays. *Educational Psychology Review, 14*, 101–120.

Schnotz, W., & Lowe, R. K. (2008). A unified view of learning from animated and static graphics. In R. K. Lowe & W. Schnotz (Eds.), *Learning with animation: Research implications for design* (pp. 304–356). New York: Cambridge University Press.

Seufert, T. (2009). Lernen mit multiplen Repräsentationen – Gestaltungs- und Verarbeitungsstrategien [Learning with multiple representations – Strategies for design and processing]. In R. Ploetzner, T. Leuders, & A. Wichert (Hrsg.), *Lernchance Computer – Strategien für das Lernen mit digitalen Medienverbünden* [The computer as a learning opportunity – Strategies for learning with digital media] (S. 45–66). Münster: Waxmann Verlag.

Stalbovs, K., Scheiter, K., & Gerjets, P. (2015). Implementation intentions during multimedia learning: Using if-then plans to facilitate cognitive processing. *Learning and Instruction, 35*, 1–15.

Stenning, K., & Oberlander, J. (1995). A cognitive theory of graphical and linguistic reasoning: Logic and implementation. *Cognitive Science, 19*, 97–140.

Stieff, M. (2017). Drawing for promoting learning and engagement with dynamic visualizations. In R. Lowe, & R. Ploetzner (Eds.), *Learning from dynamic visualization – Innovations in research and application*. Berlin: Springer (this volume).

Streblow, L., & Schiefele, U. (2006). Lernstrategien im Studium [Learning strategies in academic studying]. In H. Mandl & H. F. Friedrich (Hrsg.), *Handbuch Lernstrategien* [Handbook learning strategies] (S. 352–364). Göttingen: Hogrefe Verlag.

Svensson, L. (1977). On qualitative differences in learning. III: Study skill and learning. *British Journal of Educational Psychology, 47*, 233–243.

Thomas, E. L., & Robinson, H. A. (1972). *Improving reading in every class: A source-book for teachers*. Boston, MA: Allyn and Bacon.

Weidenmann, B. (1988). *Psychische Prozesse beim Verstehen von Bildern* [Psychological processes of picture comprehension]. Bern: Huber Verlag.

Weidenmann, B. (1989). When good pictures fail: An information processing approach to the effect of illustration. In H. Mandl & J. R. Levin (Eds.), *Knowledge acquisition from text and pictures* (pp. 157–170). Amsterdam: North-Holland.

Weidenmann, B. (1994). Informierende Bilder [Informational pictures]. In B. Weidenmann (Hg.), *Wissenserwerb mit Bildern* [Knowledge acquisition with pictures] (S. 9–58). Bern: Huber Verlag.

Weinstein, C. E., & Mayer, R. E. (1986). The teaching of learning strategies. In M. C. Wittrock (Ed.), *Handbook of research on teaching* (3rd ed., pp. 315–327). New York: MacMillan Publishers.

Winne, P. H., & Hadwin, A. F. (1998). Studying as self-regulated learning. In D. J. Hacker, J. Dunlosky, & A. C. Graesser (Eds.), *Metacognition in educational theory and practice* (pp. 277–304). Hillsdale, NJ: Lawrence Erlbaum Associates.

Wittrock, M. C. (1974). Learning as a generative process. *Educational Psychologist, 11*, 87–95.

Wittrock, M. C. (1989). Generative processes of comprehension. *Educational Psychologist, 24*, 345–376.

Zimmerman, B. J. (2002). Becoming a self-regulated learner: An overview. *Theory Into Practice, 41*, 64–70.

Chapter 16
Guiding Cognitive Processing During Learning with Animations: Commentary on Parts III and IV

Richard E. Mayer

16.1 The Potential of Animation as an Instructional Aid

Advances in low-cost, computer-based graphics over the past 20 years have allowed instructional designers to add animation to their toolbox. On the surface, it appears that animation should be a valuable aid to learning – particularly with STEM topics that involve reasoning with three-dimensional objects such as bones or chemical molecules or understanding cause-and-effect systems such as Newton's laws of motion or how a sailboat works. However, initial research evidence concerning the instructional benefits of animation has not always matched the high hopes (Lowe & Schnotz, 2008, 2014), with static diagrams sometimes being as effective or more effective than animation (Mayer, Hegarty, Mayer, & Campbell, 2005).

"Learning from Dynamic Visualizations: Innovations in Research and Application," edited by Richard Lowe and Rolf Ploetzner, offers a fresh look at the potential of animation as an instructional aid, from the vantage point of a field that is maturing in terms of empirical research depth and theoretical sophistication. In this commentary, I describe a theoretical framework for understanding how to successfully use animation in instructional messages, and I explore techniques for guiding the learner's cognitive processing during learning with animations based on the chapters in Parts III and IV of the book.

R.E. Mayer (✉)
University of California, Santa Barbara, Santa Barbara, CA, USA
e-mail: mayer@psych.ucsb.edu

R. Lowe, R. Ploetzner (eds.), *Learning from Dynamic Visualization*,
DOI 10.1007/978-3-319-56204-9_16

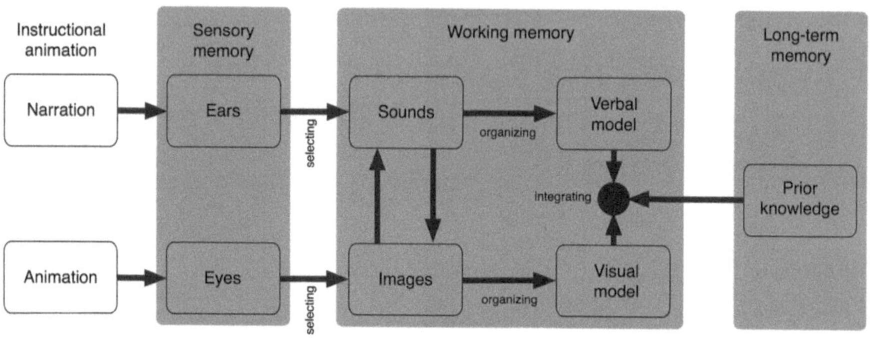

Fig. 16.1 Cognitive theory of multimedia learning applied to instructional animation

16.2 Theoretical Framework for Learning from Instructional Animations

As shown in Fig. 16.1, an instructional message enters the learner's cognitive system through the eyes and/or ears where it is held in a rapidly decaying sensory memory. If the learner pays attention to this fleeing information (indicated by the *selecting* arrows), it is transferred to working memory for further processing. In working memory, the learner can mentally organize the fragments of incoming information into a coherent structure (indicated by the *organizing* arrows) and integrate it with relevant prior knowledge activated from long-term memory (indicated by the *integrating* arrows). We begin with an information processing system that has separate channels for visual and verbal material (dual coding principle), a working memory with limited capacity (limited capacity principle), and cognitive processes (i.e., selecting, organizing, and integrating) that lead to meaningful learning (active processing principle).

As noted by the chapter authors in this book (e.g., De Koning & Jarodzka, 2017, this volume; Lowe & Mason, 2017, this volume; Ploetzner & Breyer, 2017, this volume; Stieff, 2017, this volume), the three cognitive processes required for meaningful learning – selecting, organizing, and integrating – are consistent with processing steps described in more specialized theories of learning with visualizations, such as the Animation Processing Model (Lowe & Boucheix, 2008; Lowe & Schnotz, 2014) or the Generative Theory of Drawing Construction (Van Meter & Firetto, 2013; Van Meter & Garner, 2005). Overall, the challenge of instructional design is to create instructional messages that prime the three key cognitive processes without overloading the processing capacity of working memory.

According to the Cognitive Theory of Multimedia Learning (Mayer, 2009, 2014) as reflected in Fig. 16.1 and Cognitive Load Theory (Paas & Sweller, 2014; Sweller, Ayres, & Kalyuga, 2011), instructional messages make three kinds of demands on the learner's limited cognitive processing capacity:

1. *Extraneous processing* – cognitive processing that does not serve the instructional goal and which is caused by poor instructional design.
2. *Essential processing* – cognitive processing that involves attending to relevant information and mentally representing it, and which is caused by the complexity of the material.
3. *Generative processing* – cognitive processing for making sense of the material by mentally organizing it and integrating it with relevant prior knowledge, and which is caused by the learner's motivation to engage in deep learning.

How does animation relate to this conceptualization of the human information processing system? On the positive side, animation is intended to foster essential processing (by helping make the core material visible and accessible), and animation is intended to foster generative processing (by engaging the learner). On the negative side, animation may hinder essential processing (by virtue of being transient), animation may hinder generative processing (by taking away the learner's need to mentally animate static illustrations or text), and animation may create extraneous processing (by having multiple moving elements some of which can be distracting). When animation creates a substantial amount of extraneous processing, the learner may not have sufficient remaining cognitive capacity to engage in essential and generative processing, which lead to meaningful learning. The challenge for instructional designers is to use animation in a way that minimizes extraneous processing, manages essential processing, and fosters generative processing.

16.3 Techniques for Guiding Cognitive Processing During Learning with Animations

The authors of the five papers in Parts III and IV take up this challenge by exploring techniques for guiding the learner's cognitive processing during learning with animations. There are two kinds of techniques – (1) *instructional design features*, which work by altering aspects of the instructional message, and (2) *learning strategy prompts*, which work by changing the learner's activity during learning. Examples of instructional design features are to add orientation references such as axis lines to animations of three-dimensional objects such as bones (Berney & Bétrancourt, 2017, this volume) or visual signaling such as pointing or spot lights to animations of a dynamic system (De Koning & Jarodzka, 2017, this volume). Examples of learning strategies including asking learners to generate drawings as they learn from an animated lesson (Lowe & Mason, 2017, this volume; Stieff, 2017, this volume) or answering questions as they learn from an animated lesson (Ploetzner & Breyer, 2017, this volume).

16.4 Instructional Design Features that Guide Cognitive Processing of Instructional Animations

First, let's consider instructional design features intended to guide the learner's attention during learning, which can be called the cognitive process of selecting. The goal is to help the learner focus on the essential material and thereby reduce extraneous processing. In some cases, instructional design features can also guide the process of organizing the material as well.

In "Learning Three-Dimensional Anatomical Structures with Animation: The Effect of Orientation References and Learners' Spatial Ability," (Chap. 12) Sandra Berney and Mireille Bétrancourt examine the effectiveness of adding orientation references to animations of three-dimensional objects such as bones. In an original experiment reported in the chapter, students perform faster on spatial judgment tasks involving bones if they viewed a short animation of the rotating bone that included lines indicating the two internal axes or a human-like figure facing the bone from a consistent perspective than with no added orientation references. The authors note that the orientation references mainly are intended to affect the cognitive process of selecting – that is, helping the learner attend to the relevant portions of the bone structure. These findings are consistent with previous findings showing the benefits of adding orientation references to dynamic representations of bones (Stull, Hegarty, & Mayer, 2009), and advance our understanding of how to help people reason with dynamic three-dimensional objects by adding orientation references.

In "Attention Guidance Strategies for Supporting Learning from Dynamic Visualizations," (Chap. 11) Bjorn B. De Koning and Halszka Jarodzka review research on the effectiveness of adding three types of attention guidance features (sometimes called visual signaling) to animations and video clips of dynamic systems such as the human circulatory system – cueing, eye movement modeling examples (EMME), and gesturing. Cueing involves adding features intended to draw attention to specific parts of dynamic displays, such as adding arrows, colored circles, colored lines, or spotlights that gray out the rest of the display. For example, De Koning, Tabbers, Rikers, and Paas (2007) found that adding spotlight cueing to an animation on how the heart works improved comprehension and transfer performance, but in a review, De Koning, Tabbers, Rikers, and Paas (2009) reported that visual cueing was often ineffective. Although learners spend more time looking at the signaled parts of the animation when visual cues are added – thus supporting the cognitive process of selecting – there is not strong evidence that this automatically translates into superior comprehension and transfer – suggesting that the cognitive processes of organizing and integrating also need to be primed.

The authors show that more sophisticated techniques may have potential for guiding selective attention in a way that also fosters deeper learning process – having onscreen characters point to relevant portions of an animation or video as they explain it (consistent with recent work by Fiorella & Mayer, 2016) and overlaying dots on the animation indicating the eye fixations of experts as they viewed it. Such techniques appear to be most effective when they do not distract the learner and when the material is so complex that the learner does not naturally know where to look.

Overall, these two chapters suggest that learners may need some guidance when they view dynamic visualizations – i.e., animations or video clips. In particular, these two chapters provide examples of instructional features that can guide the learner's attention to relevant parts of an animation – adding orientation references to three-dimensional animations of rotation objects such as bones, and adding various types of visual signaling to animations depicting the operation of dynamic systems such as the human heart. The authors are to be commended for showing the progress being made in finding ways to increase the instructional effectiveness of animations by guiding the learner's cognitive processing.

16.5 Learning Strategies that Guide Cognitive Processing of Instructional Animations

Second, let's consider learning strategies intended to guide how learners select and organize material in animations. A learning strategy is an activity that the learner engages in during instruction with the intention of improving learning, and research shows that some learning strategies – such as self-generated drawing or question answering during learning – can be effective under certain circumstances (Fiorella & Mayer, 2015).

In "Self-generated Drawings: A Help or Hindrance to Learning from Animation," (Chap. 13) Richard Lowe and Lucia Mason explore what research has to say about when it is useful to ask learners to produce drawings while viewing instructional animations. The authors note the benefits of asking students to generate drawings while learning from text, particularly when the demands of drawing are minimized (e.g., Schmeck, Mayer, Opfermann, Pfeiffer, & Leutner, 2014; Schwamborn, Mayer, Thillmann, Leopold, & Leutner, 2010). In a recent review, Fiorella and Mayer (2015) found that learning by drawing improved learning from text in 26 of 28 experimental comparisons, yielding a median effect size of $d = 0.40$.

When applied to the new domain of instructional animation, learning by drawing is intended to help learners select relevant aspects of the fast-moving animation and organize them into a coherent representation. For example, Mason, Lowe, & Tornatora (2013) found that learners who were asked to produce self-generated drawings while viewing an animation of a Newton's Cradle device scored better on subsequent learning outcome tests than learners who did not draw. However, the chapter authors propose that self-generated drawing is more likely to be effective with simple animations than with complex ones in which the act of drawing could require extraneous processing that overwhelms available cognitive resources. The authors suggest that the act of drawing can foster learning – by helping the learner select and organize material from the animation – as long as the drawing activity does require too much extraneous processing.

In "Drawing for Promoting Learning and Engagement with Dynamic Visualizations," (Chap. 14) Mike Stieff shows how self-drawing can be applied to learning with chemistry animations. Importantly, Stieff argues that learning by drawing can increase student engagement, which in turn leads to increased effort to engage in deep cognitive processing during learning with animations. Consistent with the chapter by Lowe and Mason (2017, this volume), Stieff shows how learners may need guidance in how to draw and adequate instructional scaffolds in order to minimize extraneous processing created by the mechanics of drawing. For example, Stieff calls for explicit instructions concerning which aspects of the animation should be in the drawing, explicit prompts for learners to reflect on their drawings, and clear feedback on the utility of their drawings. In the future, more research is needed to establish evidence-based guidelines for how best to use self-generated drawing as an aid to learning with instructional animations.

In "Strategies for Learning from Animation With and Without Narration," (Chap. 15) Rolf Ploetzner and Bianka Breyer take a look at a different set of learning strategies based on fostering the cognitive processes of selecting, organizing, and integrating. For example, Kombartzky, Ploetzner, Schlag, and Metz (2010) found better learning from students who viewed an animation on how honeybees dance while engaging in a set of learning strategies that included selecting important frames and parts of the animation, labeling them, and expressing relations among them. Ploetzner and Schlag (2013) were able to replicate the positive effects of using the learning strategy both with an animation on how honeybees dance and on how sailboats work.

The authors show how to build a set of mini-strategies aimed at crucial cognitive processes such as focusing attention on important aspects of the animation, organizing the material by specifying the key changes, and integrating the material by making global characterizations. In an original research study reported in the chapter, the authors found that asking eighth graders to study an animation on how sailboats work for up to 50 min while carrying this set of learning strategies did not result in better test performance than simply summarizing the animation in words, but did result in better being able to express more global rules of sailing, which reflects deeper understanding. Overall, the authors conclude that learners have difficulty in guiding their perceptual processing of animations, and may also had difficulty in applying learning strategies, so they either need better instructional design features (as described in the previous section) or more guidance on how to apply learning strategies.

16.6 Conclusion

The five chapters in Parts III and IV reflect advances in the study of instructional animations over earlier work and point to future directions. In terms of advances, the authors show that the initial research question of whether animations foster student learning is perhaps too broad, and a more productive path is to ask when instructional animations are most effective based on cognitive theories of learning. The authors show the benefits of shifting from a *media comparison studies* – comparing instruction based on animation versus conventional media (such as static graphics) – to *value added studies* – comparing a base version of an instructional animation with the same lesson with one feature or set of strategy prompts added. Importantly, the value added approach taken by these five chapters has yielded promising advances in what we know about the efficacy of two instructional design features (visual signaling and orientation references) and two learning strategy prompts (self-generated drawing and question answering).

The authors are to be commended for focusing on specific research issues concerning when animations are effective, extensively exploring the available evidence and theory, and in some cases even contributing to the empirical research base. In the future, what is needed is a larger research base from which to derive evidence-based guidelines for how to use instructional animations and finer tuned learning theories to ground the guidelines in an understanding of how people learn with animations. In particular, instructional design features and learning strategies for instructional animations should be grounded in an understanding of how they can guide cognitive processing during learning, such as selecting, organizing, and integrating. Instructional animations certainly belong in the instructional designer's toolbox, but researchers need to continue to determine how best to use them. In short, research and theory on instructional animations remains an important area for anyone interested in the practical issue of how to help people learn academic material and the theoretical issue of how people learn academic material.

Acknowledgements Preparation of this commentary was supported by Grant N00014-16-1-2046 from the Office of Naval Research.

References

Berney, S., & Bétrancourt, M. (2017). Learning three-dimensional anatomical structures with animation: Effects of orientation references and learners' spatial ability. In R. Lowe & R. Ploetzner (Eds.), *Learning from dynamic visualization – Innovations in research and application*. Berlin: Springer (this volume).

De Koning, B. B., & Jarodzka, H. (2017). Attention guidance strategies for supporting learning from dynamic visualizations. In R. Lowe & R. Ploetzner (Eds.), *Learning from dynamic visualization – Innovations in research and application*. Berlin: Springer (this volume).

De Koning, B. B., Tabbers, H. K., Rikers, R. M. J. P., & Paas, F. (2007). Attention cueing as a means to enhance learning from an animation. *Applied Cognitive Psychology, 21*, 731–746.

De Koning, B. B., Tabbers, H. K., Rikers, R. M. J. P., & Paas, F. (2009). Towards a framework for attention cueing in instructional animations: Guidelines for research and design. *Educational Psychology Review, 21*, 113–140.

Fiorella, L., & Mayer, R. E. (2015). *Learning as a generative activity: Eight learning strategies that promote understanding*. New York: Cambridge University Press.

Fiorella, L., & Mayer, R. E. (2016). Effects of observing the instructor draw diagrams on learning from multimedia messages. *Journal of Educational Psychology, 108*, 528–546.

Kombartzky, U., Ploetzner, R., Schlag, S., & Metz, B. (2010). Developing and evaluating a strategy for learning from animation. *Learning and Instruction, 20*, 424–433.

Lowe, R. K., & Boucheix, J.-M. (2008). Learning from animated diagrams: How are mental models built? In G. Stapleton, J. Howse, & J. Lee (Eds.), *Diagrammatic representation and inference* (pp. 266–281). Berlin: Springer.

Lowe, R., & Schnotz, W. (Eds.). (2008). *Learning with animation: Research implications for design*. New York: Cambridge University Press.

Lowe, R., & Schnotz, W. (2014). Animation principles in multimedia learning. In R. E. Mayer (Ed.), *The Cambridge handbook of multimedia learning* (2nd ed., pp. 449–463). New York: Cambridge University Press.

Mason, L., Lowe, R., & Tornatora, M. C. (2013). Self-generated drawings for supporting comprehension of a complex animation. *Contemporary Educational Psychology, 38*, 211–224.

Mayer, R. E. (2009). *Multimedia learning* (2nd ed.). New York: Cambridge University Press.

Mayer, R. E. (2014). Cognitive theory of multimedia learning. In R. E. Mayer (Ed.), *The Cambridge handbook of multimedia learning* (2nd ed., pp. 43–71). New York: Cambridge University Press.

Mayer, R. E., Hegarty, M., Mayer, S., & Campbell, J. (2005). When static media promote active learning: Annotated illustrations versus narrated animations in multimedia learning. *Journal of Experimental Psychology: Applied, 11*, 256–265.

Paas, F., & Sweller, J. (2014). Implications of cognitive load theory for multimedia learning. In R. E. Mayer (Ed.), *The Cambridge handbook of multimedia learning* (2nd ed., pp. 27–42). New York: Cambridge University Press.

Ploetzner, R., & Breyer, B. (2017). Strategies for learning from animation with and without narration. In R. Lowe & R. Ploetzner (Eds.), *Learning from dynamic visualization – Innovations in research and application*. Berlin: Springer (this volume).

Ploetzner, R., & Schlag, S. (2013). Strategic learning from expository animations: Short- and mid-term effects. *Computers & Education, 69*, 159–168.

Schmeck, A., Mayer, R. E., Opfermann, M., Pfeiffer, V., & Leutner, D. (2014). Drawing pictures during learning from scientific text: Testing the generative drawing effect and the prognostic drawing effect. *Contemporary Educational Psychology, 39*, 275–286.

Schwamborn, A., Mayer, R. E., Thillmann, H., Leopold, C., & Leutner, D. (2010). Drawing as a generative activity and drawing as a prognostic activity. *Journal of Educational Psychology, 102*, 872–879.

Stieff, M. (2017). Drawing for promoting learning and engagement with dynamic visualizations. In R. Lowe, & R. Ploetzner (Eds.), *Learning from dynamic visualization – Innovations in research and application*. Berlin: Springer (this volume).

Stull, A., Hegarty, M., & Mayer, R. E. (2009). Getting a handle on learning anatomy with interactive 3D graphics. *Journal of Educational Psychology, 101*, 803–816.

Sweller, J., Ayres, P., & Kaluga, S. (2011). *Cognitive load theory*. New York: Springer.

Van Meter, P., & Firetto, C. M. (2013). Cognitive model of drawing construction: Learning through the construction of drawings. In G. Schraw, M. McCrudden, & D. Robinson (Eds.), *Learning through visual displays* (pp. 247–280). Scottsdale, AZ: Information Age Publishing.

Van Meter, P., & Garner, J. (2005). The promise and practice of learner-generated drawing: Literature review and synthesis. *Educational Psychology Review, 17*, 285–325.

FSC
www.fsc.org
MIX
Papier | Fördert
gute Waldnutzung
FSC® C083411

Kolibri 360 GmbH
Ferdinand-Jühlke-Straße 7
99095 Erfurt, Deutschland
produktsicherheit@kolibri360.de